Whirling Disease:
Reviews and Current Topics

GW00504337

00259345

The Whirling Disease Foundation is most grateful to the Norcross Foundation for its grant in support of this publication. We also wish to thank the American Fisheries Society for its partnership in making these valuable papers available to the scientific community and the public. Our dedicated co-editors Jerri Bartholomew and Chris Wilson brought great energy and vision to this project. We also relied on the expertise of a cadre of scientists who gave of their time to serve as peer-reviewers: Mark Adkison, Thomas Baldwin, Katherine Beauchamp, Vicki Blazer, Marshall Bloom, Ken Cain, Linda Chittum, Christine Densmore, Mansour El-Matbouli, Michael Gilbert, Willard Granath, Wayne Hubert, Phil Hulbert, Keith Johnson, Deedee Kathman, Michael Kent, Jíri Lom, Harriet Lorz, Beverly Larson, Elizabeth MacConnell, Richard Mariscal, Gary Marty, Terry McDowell, Thomas McMahon, Christine Moffitt, Doug Munson, Barry Nehring, Oswaldo Palenzuela, Don Ratliff, Paul Reno, John Schachte, George Schisler, Kevin Thompson, E. Richard Vincent, Peter Walker, Glen Watson, Jim Winton, Dave Zafft, and Alexander Zale. And to you, the reader, we hope you find this to be a useful and valued reference.

Harry Piper, President
David Kumlien, Executive Director
Susan Higgins, Special Projects
Whirling Disease Foundation
Bozeman, Montana

Whirling Disease:
Reviews and
Current Topics

Edited by

Jerri L. Bartholomew
Department of Microbiology and Center for Fish Disease Research
Oregon State University, Corvallis, Oregon 97331-3804, USA

and

J. Christopher Wilson
Utah Division of Wildlife Resources, Fisheries Experiment Station
1465 West 200 North, Logan, Utah 84321-6262, USA

American Fisheries Society Symposium 29

Proceedings of the Seventh Annual Whirling Disease Symposium
held at Salt Lake City, Utah, USA
8–9 February 2001

American Fisheries Society
Bethesda, Maryland
2002

The American Fisheries Society Symposium series is a registered serial. Suggested citation formats follow:

Entire book:
Bartholomew, J. L., and J. C. Wilson, editors. 2002. Whirling disease: reviews and current topics. American Fisheries Society, Symposium 29, Bethesda, Maryland.

Chapter within the book:
Hubert, W. A., M. P. Joyce, R. Gipson, D. Zafft, D. Money, D. Hawk, and B. Taro. 2002. Whirling disease among Snake River cutthroat trout in two spring streams in Wyoming. Pages 181–193 *in* J. L. Bartholomew and J. C. Wilson, editors. Whirling disease: reviews and current topics. American Fisheries Society, Symposium 29, Bethesda, Maryland.

© 2002 by the American Fisheries Society

Printed in the United States of America on acid-free paper.

Library of Congress Control Number 2002104614
ISBN 1-888569-37-9
ISSN 0892-2284

American Fisheries Society
5410 Grosvenor Lane, Suite 110
Bethesda, Maryland 20814-2199
USA

Table of Contents

Preface

In 1996, the first whirling disease symposium was organized in response to the growing realization that this disease is a threat to trout populations in the intermountain west, and that there is an urgency to respond to the problem. Although whirling disease is not a new problem, reports of its impacts on rainbow trout populations in Colorado and Montana were the first demonstration of its effects on fish outside of hatcheries, and this became the public's first introduction to disease in wild fish. We mark a decade of research on this disease with a symposium entitled Whirling Disease: Decade of Discovery. At this meeting, we addressed familiar topics—distribution, research on the parasite and on the oligochaete and salmonid hosts, studies of ecological relationships that allow disease to occur, diagnostic methods, and options for management and control. We also solicited reviews on each of these topics to put new findings in perspective and allow a synthesis of the information. We hope that this compilation will enable you to look at whirling disease, perhaps from a slightly different angle, and to consider new interpretations.

The last ten years of field observations and laboratory investigations tell us that the interactions between the whirling disease parasite and its fish hosts are far from simple. We are constantly challenged to predict the significance of each new finding. We have learned that the host, parasite, and environment can interact in some locations to cause serious disease, and the effects on wild fish populations may be long-term in nature. Looking back on the past ten years, we are awed by the unwavering efforts and discoveries of talented researchers and field biologists studying this parasite. Few other fish pathogens have received as much attention or raised as much public awareness on the importance of healthy fisheries.

This concerted research effort is a direct result of the public awareness of the problem. In response to the declines of rainbow trout in Montana, a private foundation, the Whirling Disease Foundation, was created to raise monies for research. The result has been a renewed research effort and a massive monitoring program that encompasses nearly all of the 22 states where the parasite has been reported. Federal funds made available through the U.S. Fish and Wildlife Service for competitive research funding have generated important studies in federal, state, and academic laboratories. In the field of fish health, this is an unparalleled example of cooperation among private individuals, federal and state governments, and the scientific community. However, although this support has resulted in a tremendous increase in knowledge of all aspects of the disease, the promise of cures and solutions is not immediate, and we need to continue to examine this problem from all angles.

One of the reasons for publishing the symposium proceedings, at this time, was to make research results available to the public as well as to other researchers. The decision to solicit reviews was made not only to provide you, the reader, with the background to evaluate the new research, but also to provide a much needed reference volume. Manuscripts of all invited reviews and contributed presentations were peer-reviewed and met the standards for published AFS journals. Abstracts of posters that appear in these proceedings have also been peer-reviewed.

The editors especially wish to thank Susan Higgins for her editorial and administrative assistance in the preparation of these proceedings. We also thank the many reviewers who put aside other obligations to review these manuscripts in a timely manner. In addition, we extend our gratitude to those who participated in organizing the meeting: Eric Wagner, Utah Division of Wildlife Resources; the other session moderators who prepared review papers, Ron Hedrick, Willard Granath, Elizabeth MacConnell, Billie Kerans, and Karl Andree; and the staff of the Whirling Disease Foundation, Susan Higgins, Dave Kumlien, Wanda McCarthy, and Denise Libby. The co-sponsors of the symposium deserve special recognition, too, not only for their contributions to this event, but also for their support in creating the opportunities that make much of this research possible.

Jerri Bartholomew, Symposium Co-Chair
Oregon State University

Chris Wilson, Symposium Co-Chair
Utah Division of Wildlife Resources

Symbols & Abbreviations

The following symbols and abbreviations may be found in this book without definition. Also undefined are standard mathematical and statistical symbols given in most dictionaries.

A	ampere
AC	alternating current
Bq	becquerel
C	coulomb
°C	degrees Celsius
cal	calorie
cd	candela
cm	centimeter
Co.	Company
Corp.	Corporation
cov	covariance
DC	direct current; District of Columbia
D	dextro (as a prefix)
d	day
d	dextrorotatory
df	degrees of freedom
dL	deciliter
E	east
E	expected value
e	base of natural logarithm (2.71828…)
e.g.	(exempli gratia) for example
eq	equivalent
et al.	(et alii) and others
etc.	et cetera
eV	electron volt
F	filial generation; Farad
°F	degrees Fahrenheit
fc	footcandle (0.0929 lx)
ft	foot (30.5 cm)
ft^3/s	cubic feet per second (0.0283 m^3/s)
g	gram
G	giga (10^9, as a prefix)
gal	gallon (3.79 L)
Gy	gray
h	hour
ha	hectare (2.47 acres)
hp	horsepower (746 W)
Hz	hertz
in	inch (2.54 cm)
Inc.	Incorporated
i.e.	(id est) that is
IU	international unit
J	joule
K	Kelvin (degrees above absolute zero)
k	kilo (10^3, as a prefix)
kg	kilogram
km	kilometer
l	levorotatory
L	levo (as a prefix)
L	liter (0.264 gal, 1.06 qt)
lb	pound (0.454 kg, 454g)
lm	lumen
log	logarithm
Ltd.	Limited
M	mega (10^6, as a prefix); molar (as a suffix or by itself)
m	meter(as a suffix or by itself); milli (10^{-3}, as a prefix)
mi	mile (1.61 km)
min	minute
mol	mole
N	normal (for chemistry); north (for geography); newton
N	sample size
NS	not significant
n	ploidy; nanno (10^{-9}, as a prefix)
o	ortho (as a chemical prefix)
oz	ounce (28.4 g)
P	probability
p	para (as a chemical prefix)
p	pico (10^{-12}, as a prefix)

Pa	pascal		V	volt
pH	negative log of hydrogen ion activity		V, Var	variance (population)
ppm	parts per million		var	variance (sample)
qt	quart (0.946 L)		W	watt (for power); west (for geography)
R	multiple correlation or regression coefficient		Wb	weber
r	simple correlation or regression coefficient		yd	yard (0.914 m, 91.4 cm)
rad	radian		α	probability of type I error (false rejection of null hypothesis)
S	siemens (for electrical conductance); south (for geography)		β	probability of type II error (false acceptance of null hypothesis)
SD	standard deviation		Ω	ohm
SE	standard error		μ	micro (10^{-6}, as a prefix)
s	second		$'$	minute (angular)
T	tesla		$''$	second (angular)
tris	tris(hydroxymethyl)-aminomethane (a buffer)		$\circ$	degree (temperature as a prefix, angular as a suffix)
UK	United Kingdom		%	per cent (per hundred)
U.S.	United States (adjective)		‰	per mille (per thousand)
USA	United States of America (noun)			

Section 1

Distribution and Dissemination

American Fisheries Society Symposium 29:3–24, 2002

The History and Dissemination of Whirling Disease

JERRI L. BARTHOLOMEW*

Department of Microbiology and Center for Fish Disease Research
Oregon State University, Corvallis, Oregon 97331-3804, USA

PAUL W. RENO

Coastal Oregon Marine Experiment Station and Department of Microbiology
Hatfield Marine Science Center, Newport, Oregon 97365-5296, USA

ABSTRACT. The explosion of information on the distribution and impacts of whirling disease in the United States during the last decade has changed the way in which we view *Myxobolus cerebralis*. However, even a cursory review of whirling disease literature reveals that many of our concerns today have been expressed at some previous time in the history of our experience with this parasite. From the first description of *M. cerebralis* in Germany in 1893, it was recognized that whirling disease could severely affect the growing trout farming industry. During the first half of this century *M. cerebralis* was disseminated throughout Europe, especially following WWII when live rainbow trout *Oncorhynchus mykiss* were transferred freely. Between 1950 and 1970, the parasite began to appear at trout farms on other continents, and it was in the late 1950s that whirling disease first emerged in the United States. Nearly all reports of detection, both here and in Europe, were associated with artificial rearing facilities. Until the 1980s, the only references reporting infections in natural populations of salmonids are from Finland, Russia, and Michigan, and the reported infections were usually light. Clinical whirling disease was largely associated with culture of trout in earthen ponds, where the infective agent concentrated. In the period between 1970 and 1990, there were increasing reports of the parasite in hatcheries throughout Europe and the United States. In Europe, the perspective after many years of living with whirling disease was that eradication was not possible in most cases, but that we knew enough to reduce infection levels below the point where clinical disease occurs. In the United States, reports of whirling disease in hatcheries were often followed by destruction of any fish on the facility, but as it became apparent that proper management could reduce infection levels, and as there appeared to be no effects outside the bounds of the hatchery, these standards were relaxed. However, in the 1990s, clinical whirling disease was reported in free-ranging trout populations in Colorado and Montana, causing us, once again, to rethink how this disease can be controlled and managed.

When *Myxobolus cerebralis* was first detected in 1893 (Höfer 1903), the myxosporea were already well known fish pathogens. The first members of this taxon were described as early as 1838, from the retina of a carp (Müller 194, in Shul'man 1966). During the past 150 years, there has been considerable scientific interest in these parasites; however, because of its importance to the trout farming industry, *M. cerebralis* became a focal point upon which much of our knowledge of this group has evolved.

This review traces the history and distribution of whirling disease and will discuss prevailing perspectives on the disease, its effects, and attempts at control during different periods. Much of the information gathered here was gleaned from primary literature, previous reviews and reports, both formal and informal. In addition, personal correspondence between the authors and other scientists helped confirm previous information and update current distribution and effects. A caveat is warranted, however: while the collected data were approached conservatively, some inaccuracies with the historic data are inevitable, given the disparate nature of the sampling, monitoring, and identification of this pathogen. Thus, we have attempted to remain strictly prudent in our presentation. The information is presented chronologically, and within each timeframe, by distribution, epizootiology, and a general perspective on the disease. The information outlined below is depicted in Figure 1, for world data. Figures 2 and 3 depict parasite distribution in the United States, with Figure 2 characterizing the data by decade and Figure 3 by location and whether infected fish were cultured or wild/feral. Table 1 contains details of information on individual states.

*Corresponding author: Phone: (541) 737-1856;
Fax: (541) 737-0496; e-mail: bartholj@orst.edu

1890–1910

Distribution and Epizootiology

Germany—Höfer 1903

The first report of whirling disease, (Drehkrankheit) occurred in Germany. The infection was detected in rainbow trout *Oncorhynchus mykiss* and brook trout *Salvelinus fontinalis*. These species are not indigenous to Europe and eggs had been imported from the United States into Germany beginning in the late 1870s (Höfer 1903) to supplement the culture of the native brown trout *Salmo trutta*. The disease had a devastating effect on cultured rainbow and brook trout in the fish farms in which it was found.

Perspectives on Whirling Disease

The original report by Bruno Höfer (1903) described a myxosporidian of trout that infected the brain—hence the original designation *Myxobolus "cerebralis"*. With the recognition that this was instead a parasite of skeletal tissue, a proposal was made by Marianne Plehn to change the name to *Myxobolus chondrophagus* (Höfer 1904). This revision was not accepted due to zoological nomenclatural rules and the name "*Lentospora cerebralis*" was introduced (Höfer 1904), with the new genus name describing a species similar to *Myxobolus*, but not possessing an iodinophilic vacuole within the spore.

The first detailed descriptions of whirling disease and *Lentospora cerebralis* were by Plehn (1905). Stages of infection including initial invasion and erosion of the cartilage by the trophozoites and the subsequent formation of large lesions containing parasites and debris from cartilage degradation were described. She also speculated that because the cartilaginous region around the auditory capsule was usually invaded, either toxins released by the parasite, or simply the weakening of the capsule itself, destroyed the equilibrium of the fish and precipitated the tail-chasing behavior which is observed when fish are disturbed. Weakening of the vertebral column posterior to the 26th vertebra was proposed to exert pressure on the caudal nerves and result in loss of melanocyte control, causing black tail. The most commonly observed cranial deformities were sunken areas behind the eyes and the permanently open or twisted lower jaw, both caused by loss of cartilage during bone formation. In fish that survived to the second year, she described the formation of epitheloid granulomas, which may result in secondary damage. It was suggested that when the cartilage ossifies, the spores became encapsulated and new infections could not occur.

Infection of fish was presumed to occur by ingestion of spores, as in other myxosporidian diseases known at that time. The resilience of the spores to the effects of drying and freezing, and their ability to survive for long periods were noted (Plehn 1905), and the implications of spore stability for the eradication of the parasite were recognized.

1911–1930

Distribution and Epizootiology

Denmark—Bruhl 1926

Perspectives on Whirling Disease

The continuing spread of whirling disease throughout trout-rearing facilities in Germany caused concern and speculation on how introduction had occurred. Several reports suggested that whirling disease originated from feeding rainbow trout with raw marine fish (cod) infected with the parasite (Bergman 1922; Plehn 1924), although later investigators pointed out that introduction by this route was unlikely (Schäperclaus 1931).

Since it was recognized that the clinical signs exhibited in whirling disease were common to other fish diseases, accurate identification of the spores was necessary for confirmation. It was also recognized that because spore formation required several months to occur, diagnosis of early stages of infection was difficult. Therefore, selection of diagnostic methods depended on the age or life stage of the fish. Infection in age 0+ trout was best detected by observing external signs. In fish larger than 10 cm and beyond their first winter, spores could be detected in histological preparations or from dissection of the vestibular organ. (Plehn 1924; Schäperclaus 1931).

1931–1950

Distribution and Epizootiology

Finland (southern)—Dogel (1932, in Uspenskaya 1957)

France—Vanco (1952, in Hoffman 1970)

Italy—1950 (Scolari 1954)

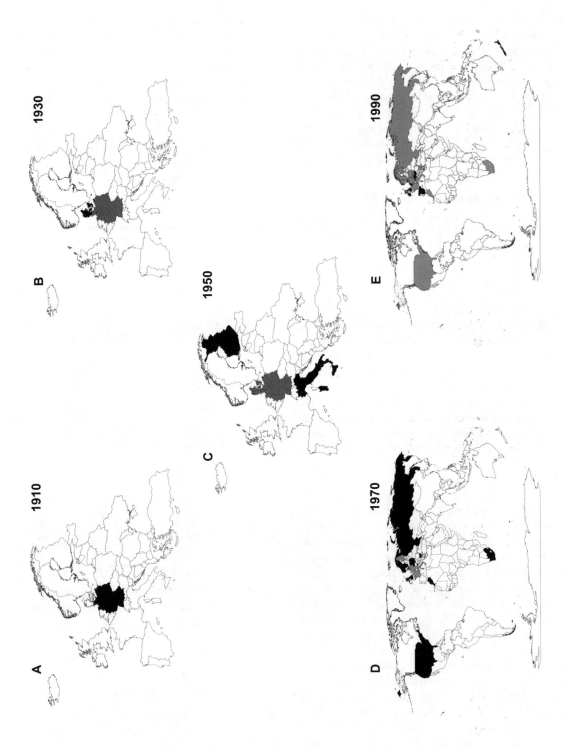

Figure 1. Geographic distribution and dispersal of *Myxobolus cerebralis* infections from its initial detection until present. Figures 1A–E represent accumulative detection for individual countries in 20-year increments. Countries in which *M. cerebralis* was detected during the period are shaded in dark grey; those in which the pathogen had been detected prior to the period are shaded in light grey. Figures 1A: 1890–1910, 1B: 1911–1930, 1C: 1931–1950, 1D: 1951–1970, 1E: 1971–2000.

In Germany, transfer of infected fish among breeding installations had led to the occurrence of whirling disease in most facilities rearing rainbow trout. The extent of the problem was evident from descriptions by Schäperclaus (1931) who indicated that trout breeders were less than forthcoming about the occurrence of whirling disease, often hiding affected fish from visitors. In areas where the parasite was established, whirling disease caused severe losses in trout pond farms and severely limited trout production.

Perspectives on Whirling Disease

Schäperclaus (1931) expanded his earlier descriptions of whirling disease. He recognized four classes of disease signs: 1) whirling movements caused by impairment of the vestibular apparatus; 2) premature death without significant signs other than loss of weight; 3) chronic disease with darkening of the tail, purportedly as a result of damage to the sympathetic nerves and; 4) chronic infection without external signs. This latter condition was particularly significant as it was recognized that these fish were likely the primary source of dissemination of the pathogen.

Other explanations for routes of parasite dissemination were proposed. Refuting earlier conclusions that raw marine fish were the primary source of infection (Plehn 1924), Schäperclaus (1931) suggested that this was unlikely because fish in hatcheries where saltwater fish were not fed also developed whirling disease. He believed that dissemination was largely a result of transfer of infected fish. The brown trout was particularly suspected as the primary source, because this species often became infected but did not show significant evidence of whirling disease. Reports of transfer of infection via shipments of eggs were noted, and were explained by contamination of the eggs with the pathogen adhered to the egg surface or on fomites such as shipping material and equipment. These observations resulted in guidelines for insuring that eggs were incubated on well water, and for disinfecting equipment. The nature of transfer in some cases suggested that eyed eggs could become infected but this was not demonstrated. Schäperclaus also investigated invertebrates as potential carriers and was especially interested in snails because of their presence in large numbers in many affected ponds. However, examination of these animals failed to support this hypothesis; other species of invertebrates were not examined.

Schäperclaus (1931) suggested guidelines for control of the disease which included setting low flow rates to avoid stirring up bottom sediment thereby making the spores available for ingestion, and keeping breeding and rearing ponds separated. After incorporating these precautions, it was recommended to disinfect unused ponds with caustic lime and calcium cyanimide, remove mud from the pond bottom and to discard the sickest fish. However, trout farmers were cautioned against seeking a panacea against whirling disease.

Continued revision of the taxonomy of the agent established the genus *Lentospora* (Plehn 1905) as synonymous with *Myxosoma* (Kudo 1933), hence the change to "*Myxosoma cerebralis*". The spores of the *Myxosoma* were differentiated from those of *Myxobolus* by their lack of an iodophorous vacuole.

1951–1970

Distribution and Epizootiology

- USSR

  Leningrad region—1952 (Uspenskaya 1955)

  Abkhazjan district—1954 (Uspenskaya 1957)

  Black River—Uspenskaya (1957)

  Sakhalin Island—1959 (Bogdanova 1960)

  Luvenga River, White Sea tributary— 1962 (Bogdanova 1964)

  Lake Issyk-Kul, south-central Russia— 1964 (Bogdanova 1968)

  Black Sea region—1964 (Bogdanova 1968)

- Europe

  Czechoslovakia—1954 (Dyk 1954)

  Poland—Kocylowski (1953)

  Bulgaria—Margaritov (1960)

  Yugoslavia—Tomasec (1960)

  Sweden—Johansson (1966)

  Scotland—Elson (1969); Roberts and Elson (1970)

  Norway—1970 (Hastein 1971)

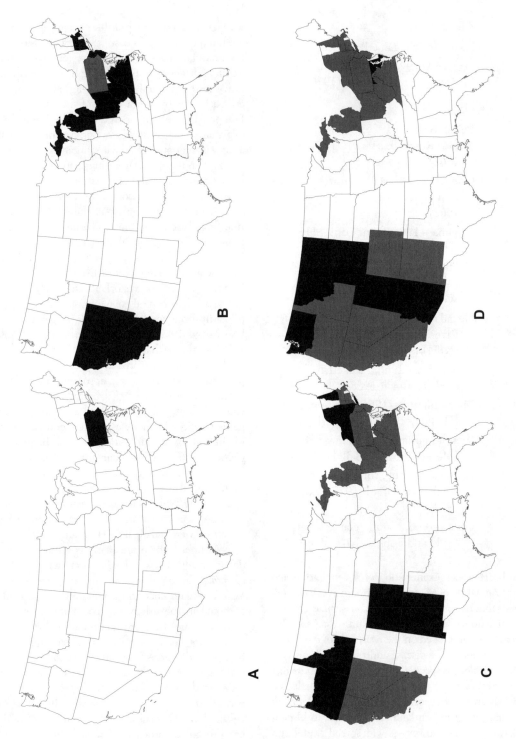

Figure 2. Geographic distribution and dispersal of *Myxobolus cerebralis* infections within the United States, from its initial detection in 1956 until present. Figures 2A–D represent accumulative detection for individual states in 10-year increments. States in which *M. cerebralis* was detected during the period are shaded in dark grey; those in which the pathogen had been detected prior to the period are shaded in light grey. The pathogen was not detected in any new states between 1971 and 1980. Figures 2A: 1951–1960, 2B: 1961–1970, 2C: 1981–1990, 2D: 1991–2000.

- Africa

  South Africa, Cape Province—van Wyk (1968)

  Morocco—Preudhomme (1970)

- USA

  Pennsylvania—1956, Benner Springs Fish Research Station (Hoffman 1962)

  Connecticut—1961, Kensington State Fish Hatchery (SFH) (Hoffman, 1962)

  Virginia—1965, Paint Bank SFH (Hoffman 1990)

  California—1965, Garrapata Creek, private hatchery (Yasutake and Wolf 1970)

  Nevada—1966, Verdi SFH, Carson River National Fish Hatchery (NFH) (Yasutake and Wolf 1970)

  New Jersey—1966, Hackettstown SFH (Hoffman 1968)

  Massachusetts—1966 private hatchery (Halliday 1976; Hoffman 1990)

  West Virginia—1968, Edray SFH (Meyers 1969, in Halliday 1976)

  Michigan—1968, three commercial hatcheries (Hnath 1970; Yoder 1972)

  Ohio—1968, commercial hatchery (Tidd and Tubb 1970)

In his thorough examination of dissemination and transfaunation of M. *cerebralis*, Hoffman (1970) suggests that the original range of M. *cerebralis* covered an area from Central Europe to Northeast Asia. This conclusion was based on the resistance of the native brown trout to the disease and the hypothesis that this resistance is specific and acquired. From its first recognition in 1903 until the mid-1930s, the disease had been reported only from Germany, Denmark (Bruhl 1926) and Finland (Dogel 1932, in Uspenskaya 1957). However, widespread detection occurred following WWII. Hoffman (1970) associated this apparent dissemination to unrestricted transfers of live rainbow trout throughout Europe. However, he also pointed out that some of the fish farms might have become infected with spores from wild native brown trout. If these subsequent reports are interpreted on this basis, then transmission was from natural sources within this region. This mechanism of dissemination would imply multiple initiation sites by "natural" routes, followed by efficient establishment in the majority of the fish farms in Europe where earthen ponds were the norm. This explanation may be viable for Europe where the parasite was enzootic in native salmonids, but would not explain the spread to regions outside Europe. In these instances, the dissemination was certain to have occurred by transport of live fish.

In Europe, the impact of the disease on wild trout populations appeared to be negligible (Christensen 1972). The hypothesis for this lack of effect is that resident species (primarily brown trout) may have acquired resistance to M. *cerebralis* and rainbow trout had not been stocked extensively. Reports of infections in natural populations of salmonids are primarily from southern Finland (Dogel 1932, in Uspenskaya 1957) and western Russia (Uspenskaya 1957). In the Sakhalin Islands, off the coast of eastern Russia, there appeared to be an enzootic focus of parasitism and the parasite apparently had a wide distribution by the late 1950s. In this region not only cultured trout, but also cultured and wild anadromous Pacific salmon were infected, enhancing the likelihood of disseminating spores. Bogdanova (1960, 1966, 1968) documented infections in salmonids from four rivers on Sahkalin Island. However, clinical disease signs were largely associated with rearing facilities using earthen ponds; clinical disease was unknown in natural populations where infections were usually light (Bogdanova 1970).

Although there are many confirmed reports documenting the worldwide dissemination of M. *cerebralis* during this time, reports from certain regions remain unconfirmed and are not generally accepted in the fish health community; for their historic value, they are mentioned below. These events are likely a result of misinterpretation because of diagnosis based on clinical signs alone or confusion of M. *cerebralis* spores with those of other myxobolid species. They may also reflect cases of parasite detection in fish that were imported for culture but where subsequent establishment of the parasite failed to occur. However, they are mentioned here as a reminder of the importance of confirming infections.

Detection of M. *cerebralis* in Korea (Tinkina 1962) and Japan (Halliday 1976) was not unexpected because of their geographic proximity to the focus

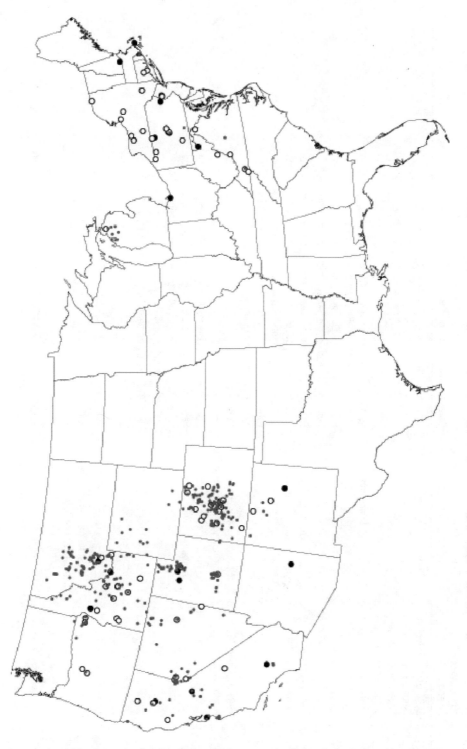

Figure 3. Locations within the United States from which *M. cerebralis* has been isolated between 1956 and present. Data were accumulated from literature citations, unpublished historical data, and personal communications; as much as possible, confirmation was acquired from telephone or electronic means. The data are grouped by type of sample collected: ● = wild or feral fish; ○ = fish in public culture/holding facilities; ● = fish in private culture/ holding facilities.

Table 1. Detection of *Myxobolus cerebralis* in hatcheries, rivers, and standing waters in states where the parasite has been detected.

State[1]	*Myxobolus cerebralis* detected[2]	Detection in culture facilities current and historical[3]	Detection in rivers/standing waters[4]
Arizona	2000 (P. Lopez, Arizona Game and Fish Department, personal communication)	None	Confirmed introduction in one private pond
California	1966 (Yasutake and Wolf 1970)	2 state hatcheries and 3 private facilities currently positive	South Fork American and Trinity Rivers; Wolf, Battle, Deer, Garrapata, and Big Creeks; Carmel River; Aptos and LaHonda Creeks and multiple points in the Owens, Lahontan, Stanislaus, Calaveras, Mokelumne, Feather, and San Lorenzo River drainages
Colorado	1987 (Barney et al. 1988; Walker and Nehring 1995)	9 state hatcheries, 1 federal hatchery and numerous private facilities positive	Enzootic in all coldwater drainages except the Animas and North Republican Rivers; populations impacted in the following rivers: Cache la Poudre, Colorado, Dolores, Fryingpan, Gunnison, Middle Fork of South Platte, South Platte, Rio Grande, Roaring Fork Rivers, as well as some smaller streams
Connecticut	1961 (Hoffman 1962; Hnath 1970)	No recent data; 2 state hatcheries historically positive	None reported
Idaho	1987 (Hauck et al. 1988)	2 of 22 state hatcheries currently positive	Coeur D' Alene, St. Joe, American, Middle Fork Salmon, Yankee Fork, East Fork, S. Fork Boise, Lemhi, Pahsimeroi, Big Lost, Big Wood, Little Lost, Henry's Fork, Teton, and Bruneau Rivers; Loving, Salmon Falls, and Jackknife Creeks and Henry's Lake
Maryland	1995 (C. Geseker, Department of Natural Resources, personal communication)	2 of 6 hatcheries positive in 1995	Jennings Reservoir on north branch of the Potomac River
Massachusetts	1966 (Hoffman 1990)	No state hatcheries currently positive; 1 commercial hatchery historically positive	Not currently detected
Michigan	1968 (Yoder 1972)	No public hatcheries currently positive, several private hatcheries positive; one state hatchery historically positive	Sturgeon, Big Manistee, Little Manistee, AuSable, Jordan, Black, and Pigeon Rivers

Table 1. Continued

State[1]	Myxobolus cerebralis detected[2]	Detection in culture facilities current and historical[3]	Detection in rivers/standing waters[4]
Montana	1994 (Vincent 1996)	None	87 waters in the following River drainages: Beaverhead, Gallatin, Madison, Bighole, Bitterroot, Blackfoot, Jefferson, Swan, Clark Fork, Missouri, Sun, Flathead, and Yellowstone
Nevada	1966 (Yasutake and Wolf 1970); 1957 (Taylor et al. 1973)[5]	1 of 4 state facilities currently positive; 4 public hatcheries historically positive (2 now closed)	Truckee, Carson, Owyhee, East Walker, and Little Humboldt Rivers; Dorsey Creek, Dry Creek, Lahontan, Wildhorse, and Wilson Sink Reservoirs; Topaz, Pyramid, and Walker Lakes, Lake Tahoe; Bilk, Clear, and Duck Creeks
New Hampshire	1981 (Jay Hendee, New Hampshire Fish and Game Department, personal communication)	No recent data; 1 private hatchery historically positive	South Merrimack River basin
New Jersey	1968 (Halliday 1976)	2 state hatcheries historically positive, no longer in operation	Not currently detected, previously stocked rivers surveyed negative
New Mexico	1987 (R. Goede, Utah Division of Wildlife Resources, personal communication)	3 of 7 state hatcheries positive	Pecos, Cebolla, San Juan, Cimarron, Red, and Canones Rivers
New York	1984 (Hoffman 1990)	1 state hatchery and 1 private facility currently positive; 6 public facilities historically positive	At least one positive population or facility in all watersheds except on Long Island
Ohio	Late 1960s (Tidd and Tubb 1970)	No recent data; 1 private hatchery in 1960s	Not currently detected
Oregon	1986 (Holt et al. 1987)	1 private facility currently positive; exposure at 2 public adult facilities from infected adult salmon; 1 state facility historically positive	Grande Ronde and Imnaha River basins
Pennsylvania	1956 (Hoffman 1962)	3 of 13 state facilities currently positve; 7 state hatcheries historically positive; no information on private hatcheries	Consider all watersheds to potentially have infected fish; infected fish detected in Lake Erie and its tributaries

Table 1. Continued

State[1]	Myxobolus cerebralis detected[2]	Detection in culture facilities current and historical[3]	Detection in rivers/standing waters[4]
Utah	1991 (Heckman 1992)	1 state hatchery previously positive, 10–12 private facilities positive	Logan, Little Bear, South Fork Ogden, Weber, Provo, Beaver, Otter, Sevier, and Fremont drainages and in some isolated ponds
Virginia	1965 (Hoffman 1970)	2 state hatcheries positive and 4 private hatcheries historically positive	Tye River, Staley Creek, Whitetop fork of Laurel River
Washington	1996 (K. Amos, Washington Department of Fish and Wildlife, personal communication)	None	Cottonwood Creek and Grande Ronde River
West Virginia	1969 (Meyers 1969)	No recent data; 1 state hatchery, 1 federal hatchery and several private facilities historically positive	Not currently detected
Wyoming	1988 (D. Money, Wyoming Game and Fish Department, personal communication)	5 of 11 state facilities positive	North Platte, S. Fork Shoshone, Salt, Yellowstone, Fire Hole, Big and Little Laramie, N. Fork Popo Agie, Jakey's Fork, E. Fork, New Fork, and Green Rivers; Fontenelle Reservoir; Yellowstone Lake

[1]Only states where M. cerebralis has been confirmed are included. Information from several states included in this table report that the parasite has not been recently detected.

[2]Dates are the first confirmed detection of the parasite in the state.

[3]Data for detection in hatcheries was acquired through a survey conducted by the Whirling Disease Foundation, personal communications with state managers, and use of historical records. It must be acknowledged that monitoring varies among states and may not have been done for private facilities. Identification of facilities as historically positive indicates that either the facility has been disinfected and not subsequently detected positive or that it has been decommissioned.

[4]Data for detection in rivers and standing waters was acquired similarly to that for hatchery data. The extent of, and methods for, monitoring of waters varies among states. The waters listed are those where M. cerebralis is known to have become established, although in most cases it was not possible to determine how widely or to what degree the parasite was established.

[5]In Nevada, the parasite was first detected in 1966; however, examination of archived samples confirmed its earlier presence.

of infection in the Sakhalin Islands. However, the original report of M. *cerebralis* in Korea (Tinkina 1962) only states that the usual preventative measures are adopted against whirling disease, and not that the parasite was detected (Halliday 1976), and subsequent detection of M. *cerebralis* has not been confirmed in Japan (Egusa 1992). Similarly, reports of clinical whirling disease from several locations in South America were not confirmed. An account of whirling disease in Venezuela was based on a misinterpretation by Bogdanova (1968, 1970) of a report by Martinez (1966) in which lentosporidiasis (whirling disease) was described as a serious problem in trout culture. However, the report did not state that the disease occurred in Venezuela. Although succeeding reports continued to describe rainbow trout with signs of whirling disease (Espinoza and Diaz-Ungria 1970; Diaz-Ungria 1970), there was no laboratory confirmation of the infection. Reports of whirling disease also appeared from Columbia (Anonymous 1972), Ecuador (Meschkat 1971) and Chile (Anonymous 1974). Conroy (1975, 1981) and Margolis et al. (1996) attempted to sort through these conflicting and unconfirmed reports. Their conclusion, based on their own observations as well as on the failure of any of the authors who observed "whirling disease" to detect M. *cerebralis* spores from the affected fish, was that the myxosporean is absent in South America. Reports of M. *cerebralis* from Mexico (Halliday 1976) and Canada (Bogdanova 1969, in Halliday 1976) are also unconfirmed. Bogdanova (1969) reported spores in wild *Salvelinus malma* from British Columbia, Canada but a study by Margolis et al. (1981) did not support this finding and there has been no subsequent evidence that the parasite is present.

In the United States, whirling disease was first diagnosed in 1956 from brook trout at the Benner Springs Fish Research Station in Bellefonte, Pennsylvania (Hoffman 1962). Imported frozen European table trout were circumstantially implicated as the source of infection as samples of these food fish were found to contain spores. It was suspected that infected fish tissues were fed to the hatchery trout, or that infected viscera were discarded in the stream (G. L. Hoffman, U. S. Fish and Wildlife Service, retired, personal communication). Another possible source of infection was live brown trout imported from Europe in the 1950s by governmental, commercial and private hatcheries. As some outbreaks in commercial hatcheries were observed in the absence of direct transfer of fish from con-

taminated hatcheries, this raises the possibility that M. *cerebralis* may have been introduced through infected fish or material at more than one site (Hnath 1996). Infected fish from the affected hatcheries were stocked throughout Pennsylvania until the early 1970s, disseminating the parasite among many hatcheries and in natural waters in the state.

In Nevada, the first reported detection was in 1966 at the Verdi State Fish Hatchery (SFH) on the Truckee River by Yasutake and Wolf (1970), but examination of preserved fish showed the parasite had been present in the state by 1957 (Taylor et al. 1973). In 1968, M. *cerebralis* was detected in wild rainbow and brown trout from the Truckee River in the Lahontan basin. Because M. *cerebralis* had been present, but undetected, since at least 1957, many of the state's waters had been stocked with infected fish. By 1969 infections had been confirmed in three state hatcheries and one federal hatchery in Nevada. Rivers that had been stocked with infected hatchery fish included the Truckee, Carson, East Fork Walker, Humboldt, and Owyhee.

In 1965 whirling disease was diagnosed in California from fish at a private hatchery on Garrapata Creek, in Monterey County (Yasutake and Wolf 1970). The hatchery had reported a history of receiving only trout eggs, and this was not believed to be the source of the infection. Imported frozen fish were again implicated (Hoffman 1990), as the affected fish had reportedly been fed frozen trout product from a Danish merchant vessel. However, because the life cycle of M. *cerebralis* was, as yet, unknown, other more likely sources of infection may not have been considered. The report of the outbreak was especially vivid. The behavior of the fish was described as "an extremely nervous twisting action as though the fish were hooked in the mouth and then got off the hook. They spiral around and around as though they were tied by the nose. This becomes so violent at times that some of them come out onto the bank." (Yasutake and Wolf 1970).

Once established at these locations in the eastern and western United States, subsequent spread of the whirling disease parasite has been attributed to transfers of live fish (Hoffman 1970, 1990). In Connecticut, whirling disease was diagnosed in 1961 from rainbow trout fingerlings at the Kensington SFH and the parasite was detected in 1966 from fish at the Burlington SFH. It was suspected that the source of the first infection was fish transferred from the infected Benner Springs facility in 1959 (Hoffman 1962). Between 1965 and 1968, infections were detected in fish from hatcheries in Virginia (Hoffman 1970), Massachusetts (Halliday 1976), West Virginia (Meyers 1969), New Jersey (Hoffman 1968: In Halliday 1976), Ohio (Tidd and Tubb 1970), and Michigan (Yoder 1972), and from additional sites in California, Pennsylvania and Nevada (Hoffman 1990). In West Virginia, introduction occurred at three separate locations in 1966, 1972, and 1984. Clinical whirling disease was diagnosed in Ohio at one location in 1967 and M. cerebralis-positive fish were reported from two additional sites (Tidd and Tubb 1970).

In Michigan, M. cerebralis was detected in 1968 at three facilities belonging to a commercial trout grower. Records showed that infection was likely a result of an importation of infected fish the previous year from a private facility in Ohio, and fish had been subsequently distributed to 159 smaller hatcheries, fee fishing ponds and private recreational ponds. The state quarantined these sites, established a permit system to prevent further spread and through voluntary agreement, treated all 159 sites to kill the fish. Subsequent monitoring at these secondary sites indicated that these actions were successful. However, action was delayed at the primary sources and the parasite became established in populations of wild brook and brown trout from the North Branch Tobacco River (Yoder 1972). The time between suspected introduction and detection in the wild populations in 1970 was 28 months. A second introduction occurred in the Sturgeon River, possibly by legal stocking of fish by a private club. The parasite became established and was eventually detected at the state hatchery on that river in 1975. All the fish at the hatchery were destroyed, and the facility was disinfected and decommissioned (J. G. Hnath, Michigan Department of Natural Resources, personal communication).

In the United States, although the pathogen was found in several regions, clinical whirling disease was only reported in hatcheries. However, there were few published studies during this time that examined infection in wild fish and low-level infections in wild fish would probably not have been evident. Exceptions to this were the survey data in Michigan demonstrating establishment of the parasite in native brook and brown trout populations below the hatchery where it had been introduced (Yoder 1972) and studies on the Truckee River and other sites in Nevada demonstrating

establishment of the pathogen in wild fish (D. Junell, Nevada Department of Conservation, Division of Wildlife, personal communication).

Perspectives on Whirling Disease

During this period whirling disease was still considered a serious problem throughout trout farming areas in Europe. Methods for control included destruction of all fish from ponds containing infected fish. It was suggested that concrete and earthen ponds should be drained and treated with calcium cyanamide, quicklime or sodium hypochlorite, with several treatments required for earthen ponds. Only fish from a known uninfected source should be used, and during the period of maximum susceptibility (the first 8 months), the young fish should be reared in spring or well water (Schäperclaus 1954; Hoffman 1962). Denmark went a step further to insure that young fish were reared on clean water supplies (Rasmussen 1967). There, they established separate facilities to rear fry to the fingerling stage, at which point they were transferred to rearing farms. This method proved successful and was implemented in other countries (Ghittino 1970). With adoption of these methods, as well as the construction of new ponds, the shift toward raising fry in troughs of concrete or synthetic fiber, and assuring that young fish were not exposed to the pathogen until their skeletons had achieved some degree of ossification, the impact of infection was reduced. However, this did not necessarily lead to a reduction of the prevalence of the pathogen in cultured or wild populations in enzootic areas.

In the United States, the discovery of M. cerebralis was considered a resource disaster because of the susceptibility of rainbow trout that are indigenous to the western United States and were widely dispersed here in the late 19th century. Efforts to restrict establishment and movement of the parasite in Michigan and in other areas where the parasite was detected were extreme, requiring destruction of fish from contaminated facilities and no transfer or stocking of fish from any infected facility.

Concerns about the effects of introduced pathogens led to development of the first national fish disease law in the United States. This legislation, a 1968 amendment to Title 50 Wildlife and Fisheries Act, attempts to exclude certain fish diseases from importation to the United States The regulations, which became effective 1 July, 1968 (Code of Federal Regulations, Title 50, Section

13.7), require that salmonids and salmonid eggs be certified free from the hemorrhagic septicemia virus (VHSV) and M. *cerebralis* before they may be imported into the United States. Suggested methods for detection of the parasite (Hoffman 1968) included 1) maceration of the head in a mortar and pestle, 2) splitting the head sagitally and scraping the cut surface with a scalpel, 3) homogenizing the head in a tissue grinder, and 4) cutting the head lengthwise and examining under a dissection microscope for granulomas or cysts, then squashing these cysts for examination. To be certified free of the pathogen, eggs and sperm were required to be from sources free of infection for at least one year before spawning, and incubation prior to shipment must have been in water from a source free of fish carrying the disease. Fish for importation were required to be from sources certified to be free of the disease during the lifetime of the fish.

Researchers continued to examine modes of transmission and the parasite's life cycle. Transovarian transmission was not documented, but it was considered possible that viable spores may be transferred with eggs (Schäperclaus 1931), although by 1954, Schäperclaus indicated that this was an unlikely means of dissemination. Schäperclaus (1954), who detected spores in the posterior gut of kingfishers, proposed the potential for transmission by avian vectors. However, he emphasized that the most common route for transmission between hatcheries was the introduction of apparently healthy carriers. It was generally accepted that spores were released when fish die and decompose in ponds; however, Uspenskaya (1955, 1957) offered evidence for live fish releasing spores. Detecting parasite spores in other tissue and organs, she hypothesized that they had been released from the destroyed cartilage and made their way via the blood or lymphatic vessels to the liver. From the liver, spores could be passed through the bile ducts to the intestinal lumen and be expelled. Rydlo (1971) supported the observations of Uspenskaya, demonstrating the presence of spores in the intestinal tract and skin of infected rainbow trout by digesting these tissues separate from cartilage.

Although it was still generally accepted that fish become infected by ingesting M. *cerebralis* spores, only Uspenskaya (1982) describes establishing infection by feeding spores. Most evidence indicated it was likely that either an intermediate host and/or certain water conditions were neces-

sary for infections to occur. The widely accepted hypothesis was that spores are infective after ageing in the presence of mud for 3–6 months (Hoffman and Putz 1969; Putz 1969).

1971–1990

Distribution and Epizootiology

- Europe

  Austria—FAO 1972 (In Halliday 1976)

  Belgium—FAO 1972 (In Halliday 1976)

  Hungary—FAO 1972 (In Halliday 1976)

  England—1981 (Hoffman 1990)

  Ireland—1971 (southern Ireland) (Halliday 1974)

  Liechtenstein—1973 (Halliday 1976)

  Luxemburg—FAO 1972 (In Halliday 1976)

  Netherlands—FAO 1972 (In Halliday 1976)

  Spain—Cordero-del-Campillo et al. 1975

  Lebanon—1973 (Halliday 1976)

  New Zealand—1971 (Hewitt and Little 1972)

- USA

  New Hampshire—1980 (Hoffman 1990)

  New York—1983 (J. Schachte, New York State Department of Environmental Conservation, personal communication), 1984 (Hoffman 1990)

  Oregon—1986 (Holt et al. 1987; Lorz et al. 1989)

  Idaho—1987 (Hauck et al. 1988)

  Colorado—1987 (Barney et al. 1988; Walker and Nehring 1995)

  Wyoming—1988, (D. Money, Wyoming Game and Fish Department, personal communication)

New Mexico—1987 (D. Eib, New
Mexico Department of Game and
Fish, personal communication)

In 1976, Halliday reviewed the status of whirling
disease in Europe. He concluded that differences in
monitoring and reporting and inconsistencies in
the literature made it difficult to determine either
the original range of M. *cerebralis* or how rapidly
the parasite was disseminated. The presence of fish
rearing facilities on the same rivers where the parasite was detected in natural populations made it
difficult to ascertain if the infection was introduced
or enzootic. By the early 1970s, M. *cerebralis* had
been reported from most European countries,
although it is certain that it existed much earlier in
some areas.

In 1971, M. *cerebralis* was reported from New
Zealand (Hewitt and Little 1972). The report
noted that heavy mortality accompanied by violent
whirling motion, a condition known locally as
"whirly-gig" disease, had occurred periodically at
the affected hatchery since 1955 and perhaps as
early as 1952. No diagnostic tests had been performed during these outbreaks; however, when an
episode occurred in 1971, spores with morphologic
characteristics that corresponded with those of M.
cerebralis were isolated from the head region of
rainbow trout. Examination of preserved specimens
demonstrated that the parasite had been present at
least five years before it was identified (Boustead
1993). At the time of the detection in 1971, there
was a complete ban on importation of salmonid
fish in any form, unless heat-treated. However, if
the earlier episodes were a result of M. *cerebralis*
infection, then introduction may have occurred
prior to the enactment of these measures in 1952.
Other suspected routes of introduction were ova
and the packing material that they are shipped in,
and live dried food for tropical fish, which may
have included infected tubificids. Import of the latter product was not prohibited until 1971.

In the northeast United States, M. *cerebralis*
was detected in New Hampshire at a private hatchery in 1980. This hatchery remained positive until
rainbow trout were removed in 1985 (J. Hendee,
New Hampshire Fish and Game Department, personal communication). The hatchery closed in
1989 and no other detection has been reported in
that state. The first documented case in New York
was at a private hatchery in 1983 (J. Schachte, New
York State Department of Environmental Conser-

vation, personal communication). The outbreak
occurred when the small rainbow trout were moved
to a pond receiving effluent from an earthen pond
where adult fish were reared, evidence that the parasite had become established some time earlier.

The immediacy and degree of active control
measures taken varied among states. In Pennsylvania, where the parasite was first detected, infected
fish were stocked until the early 1970s and the parasite became established in public and private
hatcheries and in wild populations statewide. During the 1970s earthen ponds at many state facilities
were replaced with concrete raceways, and this
resulted in disappearance of clinical disease. Infected fish were also restricted from being stocked into
waters of concern; however, monitoring of wild
populations has been limited. Other states took
more active measures and required quarantine of
affected facilities, destruction of fish, and treatment of ponds and streams with chlorine and
quicklime. In New Hampshire, West Virginia and
Ohio, where introduction was limited to a few
facilities, these measures were reported to have
been effective and the parasite has not been detect-
ed/reported since the initial introductions.

In Nevada, the perennial presence of the
pathogen, and the disease, in the two main facilities
that were sources of fish for stocking in the state led
to widespread contamination of stocked waters.
Surveys undertaken statewide during the 1970s
indicated a wide distribution of the pathogen, but
no evidence of disease outside the hatcheries. During this time no extraordinary efforts were made to
rid the facilities of the pathogen.

In California, between 1965 and 1985 five
commercial growers, one state fish hatchery and
one federal hatchery had fish diagnosed with clinical whirling disease. In each case prior to 1984, the
infected stocks at these facilities were destroyed
(Horsch 1987; Modin 1998). In response to the
control efforts, intensive monitoring was conducted during the ensuing 12 years. Results of these surveys demonstrated that the parasite had been
unknowingly dispersed, likely with transfers of subclinically infected fish or infected processed products. Infected fish were detected from the San
Lorenzo River on the central California coast in
1973 and from the Licking Fork and Middle Fork of
the Mokelumne River on the West slope of the
Sierra Nevada in 1982 and 1983, respectively. In
1984, infected fish were detected from the Owens
River in the Eastern Sierra, and in 1985 from trib-

utaries of the Sacramento River, in Northeastern California. In each of these locations, infections were detected in fish both from culture facilities and wild fish. Infections in culture facilities ranged from subclinical to epizootic, and in the wild from subclinical to observance of occasional minor skeletal anomalies (Modin 1998).

Similar surveys were conducted in Oregon, where M. cerebralis was detected in a private trout hatchery in 1986. Infected resident trout and anadromous salmonids were found in the Lostine River, in northeastern Oregon, the source of the facility's water supply (Lorz et al. 1989). Subsequent monitoring demonstrated that the parasite was established in naturally reproducing resident and anadromous salmonid populations within the Grande Ronde and Imnaha River basins. These rivers are tributaries of the Snake River, and are part of the Columbia River basin. Fish from this hatchery were distributed to more than 75 private ponds throughout the state, and fish infected with M. cerebralis were found in at least 35 of these ponds. However, subsequent surveys of wild fish from watersheds that had received infected fish failed to detect a spread of infection outside of the infected ponds (R. Holt, Oregon Department of Fish and Wildlife, personal communication). In Idaho, M. cerebralis was detected in 1987 from juvenile chinook salmon Oncorhynchus tshawytscha at two state hatcheries on the Salmon and Pahsimeroi Rivers and from rainbow trout at a private hatchery on the Lost River (Hauck et al. 1988). Myxobolus cerebralis was confirmed in rainbow trout from a state hatchery on a tributary in the Snake River system the following year. The subsequent discovery of infection in returning 3- and 4-year old chinook salmon indicated that the parasite had likely been established in the Salmon and Pahsimeroi Rivers for some time.

Elsewhere in the west, M. cerebralis was detected in Colorado, Wyoming, and New Mexico between 1987 and 1988. In Colorado, the parasite was first detected in 1987 at four facilities (one state and three private) and fish at all four locations displayed clinical signs of the disease (Barney et al. 1988). Review of shipment records indicated that the parasite had been disseminated within the state as a result of transfers and planting of infected fish from the affected state and private hatcheries. In Wyoming, M. cerebralis was first detected in 1988 from rainbow and brown trout collected from the North and South Platte River drainages; these fish

had been stocked the preceding year by a commercial grower (D. Money, Wyoming Game and Fish Department, personal communication). Introduction into New Mexico probably occurred as a result of illegal importation of trout to a private facility. Since a single focus of infection was present, an extensive effort was made to eradicate the infected fish. Subsequent sampling of these waters revealed no signs of whirling disease and monitoring was discontinued. (D. Eib, New Mexico Department of Game and Fish, personal communication).

For many states, M. cerebralis was detected only after the parasite had been introduced and had become established in populations. In these cases, control measures came too late and the parasite had already become dispersed through transfers and stocking of subclinically infected fish.

Perspectives on Whirling Disease

Legislation to control M. cerebralis was adopted in Europe in 1970 when the International Office of Epizootics accepted a proposal to include whirling disease in the "International Zoo-Sanitary Code" along with four other diseases of fish (Ghittino 1970). However, the impetus for this legislation was the fear of U.S. legislation that could effectively close U.S. markets to any country that could not guarantee absence of the parasite in exported trout/eggs. By the 1970s, whirling disease appeared to be under control in Europe and attention was turned to other pathogens. However, it was recognized that since this parasite cannot be easily eradicated, relaxing the diligence of control standards could readily lead to epizootics (Schäperclaus 1986). "Accommodation," an approach used to minimize mortality and clinical signs in facilities where the parasite is enzootic, is widely practiced in Europe where trout are used for food, and stocking of waters with rainbow trout for anglers is not a major endeavor. Although technology is available to decontaminate water supplies, the excessive cost/benefit ratio of the equipment and operation was considered counterproductive in most situations.

In the United States, methods for certification of fish for importation and transfer were updated to include the pepsin-trypsin-dextrose centrifugation method and a modified plankton centrifuge method (O'Grodnick 1975; Markiw and Wolf 1974, 1980). Although additional restrictions were placed on importation of eggs, this was because of concerns with importation of the VHS virus rather than M. cerebralis, as O'Grodnick (1975) had demonstrated that M. *cerebralis* was not transmitted vertically. However, changes in the federal regulations did result in requirements for holding sentinel fish with brood fish for detection of M. *cerebralis*.

During the 1970s and into the mid-1980s, M. *cerebralis* was not reported from any "new" states. Then, in the late 1980s, M. *cerebralis* was detected from the western states of Idaho, Oregon, Colorado, and Wyoming. Based on the experience of states that had lived with the parasite for two decades, managers looked for an approach that was less extreme, with emphasis on avoiding spread of the pathogen while allowing for management in enzootic areas. Consequently, in 1988, the Colorado River Wildlife Council, Fish Disease Subcommittee recommended that M. *cerebralis* be reclassified from prohibited to notifiable status. This reclassification continued to require inspection but did not demand depopulation and disinfection of facilities. It was felt that this gave states the flexibility to regulate the disease. Some states would remain closed; others would have more complex regulations, allowing stocking of infected fish where the parasite was already established, or where it was unlikely to become established.

Dramatic changes also occurred in our knowledge of the parasite itself. Lom and Noble (1984) synonymized *Myxosoma* with *Myxobolus* because the separating characteristic, the glycogen (iodinophilous) vacuole, is not a reliable taxonomic characteristic. Thus, after 80 years, the parasite was once again recognized as *Myxobolus cerebralis*. In the same year, Wolf and Markiw (1984) announced the discovery that a species of *Triactinomyxon* and M. *cerebralis* are alternating life stages of a single life cycle, one stage undergoing development in an aquatic worm and the other in the fish. The stunning pronouncement that the aquatic oligochaete *Tubifex tubifex* was required to complete the parasite life cycle (Markiw and Wolf 1983) lent a sense of clarity to the success of control measures like replacement of earthen bottom ponds with concrete raceways. However, at the same time, evidence of experimental infection using spores aged in mud continued to be reported (Uspenskaya 1982) and the findings of Wolf and Markiw were met with reservation. It was not until the end of the 1980s that transmission of M. *cerebralis* was independently confirmed (El-Matbouli and Hoffmann 1989). This confirmation insured that our concepts of the life cycle and taxonomy of the Myxosporea would undergo radical change. It

also opened new avenues for research on detection techniques and control methods.

1991–PRESENT

Distribution and Epizootiology

- USA
  Washington—1996 (K. Amos, Washington Dept. of Fish and Wildlife, personal communication)
  Montana—1994 (Vincent 1996)
  Maryland—1995 (C. Geseker, Department of Natural Resources, personal communication)
  Utah—1991, (Wilson 1991)
  Arizona—2000 (Patricia Lopez, Arizona Game and Fish Department, personal communication)

In the eastern United States, M. *cerebralis* was detected for the first time in Maryland (1995) at a state facility that had received fish from several sources, and in a second facility that had been stocked with fish from the first facility. The parasite was detected again in New York at the Rome SFH (1994), and subsequently at two other hatcheries. It was determined that the 1994 infection originated at the Caledonia SFH, from which it was transferred to the other facilities. Collections of wild fish from natural waters demonstrated infected fish were present, although not common, in most watersheds in the state except for Long Island. Voluntary inspections of private hatcheries demonstrated that M. *cerebralis* was present at a number of facilities (P. Hulbert, New York State Department of Environmental Conservation, personal communication). *Myxobolus cerebralis*-infected fish were removed from the four public hatcheries that were positive in 1994 and these hatcheries remain negative with the exception of the Chateaugay SFH, which tested positive in 2000 after receiving a group of infected fish from the previously uninfected Randolph SFH (J. H. Schachte, personal communication).

Re-emergence of M. *cerebralis* was also reported in Connecticut from an earthen pond at the Kensington fish hatchery, which had been contaminated in 1959. The pond had been free of fish for more than 30 yrs and was restocked with M.

cerebralis-free trout in 1994 (J. H. Schachte, personal communication). In Virginia, the parasite was detected in the early 1990s at two state fish hatcheries and four private hatcheries and is considered established in that state (J. G. Hnath, personal communication).

In California, where the range of the parasite has been monitored since its detection in 1966, M. cerebralis was reported in wild rainbow and brown trout from the Santa Ana River in the southern part of the state in 1995. It is likely that infection had been present, but undetected for a number of years, as there are records of a local trout grower making frequent purchases from parasite-positive facilities (Modin 1998). Although M. cerebralis is now enzootic in a number of rivers in California, in some waters the prevalence has declined in wild fish following closure of facilities with positive populations. The parasite can no longer be detected in wild populations from Garrapata and Tularcitos Creeks and the Carmel River, or from the Coleman National Fish Hatchery (Modin 1998).

On the heels of the accumulating evidence from California and other states that wild trout populations can exist with M. cerebralis came the first observations of overt whirling disease in free-ranging populations from Colorado and Montana in the mid-1990s. In Colorado, where M. cerebralis was first detected in 1987, whirling disease was reported in the upper Colorado River, from an area suffering recruitment failure of wild rainbow trout in 1993, without a concomitant decrease in brown trout populations (Walker and Nehring 1995). At present, the parasite is considered enzootic in all coldwater drainages of Colorado except the North Republican River. Population level reductions among wild rainbow trout have been associated with M. cerebralis in the South Platte, Dolores, Roaring Fork, Fryingpan, Gunnison, Cache la Poudre and Rio Grande as well as the Colorado River. In most of these systems, brown trout populations have been unaffected, or have increased, resulting in a stable trout biomass. However, overt whirling disease has been reported even in this resistant species (Walker and Nehring 1995).

In 1994, Montana reported whirling disease in sections of the Madison River where a precipitous decline in estimated rainbow trout numbers had occurred beginning in 1991 (Vincent 1996). In contrast to Colorado, where heavy stocking of rainbow trout had occurred, the Madison River had not been stocked with hatchery fish since the late 1970s. The origin and route of dissemination of the parasite in Montana is unclear, although it had likely been present for some time prior to its detection. Infected fish have never been detected in state, federal or private hatcheries within the state, and many affected waters are managed for wild populations. Detection of the parasite, sometimes at a high prevalence, in nine drainages in the western part of the state (Baldwin et al. 1998) indicates that the parasite has been present for a long period.

Myxobolus cerebralis was detected in Utah in 1991 at a private facility on the Fremont River, within the Colorado River drainage (Wilson 1991). The parasite was subsequently detected in fish from a pond and from two additional private facilities in the adjacent Sevier River. In 1993, M. cerebralis was detected at three private hatcheries on the Bear River drainage; in 1994, infected rainbow trout and kokanee (land-locked form of sockeye salmon O. nerka) with significant skeletal deformities were reported in Porcupine Reservoir and in 1995 the parasite was demonstrated to have extended its range within the Little Bear and Ogden Rivers. Following an extensive survey in 1996, it became clear that although the parasite distribution had expanded within the drainages where it was originally identified, it was not widespread in the state. One state hatchery and at least twelve private hatcheries have tested positive for M. cerebralis at various times. Infection at the state hatchery was detected only using molecular techniques and the facility was subsequently depopulated (C. Wilson, Utah Division of Wildlife Resources, personal communication). Population surveys have demonstrated a decrease in rainbow trout populations from 1988 levels in the Beaver River, although the causal relationship between the decline and presence of the parasite has not been established (Eric Wagner, personal communication).

Surveys conducted in Wyoming following detection of the parasite in 1988 have documented M. cerebralis from wild fish in the following river drainages: Snake, Wind/Bighorn, Madison, Yellowstone, Green, South Platte, and North Platte. Fish with gross clinical signs of whirling disease have been observed, especially from the Salt River (Snake River drainage). Infected cutthroat trout gillnetted from Yellowstone Lake in 1998 also tested positive for M. cerebralis. Five of the 11 state facilities have had infected fish on site, some the result of transfer from other facilities (D. Money, personal communication).

In New Mexico, infected fish have recently been detected in the Pecos, Cebolla, San Juan, Cimarron, Red, and Canones Rivers, and in three of the state's seven hatcheries. Coldwater streams and reservoirs in the state were inadvertently stocked with infected fish, although it is not yet known to what extent M. cerebralis has become established. Surveys of the state's trout waters for M. cerebralis are currently in progress (D. Eib, personal communication). In 2000, Arizona reported that fish had been imported to a private pond from a hatchery in Colorado where M. cerebralis was subsequently detected. These suspect fish were stocked at five locations, some consisting of multiple ponds. Fish were recovered and tested from three of the five locations, and infection was confirmed at one site. However, it is too early to determine if the parasite will become established at any of these locations (P. Lopez, Arizona Game and Fish Department, personal communication).

In the Northwest, detection of M. cerebralis infections in returning adult salmon illustrates the increased opportunities for dissemination of the parasite as these fish migrate hundreds of miles or more to and from their natal streams. Myxobolus cerebralis was confirmed in adult steelhead and chinook salmon returning to two adult capture facilities on the Deschutes River, Oregon, a tributary of the mid-Columbia River. The first record of detection was 1987, and monitoring by Oregon Department of Fish and Wildlife between 1997 and 2000 resulted in detection of infections in stray hatchery steelhead (anadromous rainbow trout), chinook and sockeye salmon. However, several years of intense monitoring of resident fish in the Deschutes River have failed to detect M. cerebralis (H. M. Engelking, this volume), and there is currently no evidence that the parasite has become established in any tributaries of the lower Columbia River basin.

Perspectives on Whirling Disease

Following the trend in the late 1980s to modify the extreme management reactions to whirling disease, measures short of eradicating the parasite were recommended. These closely followed hatchery guidelines developed in Europe. Myxobolus cerebralis was removed from the OIE list of injurious pathogens in 1993 and its presence is no longer being used in regulating imports into the United States.

This de-emphasis on M. cerebralis immediately preceded the accumulating evidence of effects on wild populations in certain states. The events that occurred in Montana and Colorado precipitated a renewed interest in whirling disease with a shift in the focus from hatcheries to impacts on wild populations. These reports also resulted in the adoption of more aggressive control policies and increased funding for surveys and research in some states. Colorado initiated an aggressive program to clean up its contaminated state hatcheries and to stop, or severely limit, stocking of infected fish. However, with the establishment of the parasite in most drainages and in a large number of private aquaculture facilities, it may be some time before the effects of these measures are evident. Montana also adopted a plan to mitigate the effects of whirling disease. Because infection has not occurred in any state or private hatcheries, this plan is directed toward educating the public and learning how to manage rivers for reduction or ultimately elimination of the parasite. It is interesting that two states with such markedly different fish management programs should share the distinction of being the most severely affected. In contrast, in California, where the parasite has been established since 1966, there are decades of data and experience to support the view that the parasite has had little impact on wild populations (Modin 1998).

SUMMARY

Although breakthroughs on the life cycle of M. cerebralis have increased our understanding of parasite biology, they have made little difference in how the disease is controlled. It is worth noting that most control methods recommended today were in effect before tubificids were implicated in the life cycle. In reviewing these historical data it has become evident that many of our hypotheses of how M. cerebralis became so broadly distributed are based on the life cycle that was known to exist prior to the implication of the alternate host. We have presented some of these "pre-triactinomyxon" views here because they were the prevailing perspectives at those times. However, it is likely that some, more probable and now obvious, routes of dissemination were dismissed at the time these hypotheses were developed.

A great deal of confusion over the present distribution of M. cerebralis in the United States stems from 1) the lack of, and inconsistencies in, survey data, and 2) the fact that many reports are a result of sampling fish known to have been infected when released. The act of stocking infected fish or the detection of infected migratory fish in a water body does not necessarily imply the establishment of the parasite life cycle capable of infecting susceptible fry or fingerlings. Confusion has also resulted from the manner in which data have been presented. Because the presence of parasite infection does not necessarily mean that disease will occur, it is important to include any data on clinical signs when sampling fish. Since clinical signs of whirling disease are not pathognomonic, and may be subtle, they might not be noted except when the disease reaches epizootic levels. This could explain why there is no differentiation between parasite presence and clinical disease in most reports. Furthermore, the presence of the pathogen and even frank disease in wild populations would be unlikely to be noticed since dead fish would be rapidly scavenged and moribund fish could be subjected to enhanced predation. In addition, most surveys have presented the distribution of M. *cerebralis* in a cumulative manner, so that locations which have tested positive at even a single point in time remain listed as such, although subsequent testing has indicated that the parasite may not have become established or may be present below detectable levels. In this review we attempted to address these considerations and portray the status of the parasite as accurately as possible. However, we acknowledge that data are continually being collected and that distribution data are constantly in flux.

Strides are being made to develop risk assessment models to determine under what circumstances the introduction of a cohort of infected fish will result in establishment of enzootic or epizootic whirling disease in wild populations. In fact, in many instances where known planting of infected fish has occurred there has been no detection of M. *cerebralis* in wild fish during subsequent surveys. This difficulty was encountered by us when drawing up Table 1, in which the simple detection of the pathogen was reported as positive or negative, depending on whether the representative considered detection or establishment as the prime criterion for reporting a positive outcome. Thus, one of the primary conclusions (and recommendations) of this manuscript is the compiling and dissemination of accurate information about this onerous pathogen and the disease it causes. This is beginning to be recognized as an important endeavor and one necessary for proper management of wild fisheries.

ACKNOWLEDGEMENTS

The authors wish to thank all of the people who contributed information that was used in this paper. Especially, we would like to thank Pete Walker, who generously loaned his library of historical manuscripts from Glenn Hoffman, and Susan Higgins, who spent many hours collecting current data from state fishery managers. Funding that supported this effort was from the Whirling Disease Foundation, the National Partnership for the Management of Wild and Native Cold Water Fisheries, Portland General Electric and from the Western Regional Aquaculture Center of the USDA. This is Oregon Agricultural Station Technical paper 11837.

REFERENCES

Anonymous. 1972. Whirling disease in Columbia. FAO Aquaculture Bulletin 1:10.

Anonymous. 1974. Situación actual de la acuicultura continental in Chile. Simposio FAO/CARPAS Sobre acuicultra en America Latina, Montevideo, Uraguay, 26 de noviembre al 2 de diciembre de 1974. CARPAS/6/74/SC8, October 1974. FAO, Rome.

Baldwin, T. J., J. E. Peterson, G. C. McGhee, K. D. Staigmiller, E. S. Motteram, C. C. Downs, and D. R. Stanek. 1998. Distribution of *Myxobolus cerebralis* in salmonid fishes in Montana. Journal of Aquatic Animal Health 10:361–371.

Barney, P., D. E. Anderson and P. G. Walker. 1988. Whirling disease identified in Colorado. American Fisheries Society, Fish Health Section Newsletter 16:3.

Bergman, A. M. 1922. Fiskarnas Sjukdomar. Stockholm.

Bogdanova, E. A. 1960. Natural habitat of the myxosporidian (*Myxosoma cerebralis*, whirling disease) at Sakhalin (S.E. Russia). Doklady Akademii Nauk SSSR 134:1501–1503.

Bogdanova, E. A. 1964. Parasites and disease of fingerling European chum, and humpback salmon in fish rearing in the Kola Peninsula. Session of the Scientific Council on the problems "Theoretical basis of rational use of the resources of the White Sea and the inland reservoirs of Karelia" pp. 2. Petrozavdsk.

Bogdanova, E. A. 1966. New data on the distribution and biology of *Myxosoma cerebralis* (Hofer). In Diseases of Fish and Treatment. Alma-Ata Agapova, A. L. et al., Izdatelbstovo Nauk Kazakhskoi SSR.

Bogdanova, E. A. 1968. Modern data on the distribution and biology of *Myxosoma cerebralis* (Protozoa, Cnidosporidia) as agent of whirling disease of salmonids. Troisiéme Symposium de la Commission de l'Office International des Épizooties pour l'Étude des Maladies des Poissons, Stockholm.

Bogdanova, E. A. 1969. New data on the distribution of *Myxosoma cerebralis* and peculiarities of its ecology depending on biotic and abiotic factors. Progress in Protozoology 2:266.

Bogdanova, E. A. 1970. On the occurrence of whirling disease of salmonids in nature in the USSR. Journal of Parasitology 56:399.

Boustead, N. C., 1993. Detection and New Zealand distribution of *Myxobolus cerebralis*, the cause of whirling disease of salmonids. New Zealand Journal of Marine and Freshwater Research 27:431–436.

Bruhl, L. 1926. Bericht über die fischereitagung in Königsberg i. Pr. Von 2. Juli bis 1. August. Fischerei-Zeitung 29:813–815.

Christensen, N. O. 1972. Panel review on Myxosomiasis (Whirling disease in salmonid fishes). FI:EIFAC 72/SC II-Symposium 8:6.

Conroy, D. A. 1975. Informe sobre las enfermedades de la trucha arco iris (*Salmo gairdneri*) in Columbia, Projecto para desarrollo de la pesca continental. INDEREN-FAO, Publication No. 1, Bogota. Pages 1–37.

Conroy, D. A. 1981. The importance of fish diseases in relation to the development of salmonid culture in South America. Revista Italiana di Piscicotura e Ittiopatologia 16:57–64.

Cordero-del-Campillo, M. A., E. Diez, M. P. Alvarez-Pellitero and F. A. Rojo Vazquez. 1975. Torneo de la trucha (Myxosomosis). Revisión. Suplemento Cientifico del Boletin Informative Consejo General de Colegio Veterinarios de España, Madrid.

Diaz-Ungria, C. 1970. Parasitología de los animals domésticos en Venezuela, Volume I. Universidad del Zulia Consejo de Desarrollo Científico y Humanístico, Maracaibo 779–783.

Dogel, V. A. 1932. Bor'ba s boleznyami ryb v prudovom khozyaistve.

Dyk, V. 1954. Nemoci y. Hasich ryb. Praha, Czechoslovakia.

Egusa, S. 1992. Infectious Diseases of Fish. Amerind Publishing Company, New Delhi. NOAA Publication 0002.

El-Matbouli, M., and R. Hoffmann. 1989. Experimental transmission of two *Myxobolus* spp. developing bisporogeny via tubificid worms. Parasitology Research. 75:461–464.

Elson, K. G. R. 1969. Whirling disease in trout. Nature 223:968.

Espinoza, V., and C. Diaz-Ungria. 1970. Presencia eo la enfermedad del 'torneo' en peces de Venequela. Proyecto MAC-FAP. Facultad de Ciencias Veterinarias de la Universidad del Zulia, Maracaibo.

FAO. 1972. International measures for the control of major communicable fish diseases. EIFAC.

Ghittino, P. 1970. Present status of whirling disease in Italian trout farms. Revista Italiana di Piscicotura e Ittiopatologia 5:89–92.

Halliday, M. M. 1974. Studies on *Myxosoma cerebralis*, a parasite of salmonids III. Some studies on the epidemiology of *Myxosoma cerebralis* in Denmark, Scotland and Ireland. Nordisk Veterinaermedicin 26:165–172.

Halliday, M. M. 1976. The biology of *Myxosoma cerebralis*: the causative organism of whirling disease of salmonids. Journal of Fish Biology 9:339–357.

Hauck, A. K., S. Landin and S. Wavra. 1988. Whirling disease in Idaho. American Fisheries Society, Fish Health Section Newsletter 16:5.

Hastein, T. 1971. The occurrence of whirling disease (Myxosomiasis) in Norway. Acta Veterinaria Scandinavica 12:297–299

Hewitt, G. C., and R. W. Little. 1972. Whirling disease in New Zealand trout caused by *Myxosoma cerebralis* (Hofer, 1903) (Protozoa:Myxosporidia). New Zealand Journal of Marine and Freshwater Research 6:1–10.

Hnath, J. G. 1970. Whirling disease in the state of Michigan. Journal of Parasitology 56:149–150.

Hnath, J. G. 1996. Whirling disease in the Midwest (Michigan, Pennsylvania, Ohio, West Virginia, Virginia, Maryland). Pages 18–22 *in* E. P. Bergerson and B. A. Knopf, editors. Whirling Disease Workshop Proceedings. Colorado Cooperative Fish and Wild-life Research Unit, Denver, Colorado.

Holt, R. A., C. R. Banner and T. O. Kreps. 1987. Whirling disease in Oregon. American Fisheries Society, Fish Health Section Newsletter 15:1.

Höfer, B. 1903. Ueber die Drehkrankheit der Regenbogenforelle. Allgemeine Fischerei Zeitschrift 28:7–8.

Höfer, B. 1904. Handbuch der Fishkrankheiten. B. Heller. Münich.

Hoffmann, G. L. 1962. Whirling disease of trout. United States Department of the Interior. Fishery Leaflet 508. Fish and Wildlife Service.

Hoffmann, G. L. 1968. Current status of whirling disease in salmonids in U. S. American Fishing U. S. Trout News 10 (Nov.–Dec.).

Hoffmann, G. L. 1970. Intercontinental and transcontinental dissemination and transfaunation of fish parasites with emphasis on whirling disease (*Myxosoma cerebralis*) and its effects on fish. Pages 69–81 *in* S. F. Snieszko, editor. Symposium on diseases of fisheries and shellfishes. American Fisheries Society, Special Publication No. 5, Bethesda, Maryland.

Hoffmann, G. L. 1990. *Myxobolus cerebralis*, a worldwide cause of salmonid whirling disease. Journal of Aquatic Animal Health 2:30–37.

Hoffmann, G. L., and R. E. Putz. 1969. Host susceptibility and the effect of aging, freezing, heat, and chemicals on spores of *Myxosoma cerebralis*. The Progressive Fish Culturist 31:35–37.

Horsch, C. M. 1987. A case history of whirling disease in a drainage system; Battle Creek drainage of the upper Sacramento River basin, California, USA. Journal of Fish Diseases 10:453–460.

Johansson, N. 1966. First identification of *Myxosoma cerebralis* in Sweden. Swedish Salmon Research Institute. Report LFI.

Kocylowski, B. 1953. Choroby nyb i ich Knalkanie. Gespodarka ryb 5:24–26.

Kudo, R. 1933. A taxonomic consideration of Myxosporidia. Transactions of the American Microscopical Society 52:195–216.

Lom, J., and E. R. Noble. 1984. Revised classification of the myxosporea Bütschli, 1881. Folia Parasitologica, Prague 31:193–205.

Lorz, H. V., A. Amandi, C. R. Banner and J. S. Rohovec. 1989. Detection of Myxobolus (Myxosoma) cerebralis in salmonid fisheries in Oregon. Journal of Aquatic Animal Health 1: 217–221.

Margaritov, N. M. 1960. Whirling disease of trout in DRS-Samokov. Rubno stopanstvo. N2 Sophia.

Margolis, L., T. E. McDonald and G. E. Hoskins. 1981. Absence of the protozoan Myxosoma cerebralis (Myxozoa: Myxosporea), the cause of whirling disease, in a survey of salmonids from British Columbia. Canadian Journal of Fisheries and Aquatic Sciences 38:996–998.

Margolis, M. L., M. L. Kent, and P. Bustos. 1996. Diseases of salmonids resembling myxosporean whirling disease, and the absence of Myxosoma cerebralis, in South America. Diseases of Aquatic Organisms 25:33–37.

Markiw, M. E., and K. Wolf. 1974. Myxosoma cerebralis: Isolation and concentration from fish skeletal elements: sequential enzymatic digestions and purification by differential centrifugation. Journal of the Fisheries Research Board of Canada 31:15–50.

Markiw, M. E., and K. Wolf. 1980. Myxosoma cerebralis: Trypsinization of plankton centrifuge harvests increases optical clarity and spore concentration. Canadian Journal of Fisheries and Aquatic Sciences 37:2225–2227.

Markiw, M. E., and K. Wolf. 1983. Myxosoma cerebralis (Myxozoa: Myxosporea) etiologic agent of salmonid whirling disease requires a tubificid worm (Annelida: Oligochaeta) in its life cycle. Journal of Protozoology 30:561–564.

Martinez, C. R. 1966. Contribución al estudio de las enfermedades de las truchas en Venezuela. Bulletin of the Office of International Epizooties 65:1135–1162.

Meyers, T. U. 1969. Whirling disease. FAO Aquaculture Bulletin 2:13.

Meschkat, A. 1971. Whirling disease in Ecuador. FAO Aquaculture Bulletin 3:11.

Modin, J. 1998. Whirling disease in California: A review of its history, distribution, and impacts, 1965–1997. Journal of Aquatic Animal Health 10:132–142.

O'Grodnick, J. J. 1975. Egg transmission of whirling disease. The Progressive Fish Culturist 37:153–154.

Plehn, M. 1905. Uber die Drehkrankheit der salmoniden [(Lentospora cerebralis) (Höfer) Plehn]. Archiv Protistenkunde 5:145–166.

Plehn, M. 1924. Praktikum der Fishkrankheiten. Stuttgart, West Germany.

Preudhomme, J. G. 1970. Whirling disease of trout in Morocco. FAO Aquaculture Bulletin 2:14.

Putz, R. E. 1969. Experimental transmission of Myxosoma cerebralis (whirling disease) and effect of freezing on the spores. Progress in Sport Fisheries Research 1970:55–57.

Rasmussen, C. J. 1967. Handbog i Orredopdraet. Rhodos, Copenhagen.

Roberts, R. J., and K. G. R. Elson. 1970. An outbreak of whirling disease in rainbow trout. Veterinary Record 86:258–259.

Rydlo, M. 1971. Evidence of Myxosoma cerebralis spores in various organs of rainbow trout afflicted with whirling disease. Österreichs Fischerei 24:97–99.

Schäperclaus, W. 1931. XXI Die Drehkrankheit in der Forrellenzucht und ihre Bekämpfung. Zeitschrift Fischerei 29:521–567.

Schäperclaus, W. 1954. Fischkrankheiten. Akademie-Verlag, Berlin. Pages 379–386.

Schäperclaus, W. 1986. Fischkrankheiten, Volume 2. Akademie-Verlag, Berlin. Pages 665–676.

Scolari, C. 1954. Sull'impiego dello stovarsolo nelle profilassi del 'capostan' o 'lentosporiasi' delle trote d'allevamento. La Clinica Veterinaria 77: 50–53.

Shul'man, S.S. 1966. Myxosporida of the USSR. Nauka Publishers, Moscow.

Taylor, R. E L., S. J. Coli and D. R. Junell. 1973. Attempts to control whirling disease by continuous drug feeding. Journal of Wildlife Diseases 9:302–305.

Tidd, W. M., and R. A. Tubb. 1970. Investigations of whirling disease in Ohio. Journal of Parasitology 56:632.

Tinkina, B. 1962. Trout hatchery in N. Korea. Rybovodstvo I Rybolovstvo 5:55–57.

Tomasec, I. 1960. Lutte contre les principles maladies infectieuses des poissons. Bulletin of the Office International des Epizootics 54:55–57.

Uspenskaya, A. V. 1955. Biology, distribution and economic importance of Myxosoma cerebralis, the causative agent of whirling disease of trout. Lectures of the Academy of Science USSR 105:1132–1135.

Uspenskaya, A. V. 1957. The ecology and spreading of the pathogen of trout whirling disease—Myxosoma cerebralis (Hofer, 1903, Plehn, 1905) in the fish ponds of the Soviet Union. Pages 47–55 in G. K. Petrushevskii, editor. Parasites and diseases of fish. Bulletin All-Union Institute of Freshwater Fisheries, Leningrad.

Uspenskaya, A. V. 1982. New data on the life cycle and biology of Myxosporidia. Archiv fur Protistenkunde 126:309–338.

Vanco, F. 1952. Contribution à l'étude de la pathologie des alevins de truits. Foulon, Paris.

van Wyk, G. F. 1968. Jonkershoek Hatchery, Division of Inland Fisheries, Department of Nature Conservation, Province of Good Hope, Republic of South Africa. Annual Report 24.

Vincent, E. R. 1996. Whirling disease and wild trout: The Montana experience. Fisheries 21(6):32–33.

Walker, P. G., and R. B. Nehring. 1995. An investigation to determine the cause(s) of the disappearance of young wild rainbow trout in the upper Colorado River, in Middle Park, Colorado. Colorado Division of Wildlife, Denver, Colorado.

Wilson, J. C. 1991. Whirling disease comes to Utah. Ichthyogram 2(3):1–2.

Wolf, K., and M. E. Markiw. 1984. Biology contravenes taxonomy in the Myxozoa: new discoveries show alternation of the invertebrate and vertebrate hosts. Science 225:1449–1452.

Yasutake, W. T., and H. Wolf. 1970. Occurrence of whirling disease of trout in the western United States. Second International Congress of Parasitology. Abstract #691. September 6–12. Washington, DC. Journal of Parasitology 56:375–376.

Yoder, W. G. 1972. The spread of Myxosoma cerebralis into natural trout populations in Michigan. The Progressive Fish-Culturist 43:103–106.

American Fisheries Society Symposium 29:25–31, 2002

Potential for Introduction of *Myxobolus cerebralis* into the Deschutes River Watershed in Central Oregon from Adult Anadromous Salmonids

H. Mark Engelking

Oregon Department of Fish and Wildlife, Fish Pathology Section
Oregon State University, Center for Salmon Disease Research
Nash Hall 220, Department of Microbiology
Corvallis, Oregon 97331-3804, USA

ABSTRACT. Anadromous fish were excluded above Pelton Round Butte Hydroelectric Project (PRB Project), located midway (RM 100) on the Deschutes River in central Oregon, beginning in 1968. Reintroduction of these fish above the PRB Project is proposed to meet conservation concerns that arise from lack of natural production and separation of populations. One consideration, when moving fish groups that have been isolated one from the other for thirty years, is that of disease. The health of the fish populations above Round Butte Dam could be seriously jeopardized by the introduction of whirling disease. Straying hatchery steelhead trout *Oncorhynchus mykiss* were detected with *Myxobolus cerebralis* spores, in 1987, at Warm Springs National Fish Hatchery, below the PRB Project. *Myxobolus cerebralis* is established in tributaries of the upper Columbia River basin and of the Snake River basin, where some of these straying hatchery and wild steelhead trout may have originated. From 1997 to 2000, fish from the Deschutes River basin have been sampled for the presence of *M. cerebralis*. The parasite has been found in both straying hatchery and unmarked adult chinook salmon *O. tshawytscha* and steelhead trout. Presently there is no evidence of infection of resident fish or in returning adult fish originating from Round Butte Hatchery, although the potential for establishment of *M. cerebralis* in the Deschutes River watershed cannot be ruled out.

The Deschutes River watershed covers an area of some 10,500 square miles, in the high desert region of north-central Oregon (Figure 1). The Deschutes River is about 200 mi in length and flows from south to north, where it joins the Columbia River. The Metolius and Crooked rivers are the two major tributaries to the Deschutes River, above the Pelton Round Butte Hydroelectric Project (PRB Project). The construction of the dams at river miles 110 (Round Butte Dam), 100 (Pelton Dam), and 97 (Re-regulating Dam) blocked migration of anadromous salmonids to the upper Deschutes River and its tributaries (Figure 1). The three dams of the PRB Project have been a complete barrier to fish migration for more than thirty years. Round Butte Dam impounds Lake Billy Chinook, a reservoir with a surface area of about 4000 acres. The lake has more than 60 mi of shoreline; at the widest, it is 3,000 ft across and has a maximum depth of 400 ft near the dam. Lake Simtustus, formed by Pelton Dam, is about 7 mi down river from Round Butte Dam. This reservoir has 540 acres of surface water, 18 mi of shoreline, and a maximum depth of 165 ft. A pool that varies in size daily is formed behind the Re-regulating Dam. Initial attempts to move fish around these barriers failed (Ratliff and Schulz 1999). A hatchery program was accepted for mitigation to replace the lost natural reproduction. Round Butte Hatchery, which began operation in 1973, was constructed immediately below Round Butte Dam and is supplied with water from Lake Billy Chinook. Hatchery summer steelhead trout *Oncorhynchus mykiss* and spring chinook salmon *O. tshawytscha* programs have achieved adult return mitigation goals, although recently the spring chinook salmon returns have fallen short of the number of adults required (Ratliff and Schulz 1999).

Although the hatchery program has met arbitrary mitigation numbers of returning adult salmonids, biological concerns have been raised, by the National Marine Fisheries Service, the U.S. Fish and Wildlife Service, other management agencies, and public interest groups, about potential genetic problems that arise from artificial propagation and lack of natural production.

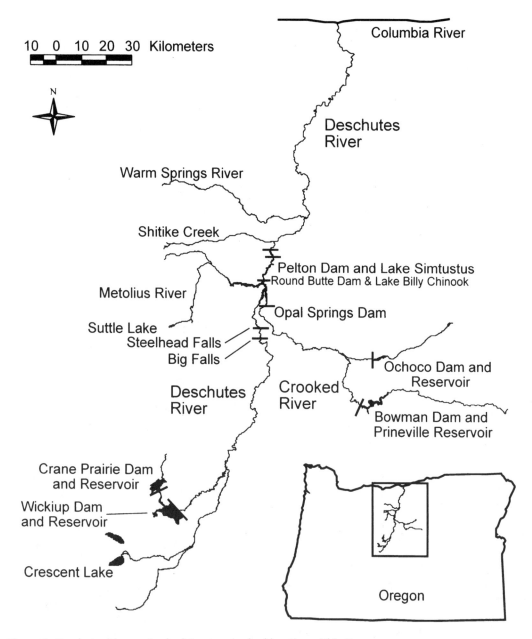

Figure 1. Deschutes River watershed, inset watershed location within Oregon.

Physical connection of the metapopulation of the bull trout *Salvelinus confluentus*, to allow gene pool mixing, is an objective to meet conservation goals (Fies et al. 1996; Stuart et al. 1996). To address these ecosystem integrity concerns, a proposal has been made to move adult anadromous fish above the dams, to spawn and provide a means for their progeny to outmigrate through

the PRB Project (Ratliff and Schulz 1999). Species involved in the proposal include sockeye *Oncorhynchus nerka* and chinook salmon, as well as steelhead *O. mykiss* and bull trout, all of which were once native in the area above the PRB hydroelectric projects (Nehlsen 1995).

Since the anadromous fish and bull trout below the PRB Project have been isolated from the upper Deschutes River watershed, several biological considerations must be addressed before any fish are moved. The survival of introduced fish species and resident populations in the area above Round Butte Dam could be seriously impacted by fish pathogens not presently found above there or from the addition and amplification of fish pathogens already present. New strains of pathogens or exotic disease organisms could reduce resident fish populations above the project. Natural production by the reintroduced anadromous fish may be unsuccessful if the juvenile fish experience excessive losses from disease. Habitat, with degraded environments, that are found above the PRB Project may favor establishment of *Myxobolus cerebralis*, by providing locations for the parasite to complete its life cycle (Engelking 2000).

Little information existed, concerning endemic fish pathogens or the presence of intermediate hosts, that could allow M. *cerebralis* to become permanently established above the PRB Project (Engelking 2000). The establishment of the parasite in new areas is dependent on (1) spores being introduced into the area, (2) the presence of *Tubifex tubifex*, the alternate host, and (3) completion of the life cycle in fish. Two factors favor the spread of whirling disease. The spore stage released from the fish is extremely resistant to inactivation from environmental extremes. Secondly, the tubificid host worm appears to be ubiquitous and likely found in all waters containing susceptible fish (Bartholomew 1998a, 1998b).

This report describes a survey for M. *cerebralis* in salmonids in the Deschutes River. If the reintroduction proposal is approved, fluctuations and changes in fish populations and the distribution of M. *cerebralis* and other fish pathogens may be more clearly understood and evaluated with this background information.

METHODS

Parasite Exams for *Myxobolus cerebralis*

All methods for the detection and identification of fish pathogens were based on standard procedures from the AFS Fish Health Blue Book (Thoesen 1994) and those of the Oregon Department of Fish and Wildlife, Fish Pathology Section laboratories. Heads from each juvenile, or small adult, fish were cut in half on the midline sagittal plane. Each large adult fish was sampled using the core method, and the core was divided into two equal parts (Lorz et al. 1989). Half of each head or core was preserved in 10% neutral buffered formalin, for histological confirmation of spores. The other half was stored frozen and later analyzed by the enzyme digest method, for the presence of myxosporean spores (Thoesen 1994; Markiw and Wolf 1974). Most samples were processed individually. Samples from some groups taken later in the study were processed with three to five fish per pool. Microscopic observation of spores that appeared morphologically similar (size 8–10 μm in diameter, round shape and polar capsules extending no more than half the length of the spore) to those of M. *cerebralis* were considered to be presumptively infected with the parasite. Two separate microscopic examinations were conducted by two fish health specialists, on each enzyme digest sample, to verify the presence or absence of these specific spores. If a sample was considered to be presumptively positive, then the portion of the sample preserved in neutral formalin was subjected to histological analysis. The slide preparations were stained with Giemsa, to enhance the appearance of internal head structures. The microscopic observation of lesions and spores, in the bone and cartilage from the head tissues, by a fish health specialist was considered to be confirmation of infection with the M. *cerebralis* (Thoesen 1994).

Fish Sampling Methods and Identification

Fish were obtained by electrofishing, by screw traps, or during stream surveys. Most anadromous fish were obtained at the Pelton Fish Trap, immediately below the Pelton Round Butte Re-regulating Dam. These fish were sampled immediately or during production spawning procedures. Knowing the origin of anadromous adult fish entering the watershed was crucial to determining if M. *cerebralis* had infected fish in the watershed. Round Butte Hatchery steelhead trout were identified by specific fin markings, in addition to removed adipose fins. Stray hatchery adult steelhead trout were identified as those with removed fins, indicating they were not of Round Butte Hatchery origin. Stray hatchery adult sockeye salmon were identified as those with fin markings because no hatchery-reared sockeye salmon are released into the Deschutes River. Round Butte Hatchery and stray hatchery chinook salmon were identified by coded wire tag data. All fish sampled for this study were given individual identification numbers and included in a relational database program (FileMaker Pro, Santa Clara, California).

RESULTS

Spores of M. *cerebralis* were detected in enzyme digest samples from anadromous adult spring chinook salmon and summer steelhead trout found below the PRB Project, but not from fish up river (Table 1). Some of these observations were confirmed by the presence of lesions and spores in cartilage from the head during histological analysis. However no Round Butte Hatchery or clearly identifiable Deschutes River fish have been confirmed to be infected by M. *cerebralis*. *Myxobolus cerebralis* enzyme digestions have been completed for more than 3,800 fish sampled throughout the Deschutes River watershed (Figure 1; Tables 1 and 2).

The stock most consistently found with M. *cerebralis* spores has been the stray hatchery steelhead trout (Table 1). In the last four years almost twenty-one percent of these fish had detectable spores, and at least five percent were histologically confirmed infections. Similarly, stray hatchery chinook salmon have been detected with M. *cerebralis* spores, with an overall infection rate of ten percent. Some of these stray hatchery chinook salmon have come from the upper Columbia (Dworshak U. S. Fish and Wildlife Service National Fish Hatchery) or lower Snake (Rapid River and McCall Idaho Fish and Game Fish hatcheries) rivers as determined by coded wire tag information. Spores, similar to those of M. *cerebralis*, have been detected in the unmarked (presumed wild) populations of steelhead trout and chinook salmon, as well. The infec-

tion rates in these populations have ranged from about five to ten percent. The origin of the unmarked M. *cerebralis* positive fish is uncertain because of the high rate of straying in the Deschutes River watershed. These unmarked fish may be wild Deschutes River fish or wild stray fish or unmarked stray hatchery fish. A single stray hatchery sockeye salmon has been confirmed to be infected with M. *cerebralis* (Table 1). Of those fish with M. *cerebralis* spores, as determined by morphology, many also had other myxosporean spores. In these presumptively positive samples, 58 were found to have other *Myxobolus* species, 10 with *Henneguya* species, and 2 with *Myxobolus insidiosus*-like spores.

The kokanee, rainbow trout, brown trout *Salmo trutta*, brook trout *Salvelinus fontinalis*, bull trout, and mountain whitefish *Prosopium williamsoni*, tested from both above and below the PRB Project, had no detectable M. *cerebralis* spores. A total of 1,469 fish above the PRB Project have been tested, and 450 resident fish from below have been examined (Table 2). No spores have been detected in wild summer steelhead smolts from the lower river. Also, no hatchery chinook or steelhead smolts have been detected with spores prior to release from the rearing ponds of Round Butte Hatchery. Other myxosporean spores have been noted in the samples, such as unidentified *Myxobolus* species, with morphology and size differences from those of M. *cerebralis* (364 fish), *Myxobolus insidiosus*-like spores (53 fish), and *Henneguya* (543 fish), among all

Table 1. Fish found to be infected with *Myxobolus cerebralis* in the lower Deschutes River basin.

Fish species stock	Years	Number of fish tested	Number of fish presumptive M. cerebralis infected	Number of fish confirmed M. cerebralis infected
Summer steelhead trout				
Unmarked[a]	1998–2000	48	5	1
Hatchery stray	1997–2000	416	87	21[b]
Spring chinook salmon				
Unmarked	1997–2000	114	6	1
Hatchery stray	1998–2000	60	6	0
Sockeye salmon				
Hatchery stray	1997–2000	4	1	1
Totals		642	105	24

[a]Unmarked fish are of unknown origin, they could be wild, stray wild, or unmarked stray hatchery fish.
[b]Most but not all the presumptive identifications of stray steelhead trout were analyzed by histology. This number therefore under represents the number of possible confirmed infections.

Table 2. Fish sampled that had no evidence of *M. cerebralis* spores from 1997 through 2000 in the Deschutes River watershed.

Species stock	Number of fish	Sampling location relative to PRB project
Spring chinook salmon		
Round Butte Hatchery stock	612	Below
Summer steelheadtrout		
Round Butte Hatchery stock	622	Below
Sockeye salmon		
Unmarked	61	Below
Mountain whitefish		
Wild Lower Deschutes River	121	Below
Rainbow trout		
Wild Lower Deschutes River	306	Below
Bull trout		
Lake Simtustus	23	Below
Total fish sampled below	1,745	
Rainbow trout		
Crooked, Metolius, Upper Deschutes rivers	611	Above
Mountain whitefish		
Crooked, Upper Deschutes rivers	186	Above
Kokanee		
Lake Billy chinook, Metolius, Upper Deschutes rivers	619	Above
Brown trout		
Metolius River	14	Above
Bull trout		
Lake Billy chinook, Metolius River	39	Above
Total fish sampled above	1,469	
Total fish sampled	3,214	

groups of fish. More than 96% of the samples have no evidence of M. *cerebralis* spores.

DISCUSSION

There is no direct evidence of M. *cerebralis* infections of fish originating from the lower Deschutes River (Table 2), although there has been at least fourteen years of exposure from stray and unmarked infected adult anadromous fish (Leek 1987). No lower river wild juveniles have been found to be infected. No smolts reared at Round Butte Hatchery have had evidence of M. *cerebralis*, prior to release. The infected unmarked chinook salmon and steelhead trout, identified in these surveys, may suggest that there is some level of infection in the lower Deschutes River. However, the high rate of adult anadromous fish straying into the Deschutes River watershed also suggests that these

infected unmarked fish are not from the Deschutes River and have come from other locations. Almost 70% of hatchery steelhead trout and up to 40% of hatchery chinook salmon entering the Deschutes River, in recent years, originate from other locations (Oregon Department of Fish and Wildlife 1996). *Myxobolus cerebralis* is established in the Grande Ronde and Imnaha rivers in Oregon, in the lower Snake River basin, in the upper Columbia River basin, and in certain Idaho streams, where some of the stray adult anadromous fish may originate. Few signs of disease and no population effects on resident fish have been noted during limited surveys in northeastern Oregon, where the parasite is found (Lorz et al. 1989; Holt 1996).

Differences in habitat are apparent above and below the PRB Project. In the lower Deschutes River basin, tributaries on the east side have seasonally varying flows, low gradients, little riparian cover, and widely varying temperatures. Most of the spawning anadromous fish are rapidly removed by predators or washed from spawning areas. This may reduce exposure of tubificid worms to M. cerebralis spores because the infected fish do not remain in the system long enough to release spores where they might infect T. tubifex. On the western side of the Deschutes River, the tributaries are less impacted, although logging and diversions for agricultural uses continue (Oregon Department of Fish and Wildlife 1996). In the upper Deschutes River basin, the Metolius River is characterized as pristine, with tributaries of low water temperatures, constant flows from spring or snow melt origins, and higher gradients. The upper Deschutes River, between Lake Billy Chinook (RM 119) and Steelhead Falls (RM 124), which is a natural barrier to anadromous fish, has relatively constant flows of cooler water temperatures in deep canyons and is fed mainly by springs. The gradients are moderate. Squaw Creek, a tributary of the upper Deschutes below Steelhead Falls, is heavily impacted by human uses in the upper reaches. Water flows are often seriously reduced by diversions for agricultural and other uses (Fies et al. 1996). The Crooked River, especially the middle river, is degraded and impacted by grazing and other agricultural uses, blocked by diversions, has widely varying temperatures, is very warm in the summer, and has a very low gradient (Dambacher and Burke 1997; Stuart et al. 1996). The alternate host for the parasite is found within all these habitats. Samples of bottom sediments, from sites above and directly below the PRB project, revealed that the oligochaete worm *Tubifex tubifex* is present in the PRB Project area (Bartholomew 1998a, 1998b). The presence of *T. tubifex* in degraded habitats may allow M. cerebralis to establish itself in some of these areas, were fish reintroduction to proceed. Specifically, the poor environments and higher water temperatures of the middle Crooked River and the upper reaches of Squaw Creek may be areas in which the parasite could establish itself because they support *Tubifex tubifex* populations. Fish could become infected in these habitats, especially if spawning occurs there. These variations may lead to local areas in which the fish populations could be reduced, from infection by the parasite. Likewise, in northeastern Oregon, in the endemic areas of M. cerebralis, there are differences among the various habitats and in the number of infected fish (Holt 1996). This suggests

habitat can affect the level of infection and possible outcomes to the fish. The upper Deschutes River watershed has differences that may be conducive to the establishment of M. cerebralis.

The presence of M. cerebralis spores in species of anadromous fish, intended for reintroduction above the PRB project, and the detection of *T. tubifex*, throughout the watershed, are cause for concern. The state agency responsible for fish health in Oregon, Oregon Department of Fish and Wildlife (ODFW) must follow fish health guidelines that do not permit the transfer of M. cerebralis infected fish to an area that is not known to have the parasite. Excluding the stray adult anadromous fish from movement above the PRB Project will be necessary to protect the upper watershed from exposure to M. cerebralis. More than 20% of the hatchery stray summer steelhead trout and about 10% of the hatchery stray spring chinook salmon carry spores of the parasite. However, methods to exclude these fish may interfere with volitional passage that is designed in the fish mitigation proposals to be submitted by Portland General Electric Company and the Confederated Tribes of the Warm Springs to the Federal Energy Regulatory Commission. This proposal, to transfer anadromous fishes above the dams, presents a dilemma. The restoration of Endangered Species Act (ESA) listed anadromous fish is considered to be dependent, in part, on conservation efforts, including the return of these fish to historic spawning areas. The potential benefits of this action must be weighed against the risks to native resident stocks, which also include ESA listed bull trout populations, within the Deschutes River watershed.

Important questions remain to be resolved. There is no direct evidence that M. cerebralis has become established in the lower Deschutes River, although there has been more than fourteen years of exposure, based on the first detection of the parasite in stray hatchery steelhead trout (Leek 1987; Holt 1996). Have the environmental factors, such as length of time of exposure and number of spores released, slowed the establishment of the parasite? If the parasite is established, why are there no signs of disease present in the fish? Has the parasite become established at a low level, and will it increase with time? Is the strain of *Tubifex* worms, found above the PRB Project, capable of supporting the production of infectious triactinomyxons? How does the number of fish transferred alter the risk of introduction and establishment of the parasite? Could M. *cerebralis* be effectively transmitted above the PRB Project by means other than moving infected fish? All these questions will need to be addressed in order to determine the feasibility of reintroduction of adult anadromous fish above the PRB Project, from a fish health perspective.

ACKNOWLEDGMENTS

This research has been funded by Portland General Electric Company, Portland, Oregon.

Robin L. Whitmore performed the pepsin trypsin digests and the initial microscopic screening. Mark Redhead took the head core samples and prepared most of the samples for fixation. The samples and fish were obtained through the cooperation and efforts of many dedicated biologists working in the Deschutes River watershed. Among these biologists are Amy Stuart, Steve Marx, Brett Hodgson, Jim Newton (retired), Steve Pribyl, Tom Nelson, Bill Nyara, former hatchery manager, and Jack Palmer, hatchery manager, and the Round Butte Hatchery crew, all of the Oregon Department of Fish and Wildlife, Mike Riehle and Brad Houslet of the U.S. Forest Service, Sisters, Oregon and Don Ratliff, Eric Schulz, Scott Lewis biologists, and the crew of Portland General Electric Co.

REFERENCES

Bartholomew, J. 1998a. Fish disease risk assessment: whirling disease and Ceratomyxosis. 1997 Annual Report. Portland General Electric Co., Portland, Oregon.

Bartholomew, J. 1998b. Fish disease risk assessment: whirling disease and Ceratomyxosis. 1998. Activity Report. Portland General Electric Co., Portland, Oregon.

Dambacher, J., and J. Burke. 1997. Stream surveys of the Pelton Round Butte drainages. Progress report of the Oregon Department of Fish and Wildlife, 1997. Portland, Oregon.

Engelking, H. M. 2000. Fish disease study associated with potential anadromous fish passage at the Pelton Round Butte Hydroelectric Project. 1999 Annual Report. Portland General Electric Co., Portland, Oregon.

Fies, T., M. Manion, B. Lewis, and S. Marx. 1996. Metolius River Subbasin fish management plan. Upper Deschutes Fish District. Oregon Department of Fish and Wildlife.

Holt, R. A. 1996. Distribution of M. *cerebralis* in Wild And Hatchery Fish in the Northwest (California, Nevada, Oregon, Idaho, and Washington). Pages 23–31 *in* Bergersen and Knopf, editors. Proceedings of the Whirling Disease Workshop. Where do we go from here? Whirling Disease Foundation, Bozeman, Colorado.

Leek, S. 1987. Monthly activity report for the month of June 1987. U.S. Fish and Wildlife Lower Columbia Fish Health Center, Underwood, Washington.

Lorz, H. V., A. Amandi, C. R. Banner, and J. S. Rohovec. 1989. Detection *Myxobolus (Myxosoma) cerebralis* in salmonid fishes in Oregon. Journal of Aquatic Animal Health 1:217–221.

Markiw, M. E., and K. Wolf. 1974. *Myxosoma cerebralis*: isolation and concentration from fish skeletal elements-sequential enzymatic digestions and purification by differential centrifugation. Journal of the Fisheries Research Board of Canada 31:1597–1600.

Nehlsen, W. 1995. Historical salmon and steelhead runs of the upper Deschutes River and their environments. Portland General Electric Co., Portland, Oregon.

Oregon Department of Fish and Wildlife. 1996. Lower Deschutes Fish Management Plan. 8 sections. Oregon Department of Fish and Wildlife, Portland, Oregon.

Ratliff, D. E., and E. E. Schulz. 1999. Fisheries Program at the Pelton Round Butte Hydroelectric Project (Oregon) 1956–1995. Portland General Electric Co., Portland, Oregon.

Stuart, A., Thiesfeld, S. L., Nelson, T. K., and T. M. Shrader. 1996. Crooked River Basin Plan. Ochoco Fish District, Oregon Department of Fish and Wildlife, Portland, Oregon.

Thoesen, John C., editor. 1994. Suggested procedures for the detection and identification of certain finfish and shellfish pathogens. 4th Edition, Version 1. Fish Health Section, American Fisheries Society.

American Fisheries Society Symposium 29:33–41, 2002
© 2002 by the American Fisheries Society

Evaluation of Risk of High Elevation Colorado Waters to the Establishment of *Myxobolus cerebralis*

GEORGE J. SCHISLER*

Colorado Division of Wildlife, 317 West Prospect Street, Fort Collins, Colorado 80526, USA

ERIC P. BERGERSEN

Colorado Cooperative Fish and Wildlife Research Unit, 203 Wagar Building
Colorado State University, Fort Collins, Colorado 80523, USA

ABSTRACT. During 1992 and 1996, fish were inadvertently stocked into 226 waters in Colorado from hatcheries that were later identified as *Myxobolus cerebralis*-positive. Seventy-two high-elevation waters, previously classified as M. *cerebralis*-negative, were sampled to determine if stocking of fish from those hatcheries contributed to the spread of M. *cerebralis* in these locations. Pepsin-trypsin digest (PTD) and polymerase chain reaction (PCR) tests were used to test for M. *cerebralis* in 1,743 fish. A total of 190 fish and 23 separate waters were identified as M. *cerebralis*-positive, with PTD. PCR identified 410 fish and 42 waters as positive for the parasite. Logistic regression and Akaiki information criterion model selection was used to identify parameters contributing to the M. *cerebralis* establishment in these high elevation waters. Within-drainage distance to the nearest known M. *cerebralis*-positive water and relative abundance of *Tubifex tubifex* habitat were found to be more important contributors to the establishment of M. *cerebralis* than the accidental stocking events.

Myxobolus cerebralis, the parasite responsible for salmonid whirling disease, was first identified in the United States in the 1950s and has been detected in at least 22 states (Bartholomew and Reno, this volume). The parasite can be potentially spread by several mechanisms. Infected fish may swim upstream or downstream from a site of initial exposure, carrying the parasite throughout a given drainage. The waterborne triactinomyxon spore (the stage infective to fish) can be carried downstream from infected areas with water currents. Mature myxospores can be passed through the digestive systems of fish-eating birds, such as great blue heron (Meyers et al. 1970), kingfishers (Schaperclause 1954), black crested night heron, and mallard ducks (Taylor and Lott 1978), facilitating transport of the organism into previously unexposed habitats. Past evaluations suggest that the most common route of transmission of M. *cerebralis* to new locations is transport of infected fish (Meyers et al. 1970). In Colorado, accidental stocking of M. *cerebralis*-suspect fish raised the concern that many high-elevation waters, previously free of the organism, may have become contaminated with the parasite. A total of 226 high-elevation waters were stocked by Roaring Judy State Fish Fatchery in 1992 and the Pitkin and Durango State Fish Hatcheries in 1996, when the facilities were considered to be free of the parasite. Most of these high-elevation stockings were aerial plants of fingerling trout. Testing of the hatcheries by the pepsin-trypsin digest method (PTD; Markiw and Wolf 1975), during annual inspections later during those years, resulted in identification of M. *cerebralis*. Additional fish were inadvertently stocked from these hatcheries the following year into some of the same or additional high-elevation habitats because they were not removed from the stocking schedules for those hatcheries. Actual M. *cerebralis* status and infection severity of the fish stocked into these high elevation waters is unknown, but it is likely that the fish were exposed to the parasite before stocking. The objective of this study was to determine the extent of the spread of M. *cerebralis* into high elevation waters presumed to be free of the parasite, identify the contribution of spread due to accidental stocking events, and determine the threat of the parasite to high-elevation trout populations in Colorado.

*Corresponding author: (970) 472-4412; george.schisler@state.co.us.

METHODS

During 1998 and 1999, resident rainbow trout *Oncorhynchus mykiss*, brown trout *Salmo trutta*, brook trout *Salvelinus fontinalis*, and cutthroat trout *Oncorhynchus clarki* were sampled from 72 waters classified as *M. cerebralis*-negative before the accidental stocking events. Sampling sites included, both, waters that had been stocked from the suspect hatcheries and those that had not been stocked from suspect hatcheries. This was done to distinguish between stocking- and nonstocking-related transfer of *M. cerebralis* to these high-elevation habitats. Waters with a wide range of elevations, 2,103–3,840 m (6,900–12,600 ft), and habitat types were sampled with gill nets, hook and line, and electrofishing. An attempt was made to collect a representative 60-fish sample from each body of water to detect *M. cerebralis* at a higher than 5% prevalence, with 95% confidence intervals (Ossiander and Wedemeyer 1973). In many cases, very few fish or only native cutthroat trout were found, so smaller samples were taken to preserve the existing populations. Sampling was focused on fish that were obviously a result of natural reproduction, to identify establishment of the parasite rather than presence of the organism in stocked fish. Wild brook, brown, and cutthroat trout, as well as younger age classes of fish, were preferentially sampled. Fish submitted for testing were at least 1 year old, to increase the likelihood that if mature myxospores were present, they would be detected using the PTD method. Parallel single-round polymerase chain reaction (PCR) and PTD tests (Schisler et al. 2001) were conducted on 1,743 fish, during the study. Myxospore burden was estimated for each fish (half-head) tested with PTD. PCR band-strengths were rated as one of five categories: negative (0), weak positive (1), positive (2), strong positive (3), and very strong positive (4). Prevalence and infection intensity were evaluated for fish in each of the waters tested. The waters were then categorized as *M. cerebralis*-positive or negative, based on the testing results.

Elevation, stocking history, presence of *Tubifex tubifex* habitat, and proximity to known infected waters were recorded for each of the waters tested. Multiple fish species were often taken from the same body of water, so waters were not classified by species. One lake was sampled in both 1998 and 1999, so data were combined for that particular lake. Samples from two other waters were incomplete, or fish were too young for valid PTD testing and were not included in the data set. The end result was 69 waters used in the analysis. The waters sampled in the study were classified into three different categories, based on stocking history. Waters never stocked with fish from suspect or positive facilities were classified as "Stock 0" waters. Waters stocked with fish from hatcheries the year before or the year of finding infected fish at the facility (suspect fish) were classified as "Stock 1" waters. Waters stocked with fish from known-positive hatcheries (year after identification of infected fish at the facility) were classified as "Stock 2" waters.

T. tubifex can be found in pristine environments and are fairly ubiquitous throughout freshwater salmonid habitat (Granath and Gilbert 2001). However, the likelihood of finding abundant *T. tubifex* is much greater in locations with fine sediments and eutrophic conditions. Sauter and Güde (1996) found that *T. tubifex* tend to prefer silt-clay substrate. Lazim and Learner (1987) found that most tubificids occurred in areas of high organic content and silt-clay sized particles. *T. tubifex* may occur in very high densities in cases of extreme organic enrichment or habitat degradation, when many other invertebrate species are eliminated (Aston 1973; Brinkhurst 1965; Lestochova 1994). Studies in Colorado have shown that eutrophic, silty locations favorable to *T. tubifex* can act as point sources of *M. cerebralis* infection (Allen 1999; Thompson and Nehring 2000). Waters were categorized based on the presence of obvious *T. tubifex* habitat. Very oligotrophic locations with cobble-boulder or bedrock substrates and minor amounts of available *T. tubifex* habitat were classified as "Tubifex 0." Waters with an intermediate level of sedimentation, gravel-cobble substrate, and a moderate amount of organic material deposition were classified as "Tubifex 1." Locations with silt-sand-clay substrate and eutrophic conditions, such as beaver ponds, were classified as "Tubifex 2." Classification into these types of habitats was not based on an absolute measure of total sedimentation, eutrophication, or total *T. tubifex* abundance. However, they are distinct enough for any layperson to identify the differences in these habitat types and rate them on a scale of 0–2.

Proximity to known *M. cerebralis* waters was evaluated by relying on past pathogen testing, conducted by the Colorado Division of Wildlife Aquatic Animal Health Laboratory in Brush, Colorado. Both straight-line distances and within-drainage distances (km) to closest known M. cerebralis-positive sites were estimated using measurements from Delorme 1:160,000 maps. Straight-line distance was simply the most direct distance to the nearest location that had previously tested positive for M. cerebralis. Within-drainage distance was the distance, following the streambed, to the nearest upstream or downstream site that had previously tested positive for M. cerebralis. Many sites in Colorado have not been tested for M. cerebralis, so the parasite is likely established in many more locations than the current data reflects, and distances recorded are based only on known positive sites. Maximum distance recorded was limited to 40.2 km (25 mi), due to the presence of some waters in dead-end drainages or in areas where limited testing had been done, which would have otherwise resulted in recording of some waters as having infinite within-drainage distances to the next known positive site.

A correlation matrix produced by PROC CORR, a procedure in SAS system software, was used to determine if significant collinearity existed between factors. Elevation was correlated with our measure of T. tubifex habitat (R^2 = 0.2628, P < 0.0001). Eutrophication and sedimentation leading to abundant T. tubifex habitat can occur at any elevation given the correct circumstances. We felt that relative amount of T. tubifex habitat was more likely to have a direct effect on establishment of M. cerebralis than elevation, and because these factors were correlated, we removed elevation as a factor in the modeling procedure. Straight-line distance was correlated with within-drainage distance to other M. cerebralis positive sites (R^2 = 0.4578, P < 0.0001). Because both of these parameters represent essentially the same factor and were correlated, we removed straight-line distance as a factor.

Apparent effects from individual factors may be confounded due to interactions, and considering each factor alone in highly complex natural systems can be misleading.

Therefore, logistic regression analysis (PROC GENMOD) in SAS system software was used, to test all factors simultaneously and to determine which (if any) of the parameters recorded contributed to M. cerebralis status of the waters tested. The major factors recorded for each water and included in the analysis as independent variables were stocking history, T. tubifex habitat, within drainage distance to the nearest known positive

water, and all interaction terms. Stocking was used as a categorical variable, with each of the three different stocking classifications as a different category. T. tubifex habitat was treated as a categorical variable in one set of models and as a continuous variable in another set of models. T. tubifex habitat was used as a categorical variable in some models to reflect the three distinct habitat types, as they were classified during sampling as minimal, moderate, or abundant habitat. T. tubifex habitat was used in other models as a continuous variable to describe a range of habitats from low to high T. tubifex habitat abundance. Treatment of the variable in this manner is useful because all waters may not always fit perfectly into one of the three categories described.

For all of the models, the sampling unit used was individual waters, and the dependent variable was the presence or absence of M. cerebralis. All of the different combinations of variables resulted in 36 different models to be evaluated. Second order Akaiki information criterion (AICc) (Burnham and Anderson 1998) was used as the model selection procedure to identify the model that best described the data without over-parameterization. AICc is much like AIC (Lebreton et al. 1992), except a small-sample bias adjustment term is added to the formula and is described as follows:

$$AICc = -2\log(L(\hat{\theta})) + 2K\left(\frac{N}{N-K-1}\right)$$

where
N = number of samples, and
K = number of parameters.

AICc values were calculated for each of the alternative logistic regression models generated. The models were then ranked based on these values, with small AICc values representing the best models and large AICc values representing poor models of the information in the data. The AICc values were then rescaled as follows to give the model with the minimum AICc a value of 0 and to allow easily comparable, rankable models:

$$\Delta_i = AICc_i - \min AICc$$

Akaiki weights (Burnham and Anderson 2001) were then calculated for each of the models, to provide "weight of evidence" for the strongest models when compared with other models evaluated. The Akaiki weight for each model is interpreted as approximately the probability that the model

is the best in the set evaluated. The weights are calculated as follows:

$$w = \frac{\exp(-\Delta_i / 2)}{\sum_{r=1}^{R} \exp(-\Delta_r / 2)}$$

This information was used to identify the best
models for describing the data from the 69 high-
elevation waters sampled. The logistic regression
and subsequent model selection allowed us to not
only identify the best models but also to quantify
relative importance and effects of the factors used
in the models.

RESULTS

Raw data suggested lower mean spore burdens, and
PCR band strengths occurred at higher elevation
waters. Mean myxospore burdens (half-heads)
increased from 1,386 per fish, at elevations greater
than 3,500 m, to 3,302 per fish, at elevations of
2,501–3,500 m. At elevations below 2,500 m,
mean myxospore burden increased to 5,727 per
fish. Corresponding mean PCR band strengths had
a similar pattern of stronger strength with lower
elevation. Prevalence of infected fish with increas-

ing elevation was fairly pronounced (Table 1).
There were some obvious exceptions, including
waters above 3,500 m that still produced high proportions of infected fish. Some low-elevation
waters also produced low proportions of infected
fish. This indicates that other factors influenced
infection prevalence besides simply elevation.

The raw data suggested that stocking history
had an effect on M. cerebralis status of the waters
tested (Table 2). A higher percentage of "Stock 2"
waters were identified as M. cerebralis-positive than
"Stock 1" or "Stock 0" waters. Only 25.0% of
"Stock 0" and only 25.7% of Stock 1 waters were
identified as M. cerebralis-positive, with PTD. PCR
testing identified 66.7% of "Stock 0" and 51.4% of
"Stock 1" waters as M. cerebralis-positive. Lower-
elevation waters adjacent to known infected waters
made up the bulk of the M. cerebralis-positive
"Stock 0" waters. By contrast, 80.0% of the "Stock
2" waters were identified as M. cerebralis-positive,
by both PCR and PTD.

Tubifex tubifex habitat was generally reduced at
higher elevations, due to the lower sediment load
and organic content of most of the higher-elevation
waters. Presence of obvious T. tubifex habitat influ-

Table 1. Mean PTD myxospore counts, mean PCR band strength ratings (half-head), and infection
prevalence for fish sampled at elevations ranging from 2,103–3,840 m (6,900–12,600 ft).

Elevation	Test	Fish	Infected (%)	Mean	St. Dev	Range
< 2,500	PTD	221	25.8	5,727	17,528	0–175,000
	PCR	221	62.0	1.71	1.57	0–4.00
2,501–3,500	PTD	965	11.4	3,302	19,538	0–291,644
	PCR	965	23.8	0.56	1.14	0–4.00
> 3,500	PTD	557	4.1	1,386	14,072	0–278,133
	PCR	557	7.7	0.17	0.67	0–4.00

Table 2. Infection prevalence of fish and percent of waters identified as M. cerebralis positive by PTD and
PCR from waters classified by stocking history. "Stock 0" waters were never stocked with fish from suspect
or known-positive facilities. "Stock 1" waters were stocked with fish from hatcheries the year before or the
year of finding infected fish at the facility (suspect fish). "Stock 2" waters were stocked with fish after
confirmation of M. cerebralis at the facility.

Stocking	Fish			Waters		
	N	%PTD+	%PCR +	N	%PTD+	%PCR+
Stock 0	751	8.3	17.6	24	25.0	66.7
Stock 1	676	5.6	14.8	35	25.7	51.4
Stock 2	316	28.5	56.3	10	80.0	80.0

enced M. *cerebralis* status of the waters tested (Table 3). PTD testing resulted in classifying 9.1%, 23.8%, and 61.5% of the populations in "Tubifex 0," "Tubifex 1," and "Tubifex 2" waters as infected with M. *cerebralis*. PCR testing identified a higher proportion of the populations as infected, but the pattern remained the same, with 45.5%, 52.4%, and 80.4% of the populations in "Tubifex 0," "Tubifex 1," and "Tubifex 2" waters identified as infected.

Average within-drainage distance to the nearest infected site was 11.2 km (N = 23; SD = 14.3) for waters found to be M. *cerebralis*-positive with PTD testing, and 16.7 km (N = 42; SD = 14.7) with PCR testing. Average within-drainage dis-

tances to the nearest infected sites were 24.9 km (N = 46; SD = 13.1) for waters found to be M. *cerebralis* negative with PTD testing, and 26.1 km (N = 27; SD = 13.6) for PCR testing.

Model Selection

The PCR data were best described using a model including T. *tubifex* habitat as a continuous variable and within-drainage distance (T_1, D). The second-choice model included T. *tubifex* habitat as a categorical variable and within-drainage distance (T_2, D). Other models with interaction terms or fewer parameters had larger AICc values and low Akaiki weights (Table 4). The PTD data were best

Table 3. Infection prevalence of fish and percent of waters identified as M. cerebralis positive by PTD and PCR for three categories of T. tubifex habitat.

Tubifex habitat	Fish			Waters		
	N	%PTD+	%PCR+	N	%PTD+	%PCR+
Minimal habitat	553	4.0	7.5	22	9.1	45.5
Moderate habitat	543	4.4	15.7	21	23.8	52.4
Abundant habitat	647	22.3	43.7	26	61.5	80.4

Table 4. Best-fitting alternative models for predicting presence of *M. cerebralis* in high elevation Colorado waters with PCR and PTD testing. N = 69 for each model.

Model	Parameters	PCR testing −2log(L(0))	AICc	Δ_i	Akaiki weights[a]
T_1, D	3	80.5109	86.8801	0.0000	0.1904
T_2, D	4	78.4234	87.0484	0.1683	0.1750
T_1, D, S, T_1*S	7	72.3160	88.1521	1.2720	0.1077
T_1, D, T_1*D	4	80.3440	88.9690	2.0889	0.0669
T_1, D, S	5	78.5539	89.5063	2.6262	0.0512

Model	Parameters	PTD testing −2log(L(0))	AICc	Δ_i	Akaiki weights[a]
T_1, D	3	60.3051	66.6743	0.0000	0.2561
T_1, D, S	5	56.6441	67.5965	0.9222	0.1615
T_2, D	4	59.2178	67.8428	1.1685	0.1428
T_1, D, T_1*D	4	59.9436	68.5686	1.8943	0.0993
T_2, D, S	6	56.3812	69.7360	3.0617	0.0554
T_1, D, S,T_1*D	6	56.4847	69.8395	3.1652	0.0526

[a]Akaiki weights for all other models were < 0.05.
T_1 = T. tubifex habitat as a continuous variable.
T_1 = T. tubifex habitat as a categorical variable.
D = Within-drainage distance to nearest infected water.
S = Stocking history as a categorical variable.
* = Interaction term.

fit with a model containing *T. tubifex* habitat as a continuous variable and within-drainage distance (T_1, D). A model with both of these variables and stocking history (T_1, D, S) was the second-best model fit. As with the PCR data, other models with interaction terms or fewer parameters had larger AICc values and low Akaiki weights.

These results indicate that a model containing within-drainage distance and *T. tubifex* habitat as a continuous variable (T_1, D) was the best overall fit for both the PTD and PCR data. Parameter estimates for models that included stocking history indicate that a within-drainage distance and *T. tubifex* habitat were the most important factors (Table 5), with stocking having a lesser influence (i.e., T_1, D, S). The best fitting model (T_1, D) pro-

vides similar results for both the PCR and PTD data, except the likelihood of finding M. *cerebralis* is greater with the PCR data (Figures 1 and 2). This is to be expected, considering the greater sensitivity of the PCR test. Increasing distance to the nearest known-infected site in the models results in lower probability of establishment of the parasite, while an increase in *T. tubifex* habitat results in greater probability of establishment.

Although the model T_1, D, S was the second-choice model for PTD testing, as identified by the AICc and Akaiki weight results, it provides some insight into the effects of stocking on M. *cerebralis* establishment in the waters in this study. Waters in the model not stocked with infected or suspect fish have the lowest overall probability of establish-

Table 5. Parameter estimates and significance levels for the two best-fitting alternative models for predicting presence of *M. cerebralis* in high elevation Colorado waters as tested by PCR and PTD.

| | | | PCR results | |
| Model = T_1, D | | | | |
Parameter		df	Estimate	Standard error
Intercept		1	0.6689	0.6061
T. tubifex		1	0.7100	0.3286
Distance		1	−0.0432	0.0195
Model = T_2, D				
Parameter		df	Estimate	Standard error
Intercept		1	2.5891	0.7806
T. tubifex	0	1	−1.4971	0.6897
T. tubifex	1	1	−1.6064	0.7804
T. tubifex	2	0	0.0000	0.0000
Distance		1	−0.0505	0.0212
			PTD results	
Model = T_1, D				
Parameter		df	Estimate	Standard error
Intercept		1	−1.2710	0.7678
T. tubifex		1	1.4944	0.4739
Distance		1	−0.0685	0.0228
Model = T_1, D, S				
Parameter		df	Estimate	Standard error
Intercept		1	5.5415	1.2434
T. tubifex		1	1.2610	0.4848
Distance		1	−0.0711	0.0245
Stocking	0	1	−1.8409	1.0676
Stocking	1	1	−1.6867	1.0089
Stocking	2	0	0.0000	0.0000

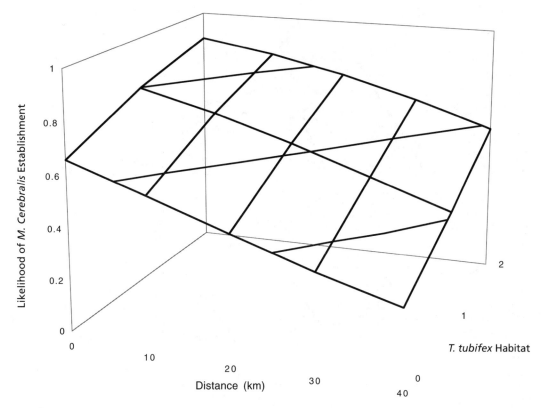

Figure 1. PCR-model results (T₁, D) for effects of *T. tubifex* habitat and distance to nearest known-positive, within-drainage site on *M. cerebralis* establishment.

ment of the parasite, followed by waters stocked with suspect fish and waters stocked from hatcheries after they were known to be contaminated with *M. cerebralis*. However, the best-fitting models indicated that the accidental stocking events had less of an effect than the other two factors.

DISCUSSION

The raw data indicated that stocking history, elevation, presence of *T. tubifex* habitat, and distance to known positive waters all contributed in some way to the *M. cerebralis* status of the waters tested. The best-fitting models identified presence of *T. tubifex* habitat and within-drainage distance to known infected waters as the most important variables in the establishment of *M. cerebralis* in high elevation waters. The importance of stocking history was low, compared with the other factors in this study. This was partially due to the small number of "Stocked 2" waters in the data set, which were the most heavily influenced by the stocking. The contribution to *M. cerebralis* establishment was not strongly influenced by the accidental

stocking that occurred in the "Stocked 1" waters. The high-elevation waters in this study were stocked only one or two times with infected or suspect fish, and lower-elevation waters with long histories of stocking of infected fish are more likely to be affected by stocking history.

Available *T. tubifex* habitat has long been known to be a contributor to the intensity and prevalence of *M. cerebralis* infection. This was the case in this study as well, with low prevalence of infection and low proportion of infected waters where little *T. tubifex* habitat was available. Infection levels may have been too low to detect in some locations with little or no obvious *T. tubifex* habitat, which would give the false impression that establishment had not occurred. However, the end result is still a reduced risk from the parasite in these locations.

The effect of distance to known positive waters on *M. cerebralis* status of the waters tested in this study suggests that transport by anglers, migrating fish, or other wildlife may be a contributing factor in the spread of the pathogen in

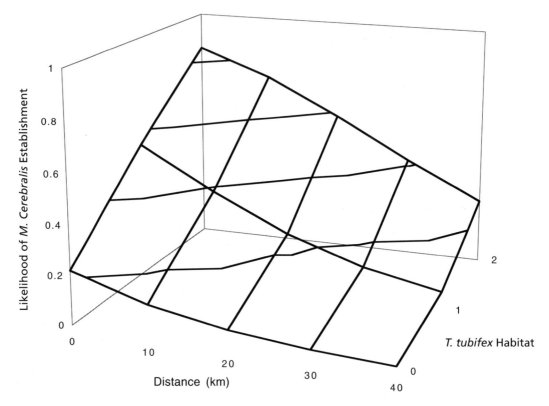

Figure 2. PTD-model results (T_1, D) for effects of *T. tubifex* habitat and distance to nearest known-positive, within-drainage site on *M. cerebralis* establishment.

Colorado. It is intuitive that waters in close proximity to other known-positive waters have a greater probability of becoming contaminated with the parasite by these means.

The models selected provide some guidelines for identifying waters at the highest risk for establishment of *M. cerebralis* and could be used to set priorities for testing or regulations. Other parameters not examined in this study may very well have an effect on presence or absence of *M. cerebralis*, and the results of this study may be an over-simplification due to exclusion of these unknown variables. The possibility exists that, with time, the range of *M. cerebralis* will expand into even the most remote locations in Colorado.

The results indicate that the accidental stocking events, occurring the year of identifying the hatcheries as *M. cerebralis*-positive, did not contribute substantially to the spread of parasite in the high-elevation waters. However, *M. cerebralis* was found in a high percentage of the few waters that were stocked the year after the hatcheries were identified as positive. Continued stocking would surely contribute more to the likelihood of *M. cerebralis* establishment. Colorado Division of Wildlife regulations will prohibit stocking of *M. cerebralis*-exposed fish into salmonid habitat by 2003. Many trout populations in Colorado are somewhat protected from *M. cerebralis*, due to their high elevations, long distances to other infected waters, lack of *T. tubifex* habitat, and preclusion of further stocking of infected fish into high-elevation waters. Populations with abundant *T. tubifex* habitat, and close proximity to known *M. cerebralis*-contaminated waters, are at high risk for establishment of *M. cerebralis* and should be closely monitored.

REFERENCES

Allen, M. B. 1999. Factors influencing the distribution of *Myxobolus cerebralis*, the causative agent of whirling disease, in the Cache la Poudre River, Colorado. Master of Science Thesis. Department of Fishery and Wildlife Biology, Colorado State University, Fort Collins, Colorado.

Aston, R. J. 1973. Field and experimental studies on the effects of a power station effluent on Tubificidae (Oligochaeta, Annelida). Hydrobiologica 42:225–242.

Brinkhurst, R. O. 1965. Observations on the recovery of a British river from gross organic pollution. Hydrobiologica 25:9–51.

Burnham, K. P., and D. R. Anderson. 1998. Model selection and inference: a practical- information-theoretic approach. Springer-Verlag Inc., New York.

Burnham, K. P., and D. R. Anderson. 2001. Kullback-Leibler information as a basis for strong inference in ecological studies. Wildlife Research 28:111–119.

Granath, Jr., W. O., and M. A. Gilbert. 2001. The role of *Tubifex tubifex* in the transmission of *Myxobolus cerebralis*. 7th Annual Whirling Disease Symposium: *a decade of discovery*. 8–9 February 2001, Salt Lake City, Utah.

Lazim, M. N., and M. A. Learner. 1987. The influence of sediment composition and leaf litter on the distribution of tubificid worms (Oligochaeta). Field and laboratory study. Oecologia 72:131–136.

Lebreton, J. D., K. P. Burnham, J. Clobert, and D. R. Anderson. 1992. Modeling survival and testing biological hypothesis using marked animals: case studies and recent advances. Ecological Monographs 62:67–118.

Lestochova, E. I. 1994. Influence of small river conditions on the abundance of Tubificidae. Hydrobiologica 278:129–131.

Markiw, M. E., and K. Wolf. 1975. *Myxosoma cerebralis*: isolation and centrifugation from skeletal elements-sequential enzymatic digestions and purification by differential centrifugation. Journal of the Fisheries Research Board of Canada 31:15–20.

Meyers, T. U., J. Scala, and E. Simmons. 1970. Modes of transmission of whirling disease of trout. Nature (London) 227:622–623.

Ossiander, F. J., and G. Wedemeyer. 1973. Computer program for sample sizes required to determine disease incidence in fish populations. Journal of the Fisheries Research Board of Canada 30:1383–1384.

Sauter, G., and H. Güde. 1996. Influence of grain size on the distribution of tubificid oligochaete species. Hydrobiologica 334:97–101.

Schaperclause, W. 1954. *Fischkrankheiten*. Akademic-Verlag, Berlin.

Schisler, G. J., E. P. Bergersen, P. G. Walker, J. Wood, and J. K. Epp. 2001. Comparison of single-round polymerase chain reaction (PCR) and pepsin-trypsin digest (PTD) methods for detection of *Myxobolus cerebralis*. Diseases of Aquatic Organisms 45:109–114.

Taylor, R. L., and M. Lott. 1978. Transmission of salmonid whirling disease by birds fed trout infected with *Myxosoma cerebralis*. Journal of Protozoology 25:105–106.

Thompson, K. G., and R. B. Nehring. 2000. A simple technique used to filter and quantify the actinospore of *Myxobolus cerebralis* and determine its seasonal abundance in the Colorado River. Journal of Aquatic Animal Health 12:316–323.

Section 2

Parasite Research

American Fisheries Society Symposium 29:45–53, 2002
© 2002 by the American Fisheries Society

Recent Advances with Taxonomy, Life Cycle, and Development of *Myxobolus cerebralis* in the Fish and Oligochaete Hosts

RONALD P. HEDRICK*
Department of Medicine and Epidemiology
School of Veterinary Medicine, University of California
Davis, California, 95616 USA

MANSOUR EL-MATBOULI
Institute of Zoology, Fish Biology and Fish Diseases, Faculty of Veterinary Medicine
Kaulbachstr. 37, 80539, University of Munich, Munich, Germany

ABSTRACT. *Myxobolus cerebralis* possesses unique phenotypic and genotypic characteristics when compared with other histozoic parasites from the phylum Myxozoa. The parasite infects the cartilage and thereby induces a serious and potentially lethal disease in salmonid fish. Comparisons of the small subunit ribosomal DNA (ssu rDNA) sequences of M. *cerebralis* to other myxozoans demonstrate that the parasite has evolved separately from other *Myxobolus* spp. that may appear in cartilage or nervous tissues of the fish host. *Myxobolus cerebralis* has a complex life cycle involving two hosts and numerous developmental stages that may divide by mitosis, endogeny, or plasmotomy, and, at one stage, by meiosis. In the salmonid host, the parasite undergoes extensive migration from initial sites of attachment to the epidermis, through the nervous system, to reach cartilage, the site where sporogenesis occurs. During this migration, parasite numbers may increase by replication. Sporogenesis is initiated by autogamy, a process typical of pansporoblastic myxosporean development that involves the union of the one cell (pericyte) with another (sporogonic). Following this union, the sporogonic cell will give rise to all subsequent cells that differentiate into the lenticular shaped spore with a diameter of approximately 10 μm. This spore or myxospore is an environmentally resistant stage characterized by two hardened valves surrounding two polar capsules with coiled filaments and a binucleate sporoplasm cell. In the fish, these spores are found only in cartilage where they reside until released from fish that die or are consumed by other fish or fish-eating animals (e.g., birds). Spores reaching the aquatic sediments can be ingested and hatch in susceptible oligochaete hosts. The released sporoplasm invades and then resides between cells of the intestinal mucosa. In contrast to the parasite in the fish host, the parasite in the oligochaete undergoes the entire developmental cycle in this location. This developmental cycle involves merogony, gametogamy or the formation of haploid gametes, and sporogony. The actinosporean spores, formed at the culmination of this development, are released into the lumen of the intestine, prior to discharging into the aquatic environment. The mechanisms underlying the complex development of M. *cerebralis*, and its interactions with both hosts, are poorly understood. Recent advances, however, are providing insights into the factors that mediate certain phases of the infection. In this review, we consider known and recently obtained information on the taxonomy, development, and life cycle of the parasite.

Myxobolus cerebralis, the parasite causing whirling disease in salmonid fish (Hofer 1903), is the most intensively studied member of the phylum Myxozoa (Hedrick et al. 1998). Recent occurrences of the disease in wild trout in North America have stimulated unprecedented efforts on behalf of federal, state, and nonprofit organizations to support the research needed to combat whirling disease. This research reveals

a complexity of interactions between the parasite, the environment, and the fish and oligochaete hosts that make controlling whirling disease a formidable challenge. In this manuscript, we examine key features of M. *cerebralis*, beginning with its relationships to other myxozoans, and proceed to an overview of the complex developmental stages in both the fish and oligochaete hosts. Where possible, we comment on potential mechanisms that the parasite employs to facilitate this life cycle.

*Corresponding author: rphedrick@ucdavis.edu

TAXONOMY

Myxobolus cerebralis is one of more than 1,350 myxozoan parasites known to infect fish (Lom and Dyková 1992). Myxozoans are a diverse group of multicellular organisms. At one time considered members of the Protozoa, more recent comparisons of ribosomal and Hox genes suggest relationships with the Bilateria or the Cnidaria (Smothers et al. 1994; Siddall et al. 1995; Schlegel et al. 1996; Anderson et al. 1998; Kent et al. 2001). If structural and functional features of the myxozoans are considered, the greatest similarity is to the Cnidaria, particularly the shared use of differentiated cells with extrusive filaments (cnidocysts) that are capable of trapping or attaching to their host or prey (Siddall et al. 1995). *Myxobolus cerebralis* falls within the order Bivalvulidae, suborder Platysporina, and finally the genus *Myxobolus* (Kent et al. 2001). *Myxobolus* is the largest of the 52 genera found in the Phylum Myxozoa but *M. cerebralis* differs from all other *Myxobolus* spp., by inducing a severe to life threatening disease in young salmonid fish (Lom and Dyková 1992). The most extensive molecular phylogenetic comparisons of the ribosomal DNA (rDNA) of myxozoans, to date, demonstrate that *M. cerebralis* branches uniquely, reflecting an evolution separate from other histozoic *Myxobolus* spp. found in fish (Figure 1). Cursory genetic studies of a limited number of *M. cerebralis* isolates from diverse geographic regions suggest little variation (e.g., in ITS regions of the rDNA), and this supports theories of a recent introduction of the parasite to North America (Andree et al. 1999). Evidence for more genetic diversity among isolates of *M. cerebralis* will likely be found among fish in Europe and Eurasia, the presumed origins of the parasite. Similarly, strains of rainbow trout *Oncorhynchus mykiss* that may have developed natural resistance to the parasite may also be present in these regions.

LIFE CYCLE

The discovery of the oligochaete host for *M. cerebralis* by Wolf and Markiw (1984) provided a model for two host life cycles for all other myxozoan parasites of fish. Although there now appears to be a number of alternate life cycle strategies, including direct cycles or involvement of other annelids or even other invertebrates (e.g., bryozoans), several members of the genus *Myxobolus* utilize oligochaetes, including *Tubifex tubifex*, as essential hosts. Our knowledge of the sequential development of *M. cerebralis*, in both the trout and oligochaete host, are from the in depth studies of El-Matbouli et al. (1995) and El-Matbouli and Hoffman (1998). These thorough studies are the basis for the following life cycle descriptions (Figure 2 and 3). Just now, however, are we beginning to understand some of the mechanisms that underlie this complex developmental cycle and how in turn the host responses to these stages differ between susceptible and more resistant fish and oligochaete hosts. Where possible, we have added evidence, either direct or indirect, on new insights into the mechanisms used by the parasite to successfully complete this two-host life cycle.

DEVELOPMENT IN THE TWO HOSTS

Rainbow Trout

Attachment and Initial Invasion

The development of *M. cerebralis* begins with the attachment and penetration of the salmonid host by the waterborne infectious triactinomyxon stages (Figure 2). These triactinomyxon stages remain viable in the water for periods from 6 to 15 d at water temperatures of 7–15°C following release from the oligochaete (Markiw 1992, El-Matbouli et al. 1999b). Upon contact with a susceptible fish host, triactinomyxons can cause significant damage to the epidermis by at least three mechanisms: (1) piercing of the epidermal cells by the extrusion of the polar filaments, (2) migrations of the sporoplasm containing germ cells between cells, a process we believe is facilitated by the release of prepackaged proteases present in the sporoplasm, and (3) the intracellular development and release of parasite daughter cells from infected host cells. Mechanisms that control extrusion, or "natural firing" of the polar filaments, appear to depend on chemical and mechanical clues, as dead or anesthetized fish or fish parts are not effective targets (El-Matbouli et al. 1999a). Subsequent damage to the epidermis, by the attachment and invasion of large numbers of triactinomyxon stages in a short but heavy exposure, can lead to death of young salmonid fish in the laboratory (Markiw 1991) and most likely in the field. Migration of the sporoplasm packet (containing up to 64 germ cells surrounded by an enveloping cell), from the intact triactinomyxon into the epidermis, may be facilitated

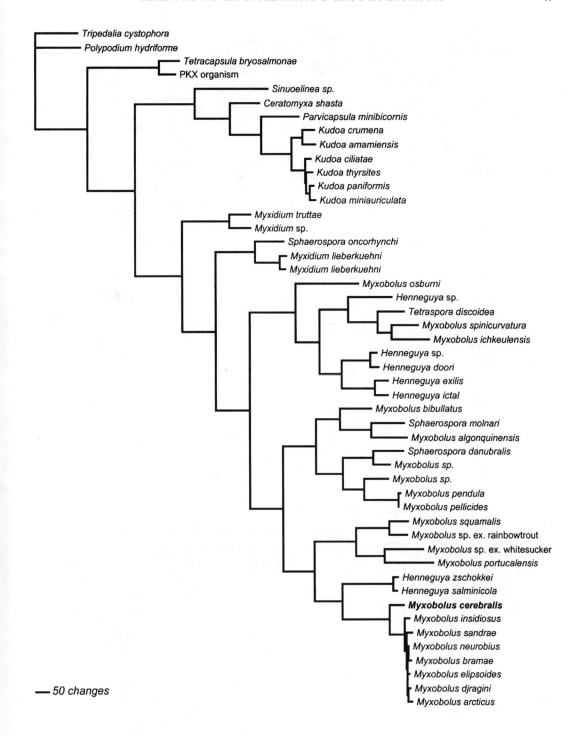

Figure 1. Phylogenetic relationships among members of the phylum Myxozoa as determined by comparisons of the small subunit ribosomal DNA (scu rDNA). Parsimony analysis using CLUSTAL as modified from Kent et al. (2001) by Dr. Mark Siddall. One tree 8951 steps, retention index = 0.64.

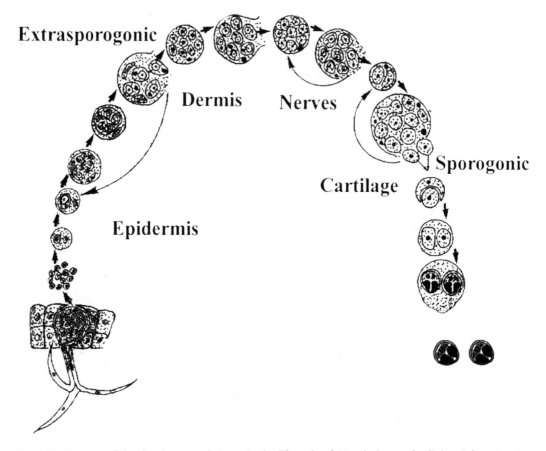

Figure 2. Diagram of the developmental stages in the life cycle of *Myxobolus cerebralis* in rainbow trout.

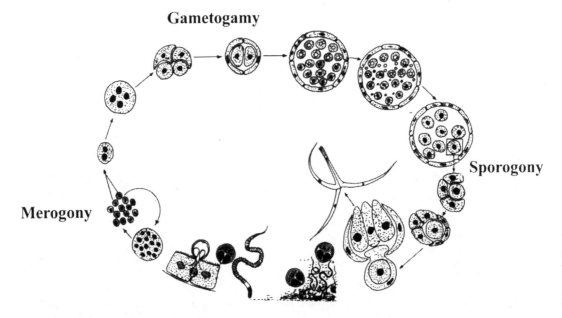

Figure 3. Diagram of the developmental stages in the life cycle of *Myxobolus cerebralis* in the oligochaete host.

by both the mechanical damage to the epidermis and the effects of released enzymes. The sporoplasm is covered with a dense coat of villi that, combined with the release of preformed proteases (M. Adkison, personal communication), facilitates entry between cells damaged by the filaments or at sites once occupied by goblet cells. A role for goblet cells as the site of entry, and perhaps stimulation of polar filament extrusion by contact with mucin or mucus from the fish, was proposed by Yokoyama et al. (1995) for *Myxobolus cultus*. There must be another mechanism with M. *cerebralis*, as trials attempting to demonstrate this response from triactinomyxons conducted in our two laboratories have failed (T. Yokoyama, University of Tokyo, personal communication). Attachment and penetration is a very rapid process, occurring within seconds of first contact between the triactionmyxon stages and the fish host. By 5 min postexposure, sporoplasms have egressed from the valves or the style of the triactinomyxon stage into the epidermis of the fish. Attachment and invasion of new triactinomyxons to the fish skin, and into the epidermis, continue in a cumulative fashion for periods up to 7 weeks (authors' unpublished data). The effects of water temperature on initial attachment and invasion have yet to be studied in detail. The knowledge of the subsequent development of the parasite is described below at a water temperature of 13–15°C (El-Matbouli et al. 1995).

Presporogonic and Sporogonic Stages

After penetration of the epidermis, and within a period as short as 1 h postexposure, individual germ cells from the sporoplasm disperse, and some begin to penetrate host cells of the epidermis. By 2 h, many parasites, and with some synchrony, undergo mitotic divisions. Subsequent stages of the parasite alternate between an intra and intercellular location and continue to multiply by endogeny (cells arising within cells), while spreading to deeper layers of the skin including the cutis and subcutis. After 24 h, there are few remaining parasites in the epidermis. By 4 d and up to 24 d postexposure, parasites are found in the nervous tissue, initially in peripheral nerves but later migrating and replicating between nerve bundles in ganglia and the central nervous system. The effects of water temperature on the early stages of the parasite in the skin and nerves are unknown, but we suspect they are somewhat delayed at colder water temperatures.

After reaching the host cartilage, the parasites undergo further replication, apparently gaining some nutritive advantage from the extracellular digestion of the cartilage matrix and chondrocytes (J. Lom, Institute of Parasitology, Czech Academy of Sciences, retired, personal communication). Enzymes that may be critical to this extracellular digestion may include serine and cystiene proteases (G. Kelley, University of California-Davis, personal communication). The appearance of the first parasite stages in the cartilage has been shown to depend upon water temperature. At 16–17°C, they appear as early as 20 d postexposure, while up to 35 d may be required at temperatures of 12–13°C (Halliday 1973). Presporogonic development ends and sporogony begins with autogamy, a process of the union of two cells to form a generative cell surrounding a sporogonic cell. This phenomenon is typical of myxozoans and signals a direct shift from a more vegetative development to sporogenesis (Lom and Dyková 1992). The sporogonic cell gives rise to all of the cells forming the multicellular myxospore stage: two cells for the polar capsules, two cells for each of the two valves, and two cells that will later fuse to form a binucleate sporoplasm. These elliptically-shaped spores, of approximately 10 μm in diameter, may exceed several million per fish and develop over a period of 52–121 d, depending on water temperature (Halliday 1973). At 16–17°C, they appear as early as 52 d postexposure, and up to 90 d may be required at temperatures of 12–13°C. At cooler water temperatures (0–7°C), spores may require up to 11 months or more to be evident. The spores, with thick protective valves, lie resident in the fish skeleton, where cartilage was once present. When the fish is either ingested by another fish, a fish-eating bird, or another animal, the spores are excreted in the feces. Alternately, severely crippled fish may die and decompose in the sediments directly. These myxospores may remain viable for periods as great as 12 years (Schaperclaus 1954). Spores surviving in sediments or excreted in feces then become available to infect the next host, the oligochaete.

Oligochaete

Attachment and Invasion

Myxospores of M. *cerebralis* ingested by susceptible aquatic oligochaetes undergo a process that may be referred to as "hatching" but that is directly analogous to the better known initial phase of the interaction between the triactinomyxon stage and the fish (Figure 3). Upon stimuli that may share characteristics with those responsible for triggering the triactinomyxon stage, the polar filaments are extruded and attach the spore to the mucosa or lining of the intestine, which is present in most segments of the worm. The mucosa of the oligochaete has an extensive microvillar network, critical to absorption of nutrients but that may function to entangle spores that attach to the mucosal epithelium (Figure 4A). Upon polar filament extrusion, the spore valves open, and the binucleate sporoplasm migrates between the mucosal epithelium, an event that can be best visualized (beginning at 24 h) by *in situ* hybridization procedures with labeled DNA probes to M. *cerebralis* (Antonio et al. 1999). In contrast to the parasite migrations in the fish host, the parasite remains in one intercellular location, between epithelial cells of the intestine, throughout all stages of development in the oligochaete host. The following developmental phases can be expected at water temperatures near 17°C and will be delayed at lower temperatures, and at temperatures that exceed 20°C, and may be arrested completely at temperatures of 30°C or greater (El-Matbouli et al. 1999b).

Merogony

Between 5 and 25 d, the binucleate amoeboid cells give rise to many uni- or bi-nucleated cells that subsequently undergo multiple divisions to produce numerous daughter cells. This cycle may continue throughout the development of the parasite, providing a reservoir of undifferentiated cells that can later contribute to gametogony and sporogony.

Gametogony

Between 25 and 46 d, oval binucleate cells give rise, through multiple divisions, to pansporocysts. The pansporocyst consists of somatic cells surrounding two generative cells. The somatic cells divide to form the pansporocyst wall, while the two generative cells further divide to form gametocytes that further differentiate into haploid alpha and beta gametocytes. The meiotic divisions that give rise to the gametes are the only phase in the parasite's life cycle where it exists in a haploid state (El-Matbouli et al. 1998). The fusion of alpha and beta gametocytes to form a zygote is the only truly sexual phase of the life cycle of M. *cerebralis* and signals the beginning of sporogony.

Sporogony

Beginning at 50 d, the fusion of gametocytes to form zygotes can be observed. Eight zygotes are formed within each pansporocyst. Each zygote, in turn, divides to eventually form four sporoblast cells, three that surround the fourth. After another division of the surrounding cells, the developing spore contains seven cells. Three cells differentiate to form the polar capsules, three other cells produce the valves, and one cell undergoes multiple divisions to give rise to an enveloping cell surrounding up to 64 internal or germ cells. The fully formed triactinomyxon spores (90 d) are released into the lumen (Figure 4B-D). The spores are folded and deflated while present in the pansporocyst. The pansporocyst wall emerges to the surface of the intestinal lumen, ruptures, and then releases the spores. Large depressions are evident in the intestinal mucosa, where spores have been released from their pansporocysts (Figure 4E). Spores in the lumen are passed towards the anus and released into the water. The space between the two delicate layers composing the valves presumably fills with water to give an inflated appearance to the fully developed triactinomyxon stage. This waterborne stage is approximately 146 µm in height (style), with tails (processes) of a 193 µm in length (El-Matbouli and Hoffmann 1998). A single oligochaete may remain infected for greater than one year and, perhaps, for life, releasing several thousand triactinomyxon stages during this period (Markiw 1986). The amount and timing of the release of triactinomyxon stages depends on many factors, two of which are clearly water temperature and the strain, or type, of worm (El-Matbouli et al. 1999b). Experimentally infected oligochaetes, held at 12.5°C, began releasing triactinomyxon stages 104 d after exposure to myxospores (Markiw 1986). Peak releases occurred at 120–170 d after exposure, with periodic releases of up to 2,090 triactinomyxons/min from certain worm groups. Total yields from worms, held for approximately 9 months, were more than 50×10^6 triactinomyxons, or a mean of 3.42×10^3 per individual worm (Markiw 1986). The effects of water temperatures of 5,10, 15, 20, 25, and 30°C on the release of triactinomyxons from oligochaetes, experimentally infected with M. *cerebralis*, were examined by El-Matbouli et al. (1999b). Releases of triactinomyxons ceased within 4 d from oligochates held at 25°C and 30°C and after 15 d at 20°C. In contrast, oligochaetes continued to release triactin-

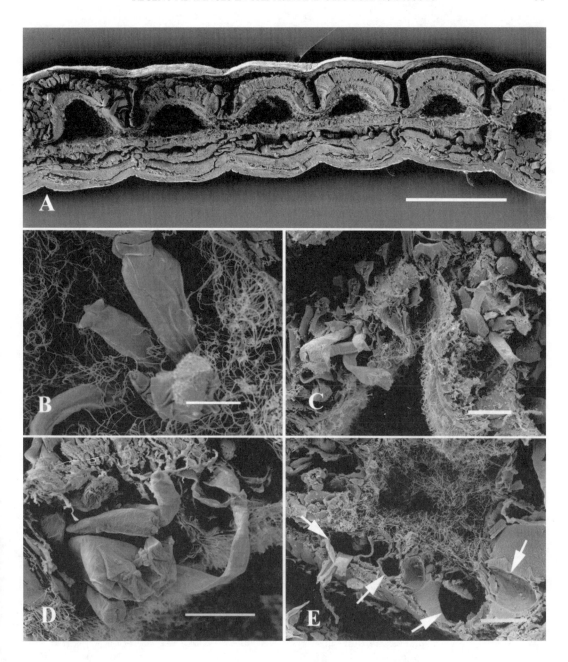

Figure 4. *Tubifex tubifex* either uninfected or with triactinomyxon stages present at 14 weeks postexposure to spores of *Myxobolus cerebralis*. Oligochaetes known to be releasing triactinomyxons were fixed in Karnovsky's solution for 12 h at 4°C and then processed by standard methods for scanning electron microscopy (El-Matbouli et al. 1999a). (A) Midline saggital section of an unexposed worm showing consecutive segments with the hollow lumen of the intestine evident in each segment, Bar = 200 μm, (B) triactinomyxon stages pushing through the heavy layer of microvilli present on the surface of the intestinal mucosa, Bar = 10 μm, (C) presence of triactinomyxon stages beneath and between cells of the mucosal epithelium, Bar = 50 μm, (D) several triactinomyxons emerging from a pansporocyst, Bar = 20 μm, and (E) cavities (arrows) left behind in the intestinal mucosa following release of the triactinomyxon stages from the pansporocyst, Bar = 20 μm.

omyxons until the end of the study at 64 d at 10°C and 15°C. At 5°C, minimal releases were detected, except for one period when the incubator temperature rose to 15°C. These results suggest that water temperatures between 10°C and 15°C are optimal for triactinomyxon release from infected worms. Certainly other factors are critical to the success of initial infections and the subsequent release of triactinomyxon stages from oligochaetes. These may include age and stage of development of the worms, water quality, and impacts of competition between susceptible and resistant oligochaetes.

CONCLUSIONS

Our studies, and those of others on M. cerebralis, reveal a complex parasite whose evolution into a serious pathogen of wild trout has had severe economic and ecological consequences in North America. In fish, the parasite may cause significant damage upon initial attachment and penetration, and subsequent migrations through nerves may also have effects on key neurological responses. Destruction of cartilage by extracellular digestive processes that continue for more than a year may severely cripple and disfigure the skeleton of young trout. Pressure on the spinal cord and hindbrain, due to the parasites and the host response to their presence, are thought to induce significant swimming disorders (Rose et al. 2000). In the oligochaete, there is also evidence that parasites resident between cells in the intestinal mucosa may compromise the host. Pressure atrophy of the host cells alone, during parasite development, would decrease the absorptive surface of the intestinal mucosa. Studies examining growth and reproduction of infected oligochaetes are needed to confirm these suspected negative effects.

While certain of the mechanisms underlying functions of the parasite at each developmental stage are becoming clear, many or most remain unknown. Attempts to understand these mechanisms, by studying resistant and susceptible species of fish and oligochaetes, are providing additional insights, some of which in part should explain population level impacts of the disease in certain geographic regions. Exploiting this new information to develop new and innovative control measures are goals of these current studies.

ACKNOWLEDGMENTS

This work was supported in part by the U.S. Fish and Wildlife Service and the Whirling Disease Foundation. We thank Mark Siddall, Division of Invertebrate Zoology, American Museum of Natural History, New York for the preparation of the phylogenetic comparison as modified from that of Kent et al. (2001).

REFERENCES

Anderson, C. L., E. U. Canning, B. Okamura. 1998. A triploblast origin for myxozoa? Nature 392:346–347.

Andree, K. B., M. El-Matbouli, R. W. Hoffmann, and R. P. Hedrick. 1999. Comparison of 18S and ITS-1 rDNA sequences of selected geographic isolates of Myxobolus cerebralis. International Journal for Parasitology 29:771–775.

Antonio, D. B., M. El-Matbouli, and R. P. Hedrick. 1999. Detection of early developmental stages of Myxobolus cerebralis in fish and tubificid oligochaete hosts by in situ hybridization. Parasitological Research 85:942–944.

El-Matbouli, M., R. W. Hoffmann H. Shoel, T. S. McDowell, and R. P. Hedrick. 1999a. Whirling disease: host specificity and interaction between the actinosporean stage of Myxobolus cerebralis and rainbow trout (Oncorhynchus mykiss) cartilage. Disease of Aquatic Organisms 35:1–12.

El-Matbouli, M., T. S. McDowell, D. B. Antonio, K. B. Andree, and R. P. Hedrick. 1999b. Effect of water temperature on the development, release and survival of the triactinomyxon stage of Myxobolus cerebralis in its oligochaete host. International Journal for Parasitology 29:627–641.

El-Matbouli M., and R. W. Hoffman. 1998. Light and electron microscopic study on the chronological development of Myxobolus cerebralis in Tubifex tubifex to the actinosporean stage triactinomyxon. International Journal of Parasitology 28:195–217.

El-Matbouli, M., T. W. Holstein, and R. W. Hoffmann. 1998. Determination of the nuclear DNA concentration in cells of Myxobolus cerebralis and triactinomyxon spores, the causative agent of whirling disease. Parasitological Research 84:694–699.

El-Matbouli M., R. W. Hoffmann, and C. Mandok. 1995. Light and electron microscopic observations on the route of the triactinomyxon-sporoplasm of Myxobolus cerebralis from epidermis into rainbow trout (Oncorhynchus mykiss) cartilage. Journal of Fish Biology 46:919–935.

Halliday, M. M. 1973. Studies on Myxosoma cerebralis, a parasite of salmonids II. The development and pathology of Myxosoma cerebralisi in experimentally infected rainbow trout (Salmo gairdneri) fry reared at different temperatures. Nordic Veterinary Medicine 25:349–358.

Hedrick, R. P., M. El-Matbouli, M. Adkison, and E. Mac-Connell. 1998. Whirling disease: re-emergence among wild trout. Immunology Reviews 166:365–376.

Hofer, B. 1903. Ueber die Drehkrankheit der Regenbogenforelle. Allgemeinen Fischerei-Zietung 8(1):7–8.

Kent, M. L., K. B. Andree, J. L. Bartholomew, M. El-Matbouli, S. Desser, C. Xiao, R. H. Devlin, R. P. Hedrick, L. Khattra, O. Palenzuela, and M. Siddall. 2001. Recent advances in our knowledge of the Myxozoa. Journal of Eukaryotic Microbiology 48:395–413.

Lom J., and I. Dyková. 1992. Protozoan parasites of fishes. Elsevier, Amsterdam.

Markiw, M. E. 1992. Experimentally induced whirling disease II: determination of longevity of the infective triactinomyxon stage of Myxobolus cerebralis by vital staining. Journal of Aquatic Animal Health 4:44–47.

Markiw, M. E. 1991. Whirling disease: earliest susceptible age of rainbow trout to the triactinomyxid of Myxobolus cerebralis. Aquaculture 92:1–6.

Markiw, M. E. 1986. Salmonid whirling disease: dynamics of experimental production of the infective stage—the triactinomyxon spore. Canadian Journal of Fisheries and Aquatic Sciences 43:521–526.

Rose, J. D., G. S. Marrs, C. Lewis, and G. Schisler. 2000. Whirling disease behavior and its relation to pathology of the brain stem and spinal cord in rainbow trout. Journal of Aquatic Animal Health 12:107–118.

Schaperclaus, W. 1954. Fsich-Krankheiten. Akademie-Verlag, Berlin.

Schlegel, M., J. Lom, A. Stechmann, D. Bernhard, D. Leipe, I. Dyková, M. L. Sogin. 1996. Phylogenetic analysis of complete small subunit ribosomal RNA coding region of Myxidium lieberkuehni: evidence that Myxozoa are metazoa related to the bilateria. Archives Fur Protistende Kunde 147:1–9.

Siddall, M. E., D. S. Martin, D. Bridge, S. S. Desser, and D. K. Cone. 1995. The demise of a phylum of protists: phylogeny of Myxozoa and other parasitic cnidaria. Journal of Parasitology 8:961–967.

Smothers, J. F., C. D. von Dohlen, L. H. Smith, and R. D. Spall. 1994. Molecular evidence that the Myxozoan protists are metazoans. Science 265:1719–1721.

Wolf, K., and M. E. Markiw. 1984. Biology contravenes taxonomy in the Myxozoa: new discoveries show alternation of invertebrate and vertebrate hosts. Science 225:1449–1452.

Yokoyama, H., Ogawa, K., and H. Wakabayashi. 1995. Chemoresponse of actinosporean spores of Myxobolus cultus to skin mucus of goldfish Carassius auratus. Diseases of Aquatic Organisms 21:7–11.

American Fisheries Society Symposium 29:55–59, 2002

Detection of Glycoconjugates and Lectin-like Activities in the Causative Agent of Whirling Disease

Mansour El-Matbouli*, Christiane Anders, and Rudolf Hoffmann

Institute of Zoology, Fish Biology and Fish Diseases, Faculty of Veterinary Medicine
Kaulbachstr. 37, 80539, University of Munich, Munich, Germany

Hans-Joachim Gabius and Herbert Kaltner

Institute of Physiological Chemistry, Faculty of Veterinary Medicine, Veterinärstr. 13,
80539 Munich, University of Munich, Germany

ABSTRACT. Attachment of triactinomyxon spores of *Myxobolus cerebralis* to the epidermis of the rainbow trout (*Oncorhynchus mykiss*) host and the migration of sporoplasm cells to their target tissues are the first steps in the development of whirling disease. Morphologically these processes are well-documented at the level of light and electron microscopy. In order to address the question of whether interactions of lectins and carbohydrate structures are involved in these processes, biotin-labelled *Viscum album* agglutinin (VAA) and the neoglycoprotein mannose-BSA (bovine serum albumin) biotin were used. These markers enabled the detection of galactoside-containing glyco-conjugates and binding sites with specificity to D-mannose. VAA-reactive sites are distributed in all compartments of the triactinomyxon stages and in the polar capsules of *M. cerebralis*. Mannose-BSA shows weak to moderate binding to the polar capsules of the triactinomyxon spores. This indicates the presence of lectin-like activity, which may imply that a lectin is involved in the adhesion of the triactinomyxon spores to rainbow trout.

Whirling disease is a disorder of salmonid fish caused by the microscopic myxozoan parasite *Myxobolus cerebralis*. The disease, initially thought to be a problem only of trout hatcheries, has caused catastrophic losses to rainbow trout populations in several of the western North American waters. Currently, it is one of the most widely discussed fish diseases in the United States (Hedrick et al. 1998). It is well documented at the electron microscope level that the triactinomyxon spores of *Myxobolus cerebralis* interact with the epidermis of the salmonid host and that the sporoplasm cells migrate to the target tissue (El-Matbouli et al. 1995, 1999; El-Matbouli and Hoffmann 1998). However, the molecular mechanisms responsible for these processes and the ensuing pathogenesis are not known in detail.

Obligatory parasites must enter target cells for infection and proliferation. This phenomenon implies that the parasites must recognize, and bind to, the host cell surface. This recognition can be accomplished by the interaction of carbohydrate-binding proteins on the parasites and complementary sugars on the host cells, or vice versa (Sharon

1984). In fact, oligosaccharides are now known to store and to transmit biological information (Laine 1997; Gabius 2000). To decode the informational content of these moieties, endogenous lectins are present in organisms from bacteria or viruses up to mammals (Gabius 1997; Kaltner and Stierstorfer 1998; Rüdiger et al. 2000). While plant lectins are frequently used to detect carbohydrate determinants by histochemistry (Spicer and Schulte 1992; Gabius and Gabius 1993; Rüdiger 1998), a different class of tools is required to detect endogenous lectins. The characteristic attribute of lectins is the binding of a carbohydrate moiety. By covalently linking sugar-ligands to an inert carrier, a neoglycoprotein is prepared that presents its carbohydrate part to respective binding sites in the tissue. Initially used exclusively to raise sugar-specific antibodies, these neoglycoproteins have been applied to localize endogenous lectins by histochemistry (Gabius and Bardosi 1991; Gabius et al. 1993, 1994; Bovin and Gabius 1995; Gabius, 2001).

Herein, we introduce monitoring with a labeled neoglycoprotein to the whirling disease infection model, in the framework that protein-carbohydrate interactions are important for steps within the infectious process. Studies of carbohydrate

*Corresponding author:
Elmatbouli@zoofisch.vetmed.uni-muenchen.de

residues of myxosporean parasites are very limited (Hedrick et al. 1992; Muñoz et al. 1999). In this study, triactinomyxon spores in infected *Tubifex tubifex*, and M. *cerebralis* in infected rainbow trout, were probed for glycans and, for the first time, for carbohydrate-binding proteins, which could be involved in host-parasite interaction.

MATERIAL AND METHODS

Actinosporea-free T. *tubifex* (10 g) were exposed in our laboratory to M. *cerebralis* with a dose of 400 spores per tubificid, as described by El-Matbouli and Hoffmann (1998). The infected tubificid used in the present work were sampled at 95 d postexposure, fixed in 5% buffered formalin, and embedded in paraffin. Tissue sections 4–5 μm thick were cut and transferred to microscope glass slides.

Rainbow trout, experimentally infected with the triactinomyxon stages of M. *cerebralis*, were also sampled at 120 d postexposure. Fish were killed by overdose with the anesthetic chlorobutanol (1.1.1-trichlor-methylpropanol, 0.1 g l⁻¹ water), necropsied, and divided sagitally. They were then fixed either in 5% buffered formalin or in Bouin's solution. Sections were prepared, as described above, for the tubificid.

Histochemistry

For detection of carbohydrate moieties, the following biotinylated lectins were used: *Viscum album* L. agglutinin (VAA) and the endogenous mammalian lectins galectin 1 and –3. VAA is specific for terminal ∝/β-D-galactose residues and the mammalian lectins for terminal β-D-galactose and internal [Galβ1–4GlcNAc]$_n$ poly-N-acetyllactosamine sequences. VAA was purified from extracts of dried leaves of the mistletoe plant, whereas recombinant galectins were expressed from the plasmids pH14Gal (human galectin-1) and prCBP35 (murine galectin-3) in E. *coli* JA221 cells (Hirabayashi et al. 1989; Agrwal et al. 1993). Final purification was performed using Sepharose 4B (Pharmacia, Freiburg, Germany) to which lactose had been coupled after activation with divinyl sulfone, as described by Gabius (1990). The neoglycoproteins D-mannose-bovine serum albumin (BSA)-biotin, lactose-BSA-biotin and L-Fucose-BSA-biotin were chemically synthesized by coupling the saccharide derivative to the carrier molecule BSA according to standard procedures (Gabius and Gabius 1993; Gabius et al. 1993). The concentrations of the lectins were 2.5 μg, and the neoglycoproteins were 5.0 μg per mL incu-

bation solution (PBS, pH 7.4). Histological sections were dewaxed using xylene, rehydrated in graded alcohol solutions, and incubated with 0.4% hydrogen peroxide for 30 min at room temperature, to inhibit the endogenous peroxidase activity. After rinsing in phosphate-buffered saline (PBS, pH 7.4), the sections were preincubated with 0.1% carbohydrate-free BSA for 30 min, to block nonspecific binding of BSA-biotin derivatives used in the following steps. Residual solution was blotted from the slides, and the sections were incubated, with biotinylated VAA (2.5 μg/ml PBS, pH 7.4) or biotinylated neoglycoprotein (5.0 μg/mL PBS, pH 7.4), overnight at room temperature. After rinsing, the sections were incubated with avidin-biotin-peroxidase complex (ABC) (Vector Laboratory, Burlingame, California, USA) for 30 min at room temperature. Rinsing was performed again three times in PBS, and bound peroxidase was detected by adding the chromogen 3,3'-diaminobenzidine tetrahydrochloride (DAB) (Sigma) for 10 min. The reaction was stopped by washing with PBS, and the sections were counterstained, using hematoxylin, dehydrated in ethanol solutions, cleared in xylene, and finally mounted in eukitt. Incubation of ABC reagents with the tissue sections alone served as control to exclude the presence of endogenous biotin-binding proteins. Controls to verify the specificity of protein-carbohydrate interaction included preincubation of the labeled lectins with 0.1 M lactose and, when neoglycoproteins were applied, incubation of the sections in the presence of the haptenic monosaccharide (0.1 M).

RESULTS AND DISCUSSION

Histochemistry with *Viscum album* agglutinin (VAA) resulted in strong staining of the triactinomyxon spores in the gut of infected T. *tubifex*. VAA bound strongly to the sporoplasm cells, weakly to moderately to the valvogenic cells and the capsulogenic cells of the triactinomyxon (Figure 1a). The polar capsules of M. *cerebralis* myxospores were strongly reactive and the valves weakly to moderately with VAA (Figure 1c). Both galectins indicative for N-acetyllactosamine moieties showed no reactivity. The neoglycoprotein D-Mannose-BSA-biotin revealed a weak to moderate specific binding to the polar capsules of the triactinomyxon spores (Figure 1e). No reaction could be seen in all other spore components, thus serving as further internal specificity control. The other neoglycoproteins applied, L-Fucose-BSA-

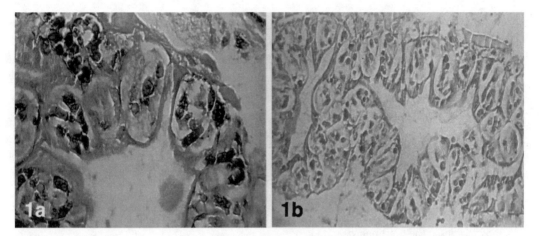

Figure 1a–1b: Histochemical detection of glycan structures of *Myxobolus cerebralis* spores and their triactinomyxon stages with biotin-labeled *Viscum album* agglutinin (VAA) in longitudinal sections of infected *Tubifex tubifex*. 1a) Note the strong positive staining of the sporoplasm cells of the triactinomyxon. 20d post exposure, 1000X. 1b) Control section to verify the carbohydrate-specific binding of VAA. Biotin-labelled VAA was applied in the presence of the respective inhibitory sugar (0.1 M lactose). 600X.

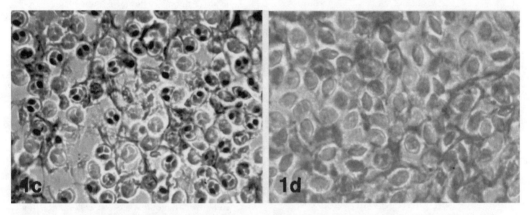

Figure 1c–1d: Histochemical detection of glycan structures of *Myxobolus cerebralis* spores and their triactinomyxon stages with biotin-labelled *Viscum album* agglutinin (VAA) in sections of the head of rainbow trout infected with *M. cerebralis*. 1c) Note the strong positive staining of the polar capsules. 150d post infection, 1500X. 1d) Control section to verify the carbohydrate-specific binding of VAA. Biotin-labelled VAA was applied in the presence of the respective inhibitory sugar (0.1 M lactose). 1500X.

biotin and lactose-BSA-biotin, showed a weak staining but no specific binding pattern. The early developmental stages of the parasite, in both hosts, did not show any binding with any applied lectins or with the neoglycoproteins. Binding to the host tissues could be detected only with VAA, which reacted weakly to the gut epithelial cells of the *T. tubifex* and moderately to strongly to the epidermis and bone of rainbow trout. Lectin staining was inhibited in all cases after blocking with specific sugar (Figure 1b, d).

The ability of parasitic organisms and their developmental stages to adhere to the surface of the host cell is the first step for invasion and infectivity. Jacobson and Doyle (1996), Ward (1997) and Cummings and Nyame (1999) reported that the parasite surface coat is involved in many host-parasite interactions. It is, among other components, constituted by glycan chains of glycoproteins and glycolipids. As outlined in the introduction, it is reasonable to assume functional implications of these chains. Therefore, the carbohydrate

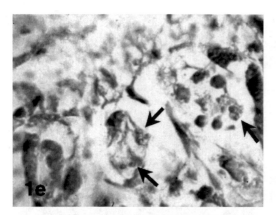

Figure 1e: Histochemical detection of carbohydrate-binding activities in the triactinomyxon stage of *Myxobolus cerebralis* using the biotin-labeled neoglycoprotein, mannose-BSA, in longitudinal sections of infected *Tubifex tubifex*. Note the positive staining in the polar capsules of the triactinomyxon spores (arrows). 2000X.

composition of the polar capsules and valves of both parasitic stages, M. *cerebralis* myxospore and the triactinomyxon spore, is proposed to be important for understanding the host-parasite interaction of this parasite. Besides profiling glycan structures, it will be pertinent to pursue our initial detection of a carbohydrate-binding activity by using the labeled neoglycoprotein. Consequently, our glycohistochemical analysis will be extended.

ACKNOWLEDGMENTS

This work has been supported by the Whirling Disease Foundation.

REFERENCES

Agrwal N., Q. Sun, S. Y. Wang, and J. L. Wang. 1993. Carbohydrate-binding protein 35. I. Properties of the recombinant polypeptide and the individuality of the domains. Journal of Biological Chemistry 268:14932–14939.

Bovin N. V., and H. J. Gabius. 1995. Polymer-immobilized carbohydrate ligands: versatile chemical tools for biochemistry and medical sciences. Chemical Society Reviews 24:413–421.

Cummings, R. D., and A. K. Nyame. 1999. Schistosome glycoconjugates. Biochimica et Biophysica Acta 1455:363–374.

El-Matbouli, M., and R. W. Hoffmann. 1998. Light and electron microscopic study on the chronological development of *Myxobolus cerebralis* in *Tubifex tubifex* to the actinosporean stage triactinomyxon. International Journal of Parasitology 28:195–217.

El-Matbouli, M., R. W. Hoffmann, and C. Mandok. 1995. Light and electron microscopic observations on the route of the triactinomyxon-sporoplasm of *Myxobolus cerebralis* from epidermis into the rainbow trout (*Oncorhynchus mykiss*) cartilage. Journal of Fish Biology 46:919–935.

El-Matbouli, M., R. W. Hoffmann, H. Schoel, T. S. McDowell, and R. P. Hedrick. 1999. Whirling disease: host specificity and interaction between the actinosporean stage of *Myxobolus cerebralis* and rainbow trout (*Oncorhynchus mykiss*). Diseases of Aquatic Organisms 35:1–12.

Gabius, H. J. 1990. Influence of type of linkage and spacer on the interaction of β-galactoside-binding proteins with immobilized affinity ligands. Analytical Biochemistry 189:91–94.

Gabius, H. J. 1997. Animal lectins. European Journal of Biochemistry 243:543–576.

Gabius, H. J. 2000. Biological information transfer beyond the genetic code: the sugar code. Naturwissenschaften 87:108–121.

Gabius, H. J. 2001. Glycohistochemistry: The why and how of detection and localization of endogenous lectin. Anatomia Histologia Embryologia 30:3–31.

Gabius, H. J., and A. Bardosi. 1991. Neoglycoproteins as tools in glycohistochemistry. Progress in Histochemistry and Cytochemistry 22(3):1–66.

Gabius, H-J., and S. Gabius, editors. 1993. *Lectins and glycobiology*. Springer, Berlin.

Gabius H. J., S. Gabius, T. V. Zemlyanukhina, N. V. Bovin, U. Brinck, A. Danguy, S. S. Joshi, K. Kayser, J. Schottelius, F. Sinowatz, L. F. Tietze, F. Vidal-Vanaclocha, and J. P. Zanetta. 1993. Reverse lectin histochemistry: design and application of glycoligands for detection of cell and tissue lectins. Histology and Histopathology 8:369–383.

Gabius H. J., S. André, A. Danguy, K. Kayser, and S. Gabius. 1994. Detection and quantification of carbohydrate-binding sites on cell surfaces and in tissue sections by neoglycoproteins. Methods in Enzymology 242:37–46.

Hedrick, R. P., M. El-Matbouli, M. Adkison, and E. MacConnell. 1998. Whirling disease: re-emergence among wild trout. Immunological Reviews 166:365–376.

Hedrick, R. P., M. Marin De Mateo, M. Castagnaro, D. Monge, and P. Kinkelin. 1992. Rapid lectin-based staining procedure for detection of the myxosporean causing proliferative kidney disease in salmonid fish. Diseases of Aquatic Organisms 13:129–132.

Hirabayashi J., H. Ayaki, G. Soma, and K. Kasai. Production, and purification of a recombinant human 14 kDa beta-galactoside-binding lectin. FEBS Letters 250:161–165.

Jacobson, R. L., and R. J. Doyle. 1996. Lectin-parasite interactions. Parasitology Today 12:55–61.

Kaltner, H., and B. Stierstorfer. 1998. Animal lectins as cell adhesion molecules. Acta Anatomica 161:162–179.

Laine, R. A. 1997. The information-storing potential of the sugar code. Pages 1–14 *in* H-J. Gabius and

S. Gabius, editors. Glycosciences: status and per-spectives. Chapman and Hall, London-Weinheim.

Muñoz P., O. Palenzuela, P. Alvarez-Pellitero, and A. Sitja-Bobadilla. 1999. A comparative studies on carbohydrates of several myxosporean parasites of fish using lectin histochemical methods. Folia Para-sitologica 46:241–347.

Rüdiger, H. 1998. Plant lectins—more than just tools for glycoscientists. Occurrence, structure and possible functions of plant lectins. Acta Anatomica 161:130–152.

Rüdiger, H., H. C. Siebert, D. Solís, J. Jimenez-Barbero, A. Romero, C. W. von der Lieth, T. Diaz-Mauriño, and H.-J. Gabius. 2000. Medicinal chemistry based on the sugar code: fundamentals of lectinology and experimental strategies with lectins as targets. Cur-rent Medicinal Chemistry 7:389–416.

Sharon, N. 1984. Surface carbohydrates and surface lectins are recognition determinants in phagocyto-sis. Immunology Today 5:143–147.

Spicer, S. S., and B. A. Schulte. 1992. Diversity of cell glyco-conjugates shown histochemically: a perspective. Jour-nal of Histochemistry and Cytochemistry 40:1–38.

Ward, H. D. 1997. Glycobiology of parasites: role of carbo-hydrate-binding proteins and their ligands in the host-parasite interaction. Pages 399–413 in H-J. Gabius and S. Gabius, editors. Glycosciences: status and perspec-tives. Chapman and Hall, London-Weinheim.

American Fisheries Society Symposium 29:61–76, 2002
© 2002 by the American Fisheries Society

Extrusion of Polar Filaments of the *Myxobolus cerebralis* Triactinomyxon by Salts, Electricity, and Other Agents

ERIC J. WAGNER*, QUINN CANNON, MARK SMITH, RYAN HILLYARD, AND RONNEY ARNDT
Fisheries Experiment Station, Utah Division of Wildlife
1465 West 200 North, Logan, Utah 84321, USA

ABSTRACT. The ability of several compounds to discharge the polar filaments of polar capsules of the triactinomyxon stage of *Myxobolus cerebralis* was tested. Premature polar filament discharge may provide a means for preventing the infective stage of myxozoan parasites from attaching to fish hosts. The discharge regimes evaluated included high and low pH, chloride and phosphate salts, calcium chelators, direct current, mucus, tricaine methanesulfonate anesthetic, neurochemicals, and chemosensitizing agents that are effective discharge agents for members of the phylum Cnidaria. Polar filament discharge, in response to HCl or NaOH, did not differ from controls until pH levels dropped to 1.1 or increased to 11.7. Among the chloride salts tested (NaCl, KCl, CaCl₂, NH₄Cl, MgCl₂), discharge increased at concentrations ranging from 3.1 to 100‰. Discharge varied among the salts tested, peaking at 71% for 100‰ KCl; however, the phosphate salts K⁺ and Na⁺ did not differ in discharge ability. Comparison among KCl, KI, and KPO₄ indicated that Cl⁻ was significantly more effective at both 6.2‰ (45.6% discharge) and 12.5‰ (57.8%) than the other anions. The calcium chelators sodium citrate and EGTA did not induce any significant increase in discharge, nor did the neurochemicals angiotensin, bradykinin, and acetylcholine chloride. Compounds, such as N-acetyl neuraminic acid, proline, and glutathione, that have been reported as chemosensitizers for cnidae discharge among cnidarians, were ineffective discharge agents for triactinomyxon polar capsules. Mucus from rainbow trout or bovine submaxillary gland failed to significantly increase discharge. Attempts to combine mucus with force (stirring rod) or a 0.45 Gauss magnetic field did not increase discharge rates. However, using an electroporator to administer direct current, the discharge rate increased with pulse length (up to 99 μsec) and the number of pulses (0–25). Maximum discharge (98%) and mortality (100%) was observed after 25 99-μsec pulses of 3 kV. Results with electricity indicate a potential for using direct current as a means of disinfection. The data suggest some similarities and differences with similar research on Cnidaria that is discussed.

Myxozoa is the animal phylum characterized by spores made of several cells forming 1–7 shell valves, 1–2 amoeboid infective sporoplasms, and 2–7 polar capsules (Lom and Dyková 1992). The polar capsules each contain an eversible polar filament that serves to anchor the parasite to the host tissue. The potential for premature polar filament discharge to disarm Myxozoan parasites has been recognized by previous researchers (Hoffman et al. 1965).

The myxozoan parasite *Myxobolus cerebralis* causes whirling disease, an important disease of salmonid fishes. The infective stage of the parasite to fish is known as the triactinomyxon (Markiw 1992a). This stage is also broadly classified as an actinospore, a group of organisms once thought to be separate species but now considered alternate stages of myxozoan parasites (Kent et al. 1994). The triactinomyx-

on has 3 polar capsules, with eversible polar filaments used to attach to the fish host. Research on the discharge mechanisms of the polar capsules could potentially lead to methods for control of whirling disease and other myxosporean diseases.

Research on polar filament discharge in Myxozoa has primarily been conducted on the myxospore stage of the parasite. A number of agents have been tested, but only a few have been successful. These discharge agents include strong bases, such as KOH (Yasutake and Wood 1957; Uspenskaya 1957; Lewis and Summerfelt 1964), saturated urea (Lom 1964), hydrogen peroxide (Kudo 1918; Lom 1964), and direct pressure on wet mounts (Herrick 1941; Guilford 1963; Iversen 1954).

Discharge of the polar filaments of the actinospore stage of myxozoan parasites has been investigated for a few species. Yokoyama et al. (1995) were able to discharge filaments of raabeia-type

*Corresponding author: nrdwr.ewagner@state.ut.us

actinospores of *Myxobolus cultus*, using mucus from a variety of fish species or bovine mucin. Mucus has also been effective for discharge of *Zschokkella nova* actinospores (Uspenskaya 1995) and aurantiactinomyxon, neoactinomyxum, echinactinomyxon, and raabeia forms of actinospore (Xiao and Desser 2000). However, for M. *cerebralis* triactinomyxon actinospore, rainbow trout mucus was ineffective as a discharge agent (El-Matbouli et al. 1999).

The natural discharge mechanisms for both the myxospore and actinospore stages in Myxozoa have yet to be described. Much more research on discharge has been conducted on cnidae, specialized structures with an eversible filament characteristic of the phylum Cnidaria (Mariscal 1974). The development of polar capsules and cnidae are very similar (Lom and de Puytorac 1965), so knowledge gained by researchers studying cnidae may apply to polar capsule discharge. The literature on cnidae discharge is vast (see reviews by Pantin 1942; Picken and Skaer 1966; Mariscal 1974; Tardent 1988; Thorington and Hessinger 1988). Numerous discharge agents have been tested, including various salts, calcium chelating agents, electricity, and strong acids and bases that are the focus of this article (Pantin 1942; Yanagita and Wada 1953; Salleo et al. 1983b). Effects of discharge agents may vary depending upon whether the cnida is still within the cnidocyte (in-situ) or isolated after being artificially expelled from the cnidocyte. For example, in-situ cnidae discharged at pH values below 4 and above 11, whereas isolated cnidae discharged only at pH values of 2 and below (Blanquet 1970). Sodium citrate and EGTA are calcium chelators that have been effective in discharging isolated cnidae (Hidaka and Mariscal 1988). Pantin (1942) found that electrical shocks (240 V from a 2 μF condenser and passed through a 400 Ω potentiometer) stimulated cnidae discharge. Among anions tested by Salleo et al. (1983b), discharge potency of sodium salts varied as follows: SO_4^{2-} < CH_3COO^- < F^- < Cl^- < Br^- < NO_3^- < I^- < ClO_4^- < SCN^-. Santoro and Salleo (1991) found that high K^+ concentrations induced massive discharge of cnidae.

Based upon the previous research, discharge of the polar filaments of *Myxobolus cerebralis* triactinomyxons was attempted in a series of tests. The effects of pH, various salts, calcium chelating agents, chemosensitizing agents, neurochemicals, and direct electrical current were evaluated as potential discharge agents. Survival of discharged

sporozoites was also of interest. If the sporozoites were still alive, this would be useful for in-vitro tissue culture and research on the sporozoite. If dead, the treatment could potentially be applied to control the parasite.

METHODS

The triactinomyxon stock solutions were made by filtering the supernatant from worm cultures through a 20 μm mesh screen and rinsing the retentate into a vial with hatchery well water. The hatchery well water had a pH of 7.5–8.0, total hardness of 222 mg/L as $CaCO_3$, and total alkalinity of 222 mg/L. For each test, the triactinomyxons were categorized after treatment, as fired, unfired, or empty. Triactinomyxons with at least one of the three polar filaments discharged were categorized as fired. Unfired triactinomyxons were those that retained all their filaments within the polar capsules. The 'empty' category was for triactinomyxons without a spore body. Triactinomyxons recorded as empty were included in the total for percentage calculations. To adequately reflect the effect of pH on discharge, the empty and fired categories were combined for analysis and graphing. Percentage data were arc-sine transformed before analysis, with either *t*-tests or analysis of variance (ANOVA).

Vital staining with propidium iodide (PI) and fluorescein diacetate (FDA) was used to assess viability (Jones and Senft 1985; Markiw 1992b) of triactinomyxons exposed to KCl or direct current. An aqueous stock solution of PI (52 mg/mL) was made and kept frozen (-40°C) in 1-mL aliquots. Vials were then thawed and ready for use on the test day. A concentrated stock solution of FDA (5 mg/mL acetone) was also kept at -40°C. On the day of an experiment, 100 μL of the concentrated FDA solution was added to 8 mL of well water to make a working stock solution of FDA stain.

Effect of pH on Discharge

Acid pH was achieved using HCl and basic pH using KOH. The ranges of pH tested were 3.8–1.1 and 11.7–12.9. Acids and bases were diluted with de-ionized water to achieve the desired pH. These solutions were combined with equal volumes of triactinomyxon stock solution (pH 7.5–8.0) on slides. The resulting pH experienced by triactinomyxons was measured separately in test tubes with a freshly calibrated pH meter (Orion, model SA 720). After mixing triactinomyxon and pH solutions on a slide, the slide was immediately cover slipped and observed. Time required for reading one slide (i.e., the maximum exposure time) averaged 3.5 min.

Effect of Salts on Discharge

The effects of various chloride salts (NaCl, KCl, CaCl$_2$, NH$_4$Cl, MgCl$_2$) on polar filament discharge were tested at concentrations ranging from 3.1 to 100‰. In addition, other salts (NaPO$_4$, KPO$_4$, and KI) were evaluated for a narrower range of concentrations, to determine if certain anions or cations were better discharge agents. Salts were evaluated by combining a salt stock solution of known concentration with equal volumes of triactinomyxon stock solution on glass slides.

For the viability test, 30 µL of triactinomyxon stock solution, 30 µL each of FDA and PI and 90 µL of 20‰ KCl, were added to a microscope slide. This solution was mixed carefully using the edge of a cover slip, producing a 10‰ KCl exposure. The slides had cover slips placed on them and were incubated in an opaque plastic chamber in a refrigerator for 30 min prior to reading.

To determine the effect of calcium ion removal on discharge, two calcium chelators were tested. These were sodium citrate and EGTA (ethelenebis [oxyethylenenitrilo] tetra-acetic acid). A 50-mM solution of sodium citrate was put on three slides with equal volumes of triactinomyxon solution, yielding 25-mM exposure concentrations. Slides were then incubated for 60 min, and discharge was recorded as noted above. In addition to controls (triactinomyxon stock solution only), slides were made in which a NaCl solution of the same molarity was mixed with the triactinomyxon stock solution. This was to compare the discharge effect of sodium between the two solutions. EGTA is only soluble in solutions with relatively high pH, so it was placed in de-ionized water into which NaOH was titrated until all the EGTA dissolved. The resulting pH was 9.4. The EGTA concentration, after dilution with an equal volume of triactinomyxon stock solution, was 12.5 mM. To rule out pH as the factor causing firing, a solution of NaOH and de-ionized water of pH 9.4 was used as a control. Slides were made, incubated, read, and recorded, as in the sodium citrate test.

Electricity and Discharge

Electrical pulses of direct current were also examined as a method for firing triactinomyxon polar filaments using an electroporator (model T 820, BTX, San Diego, California). Electroporation involves administering short, high-voltage pulses of electricity to a cell (Palaniappan et al. 1990). We put 100-µL samples of triactinomyxon stock solution into a 1 mm wide chamber, through which the electricity was pulsed. Voltage (1 or 3 kV), pulse length (0–99 µsec), and the number of pulses (0–25) were manipulated to evaluate effects on triactinomyxon polar filament discharge. Two slides of 30 triactinomyxons each were counted for each treatment combination, but empty triactinomyxons were not included in the total or percentage calculations.

Viability of triactinomyxons was assessed after treating 100 µL of triactinomyxon stock solution to a varying number (1–25) of 3 kV, 99 µsec pulses of direct current. After electrical exposure, the treated solution was put on a slide and stained with 50 µL each of FDA and PI. Two slides of 30 triactinomyxons each were made for each treatment. Cover slips were placed on the slides, which were incubated in a dark chamber in a refrigerator for at least 30 min before reading.

Effects of Mucus and Tricaine Methanesulfonate on Discharge

Three tests were conducted in which mucus from rainbow trout *Oncorhynchus mykiss* was evaluated as a discharge agent. Additional tests were conducted to evaluate the effect of tricaine methanesulfonate (MS-222) or bovine mucin on discharge of triactinomyxon polar filaments. In the first test, mucus was scraped from a 250-mm fish using a microscope slide. A drop of the mucus, applied to a microscope slide, was mixed with 50 µL of triactinomyxon stock solution harvested the same day. The corner of a cover slip was used to gently mix the two solutions on the slide. The slide was cover slipped and viewed at 400×, either immediately or after 30 min. Triactinomyxon discharge was categorized as noted above.

A subsequent test was conducted to determine if mucus in combination with mechanical contact would trigger polar filament extrusion. Mucus, scraped with a glass slide from a rainbow trout, was applied to a glass rod. The rod was stirred for 1 min, within a 1-mL suspension of triactinomyxons in well water. The mucus on the rod was transferred to a microscope slide for examination, and 200 µL of the triactinomyxon solution remaining in the test tube was added to the slide as well. Slides were examined at 400×, and discharge was categorized as noted above.

In the third test, the effect of a refrigerator magnet on discharge was tested with or without mucus. The magnet strength was 0.45 Gauss, as measured by a gaussmeter. Mucus, scraped from the side of a rainbow trout (250 mm) anesthetized with tricaine, was transferred to each of 3 slides. Triactinomyxons, from a stock solution in hatchery well water, were added (100 uL) on top of the mucus. These slides were exposed to the magnetic field for 1 min, by placing the slide directly upon the magnet. Control slides, of triactinomyxon stock only, were not exposed to the magnetic field. Three slides of mucus and triactinomyxons were not exposed to the magnetic field. Three slides were exposed to the magnetic field for 1 min without any mucus present. A total of 50–100 triactinomyxons were viewed on each slide.

Due to contradictory results from the previous tests, the potential for tricaine methanesulfonate to act as a discharge agent was investigated in two series of tests. The first series evaluated discharge at concentrations ranging from 0.091 to 8,182 mg/L (after mixing with triactinomyxon stock). In the second series, tricaine concentrations of 50, 100, 200, and 300 mg/L were evaluated. These solutions were made by diluting appropriate amounts of a 100 mg/mL stock solution with de-ionized water. For each concentration, tricaine and triactinomyxon stock were mixed on each of 3 slides, and 100 triactinomyxons were observed on each slide. Control slides were made by mixing equal amounts of triactinomyxon stock and de-ionized water on each of 3 microscope slides.

Bovine submaxillary mucin (Type 1-S) diluted with de-ionized water was mixed with triactinomyxon stock solution (50 uL of each) to achieve concentrations of 5, 50, 500, and 1000 mg/L. Three slides were observed for each concentration, categorizing 100 triactinomyxons per slide as noted above. Three control slides were made directly from the triactinomyxon stock solution.

Other Discharge Agents

Glutathione was mixed with an equal amount (200 μL) of freshly-harvested triactinomyxons in two microcentrifuge tubes. The glutathione stock solutions were prepared with de-ionized water. After mixing in the tubes, glutathione concentrations were either 0.5 or 0.05 M. After 10–15 min in the tubes, 180 μL of the mix was transferred to a slide. Discharge was categorized as noted above, using 2 slides per concentration. An additional pair of slides was prepared, on which either 1.0 or 0.1 M glutathione was dried on each slide. To each slide, 180 μL of triactinomyxons was added and incubated for 1–3.5 h. Triactinomyxons on the slide were categorized as noted above.

Serial dilutions of a 0.01 M proline solution were prepared using de-ionized water. Equal quantities of proline and triactinomyxon stock solution (50 μL) were mixed on a microscope slide, covered with a cover slip, and immediately examined. The proline concentrations after mixing were 5×10^{-3}, 5×10^{-4}, 5×10^{-5}, and 5×10^{-6} M. Three slides, with a total of 92–100 triactinomyxons per slide, were observed for each concentration.

NANA (N-acetyl neuraminic acid, type VI from *E. coli*) was tested at concentrations (after mixing) of 0.5×10^{-7}, 1.0×10^{-7}, 2.0×10^{-7}, and 4.0×10^{-7} M. On each of 3 slides per concentration, 100 triactinomyxons were observed. For the NANA and proline tests, control slides were made by mixing 50-μL triactinomyxon stock solution with 50-μL de-ionized water. Discharge was categorized as noted above, for replicate slides.

Other agents evaluated included three neurochemicals. A 0.79-mM stock solution of synthetic salmon angiotensin-1 was made with de-ionized water. Serial dilutions with de-ionized water provided test concentrations of 0.395, 0.0395, and 3.95×10^{-3} mM, after mixing equal amounts (50 μL) of angiotensin and freshly harvested triactinomyxon stock on a microscope slide. Two slides, with 32–71 triactinomyxons on each, were observed for each concentration. Acetylcholine chloride was evaluated at concentrations (after mixing with triactinomyxons) of 1.13, 11.3, and 113 mg/L. Bradykinin concentrations (after mixing with triactinomyxons) tested were 5×10^{-4}, 5×10^{-5}, 5×10^{-6}, 5×10^{-7}, and 5×10^{-8} M. For both the bradykinin and acetylcholine chloride tests, 3 slides were observed per concentration, categorizing up to 100 triactinomyxons per slide as noted above. For each of the neurochemical tests, 3 control slides were made in which 50-μL de-ionized water was mixed with triactinomyxon stock.

RESULTS

Effect of pH on Discharge

Discharge of triactinomyxon polar filaments increased toward both extremes of the pH range (Figure 1). However, the percentage discharged

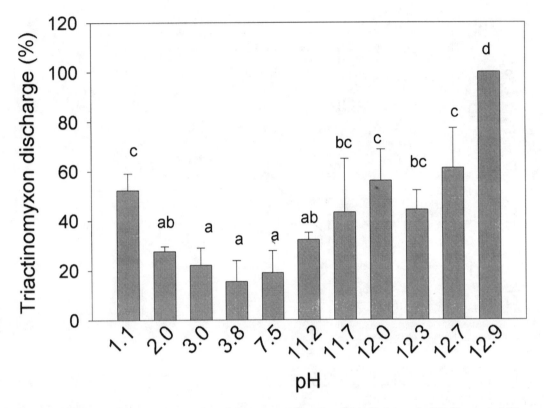

Figure 1. Mean percentage of *Myxobolus cerebralis* triactinomyxons with discharged polar filaments (±SD; $N = 3$ slides of 30 triactinomyxons each, fired and empty categories combined) after exposure to high and low pH. Significant differences ($P \leq 0.050$) between pH treatments are noted by a different letter.

was greater at the basic end of the pH spectrum. A significant difference ($P \leq 0.050$) from controls was not found until the pH was at or below 1.1 or above 11.7. A pH of 12.9 was needed to discharge 100% of polar filaments.

Effect of Salts on Discharge and Viability

Comparison of discharge among the chloride salts indicated that some cations were significantly more effective than others (Figure 2). K^+ tended to have higher discharge percentages than the other ions, though the difference was significant only at 50‰ (ANOVA, $P = 0.039$). NH_4^+ had significantly lower discharge percentages at 6.2‰ and 12.5‰ than for K^+, Mg^{2+}, or Ca^{2+}. There was no significant difference in the percent discharged (t-test, $P = 0.28$) between $NaPO_4$ (21.1 ± 5.1%, SD) and KPO_4 (26.7 ± 5.8%) at 6.2‰. Similarly, at 12.5‰, there were no significant differences in discharge between $NaPO_4$ (30.0 ± 6.7%) and KPO_4 (42.2 ± 8.4%, $P = 0.12$). Generally, as salt concentration increased, the discharge percentage increased, peak-

ing at 71% for 100‰ KCl. Most of the salts induced discharge, indicating that high osmolarity in general can induce discharge. Survival was compromised by the use of KCl. After exposure to 10‰ KCl for 30 min, only 38% of triactinomyxons remained viable, compared with 74% with controls.

Discharge differences among anions were also significant. Comparison of KCl, KI, and KPO_4 indicated that Cl^- was significantly more effective ($P \leq 0.02$) at both 6.2 (45.6% discharge) and 12.5‰ (57.8%) than the other anions. Comparison of NaCl and $NaPO_4$, at 12.5‰, also indicated chloride ion was more effective (44.4 versus 30.0% discharged, respectively; t-test, $P = 0.02$).

Both sodium citrate and NaCl, at tested concentrations, had significantly more fired polar filaments ($P = 0.010, 0.015$) than did controls (Table 1). However, there was no significant difference between NaCl and sodium citrate treatments ($P = 0.336$). After 1 h, both the EGTA and pH 9.4 solutions produced significantly more triactinomyxons

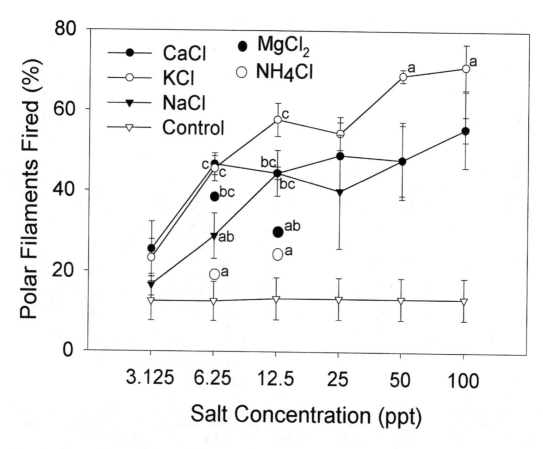

Figure 2. Mean percentage of *M. cerebralis* triactinomyxons with discharged polar filaments (±SD; *N* = 3 slides of 30 triactinomyxons each) after exposure to various concentrations of chloride salts. Within a given concentration, means that are significantly different (ANOVA, $P \leq 0.050$) among the salts are noted by a different letter. Salt concentrations reflect actual concentration after mixing on the slide.

Table 1. The effect of sodium citrate and EGTA on the percentage (Mean ± SD; *N* = 3) of discharged polar filaments of *Myxobolus cerebralis* triactinomyxons after 1 h incubation. Data for controls for the cation and pH effects are shown as well.

Treatment	Polar filament discharge (% ± SD)
25 mM NaCl	32.22 ± 10.99*
25 mM sodium citrate	22.22 ± 6.84*
12.5 mM EGTA (pH 9.4)	35.55 ± 12.27*
NaOH (pH 9.4)	16.66 ± 7.20*
Control	0.33 ± 0.47

*Denotes treatments that are significantly different ($P \geq 0.050$) from control.

with fired polar filaments (P = 0.015, 0.033) than controls (Table 1). EGTA treatments extruded more polar filaments (36%) than did the pH 9.4 NaOH solution (17%), but the difference was not significant (P = 0.134). It appears that most of the firing observed with EGTA can be explained by the pH of the solution as opposed to the chelating action of EGTA.

Effect of Electricity on Discharge

Polar filament discharge varied with electrical pulse length. At 1 kV, a 1-μsec pulse induced 18% of triactinomyxons to fire polar filaments. Pulse lengths of greater than or equal to 5 μsec significantly increased the percentage of polar filaments fired (Figure 3). When the pulse length was increased to 99 μsec (electroporator maximum), 76.7% of triactinomyxons had discharged filaments after a single pulse. There was no significant difference in the percent discharged between a single 99-μsec pulse of 1 kV (74.0%) and the machine maximum of 3 kV (76.7%; P = 0.877). At a con-

stant pulse length of 99 μsec and a voltage of 3 kV, the number of pulses significantly increased the number of triactinomyxons firing polar filaments (Figure 4). There was no significant increase in discharge if the number of pulses increased from 1 to 20 (P = 0.737). However, after 25 pulses, 98% of triactinomyxons had fired filaments; this was a significant increase from a single pulse (P = 0.028). The correlation between the number of pulses and the percentage fired was significant (P = 0.002, r^2 = 0.50, ordinary least-squares regression). The correlation between pulse length and polar filament discharge was also significant (P = 0.015, r^2 = 0.24).

Viability was also affected by electricity. After a single 3-kV, 99-μsec pulse, 83% of triactinomyxons were dead, a significant increase from controls. After treatment with various numbers of pulses of direct current, a significant correlation (r^2 = 0.914, P < 0.001) was found between the percentage of triactinomyxons with fired polar filaments and the percentage of dead triactinomyxons (Figure 4). As

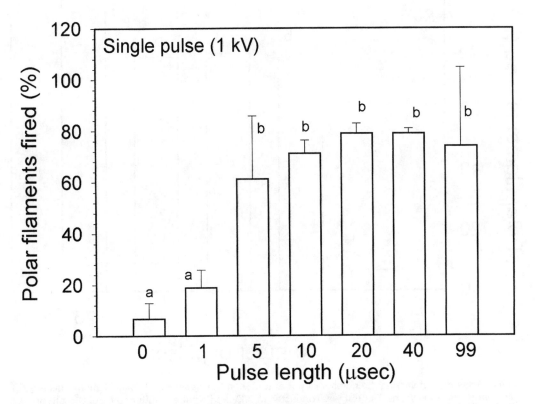

Figure 3. The mean percentage of *M. cerebralis* triactinomyxons with discharged polar filaments (±SD) after exposure to a single 1kV pulse at various pulse lengths. Means for each pulse length (N = 3) that are significantly different from each other ($P \leq 0.05$) have different letters.

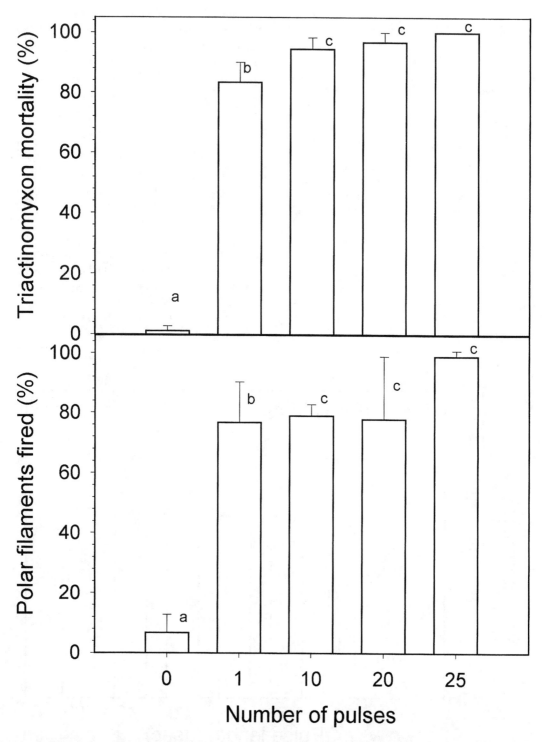

Figure 4. The effect of varying the number of pulses (0–25) on the viability and polar filament discharge of *M. cerebralis* triactinomyxons. The test was conducted at 3 kV and a pulse length of 99 μsec. Results are expressed as mean percentage of dead triactinomyxons ± SD (top) and mean (±SD, *N* = 3) percentage of triactinomyxons with fired polar filaments (bottom). Significant differences (*P* ≤ 0.050) between the number of pulses are noted by a different letter.

the number of pulses increased, more triactinomyxons were killed. However, the differences in percent
mortality did not differ among triactinomyxons
exposed to 10, 20, and 25 pulses. To achieve consistent 100% mortality of triactinomyxons, 25 pulses
of 3 kV for 99 μsec were required.

Effect of Mucus and Tricaine Methanesulfonate on Discharge

In the first test, no discharge was observed among
triactinomyxons that had been exposed to mucus
(N = 29–33 triactinomyxons). If mechanical agita-
tion was used in combination with the mucus, no
extruded polar filaments were seen (N = 116 triactinomyxons). Bovine mucin failed to induce
higher rates of polar filament extrusion than
observed for controls (Table 2). Concentrations of
mucin from 5 to 1000 mg/L resulted in only
13.3–22.0% discharge.

The magnetic field test indicated that the
percentage of polar filaments discharged in a magnetic field alone (9.0 ± 2.0% discharge) did not
differ significantly from controls (4.3 ± 2.3%).
However, rainbow trout mucus induced significantly higher (ANOVA, P < 0.001) discharge
rates, either with (39.3 ± 3.0%) or without the
magnetic field present (31.3 ± 16.3%). The effect
of mucus in this test was notably different from the
previous experience, so use of MS-222 anesthetic
in the second trial was suspected.

Further testing indicated that the anesthetic
had a significant effect on discharge, although the
doses required for discharge were higher than nor-
mally used for anesthesia. Tricaine caused significantly higher rates of discharge than observed in
controls, but only at concentrations at 909 mg/L
and above (Figure 5). Maximum mean discharge
was observed at 6,364-mg/L tricaine (29.4 ±

Table 2. Effect of mucin from bovine submaxillary gland on the mean percent discharge (± SD) of polar
filaments of the *Myxobolus cerebralis* triactinomyxon.

		Concentration (mg/L)		
0.0	5.0	50	500	1,000
13.4 ± 5.6	22.0 ± 6.1	14.0 ± 2.6	13.3 ± 2.9	14.7 ± 8.3

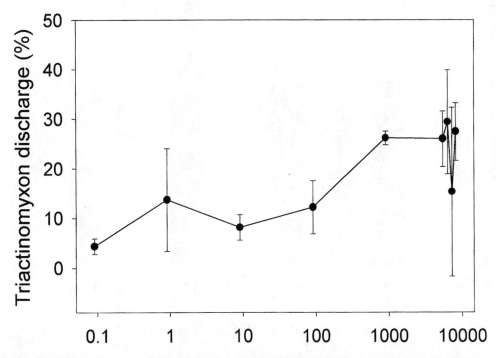

Figure 5. The effect of tricaine methanesulfonate anesthetic (MS-222) on the mean percentage of *M.
cerebralis* triactinomyxons with discharged polar filaments (±SD, N = 3 slides of 38–395 triactinomyxons each).

10.5%). At 90.9 mg/L of MS-222, discharge was about 12.2% but did not significantly differ from controls (1.9%). In the second series of anesthetic tests, the percent discharge at MS-222 concentrations, ranging from 50 to 300 mg/L, varied from 16.3 to 32.3%, and did not differ significantly from controls (26.1%; Figure 6).

Other Agents

Glutathione discharge in the 0.05 M treatment averaged 34.2% (N = 24, 37 triactinomyxons) and 26.3% (N = 39, 46 triactinomyxons) in the 0.5-M treatment. Controls averaged 22.5% discharge (N = 20, 25 triactinomyxons). Additional tests with 0.1-M gluthione dried on a slide resulted in an average of 43.1% discharge. Slides coated with1.0 M glutathione were too thick with crystals to adequately observe the triactinomyxons.

Polar filament discharge after exposure to proline concentrations of 5×10^{-6} to 5×10^{-3} M did not significantly differ from controls (Table 3). Similar-ly, NANA failed to significantly increase the percentage of discharged triactinomyxons (Table 3), varying from only 10.6% to 18.1% discharge among the concentrations.

The discharge percentages among controls varied from 7.7% to 17.7%. The neurochemicals tested did not induce any significantly higher discharge of polar filaments in the ranges of concentration tested. Mean discharge percentages for acetylcholine chloride ranged from only 15.0% to 19.1% (Table 4). For angiotensin and bradykinin, mean discharge percentages were also low, ranging from 5.5% to 15.0% and 8.4% to 10.0%, respectively.

DISCUSSION

pH

In this study, pH effects on triactinomyxon polar filament extrusion were observed only at extreme values. Acid-induced discharge was only signifi-

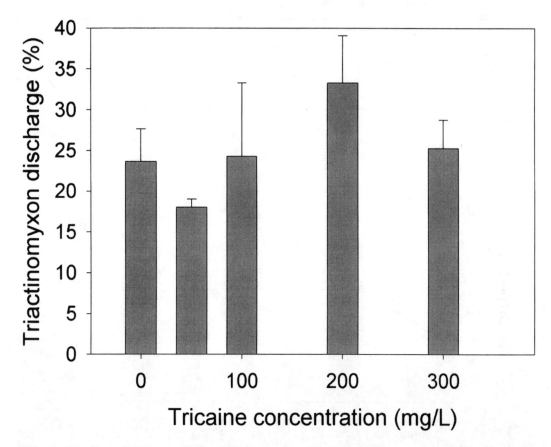

Figure 6. The effect of tricaine methanesulfonate concentrations of 50–300 mg/L on the mean percentage of *M. cerebralis* triactinomyxons with discharged polar filaments (±SD, N = 3 slides of 100 triactinomyxons each).

Table 3. Effect of proline or NANA on the percentage (± SD, N = 3) of triactinomyxons (tams) of *Myxobolus cerebralis* that had at least 1 polar filament discharged or were lacking a spore body (empty). The range in the number of triactinomyxons per slide is also presented.

Chemical concentration	Discharged (%)	Empty (%)	Tams/slide (range)
Proline			
0.0 M	5.5 ± 1.5	4.0 ± 1.7	100
5×10^{-6} M	5.4 ± 2.1	6.4 ± 1.1	98–100
5×10^{-5} M	6.2 ± 1.6	5.2 ± 0.8	93–100
5×10^{-4} M	5.1 ± 2.0	2.3 ± 1.1	100
5×10^{-3} M	6.5 ± 2.1	3.8 ± 2.5	92–100
NANA			
0.0 M	7.0 ± 1.0	9.3 ± 2.5	100
1.0×10^{-7} M	17.0 ± 3.5	6.3 ± 2.1	100
2.0×10^{-7} M	13.0 ± 6.6	9.0 ± 2.6	100
4.0×10^{-7} M	14.7 ± 2.9	7.7 ± 1.5	100
8.0×10^{-7} M	9.7 ± 4.0	7.7 ± 4.0	100

Table 4. Effect of various neurochemicals on the percentage (± SD, N = 3) of triactinomyxons (tams) of *Myxobolus cerebralis* that had at least 1 polar filament discharged or were lacking a spore body (empty). The range in the number of triactinomyxons per slide is also presented.

Chemical concentration	Discharged (%)	Empty (%)	Tams/slide (range)
Acetylcholine chloride			
0.0 mg/L	12.0 ± 6.0	4.0 ± 1.7	100
1.13 mg/L	18.0 ± 5.0	5.3 ± 2.5	100
11.3 mg/L	15.3 ± 4.0	6.0 ± 3.6	100
113.0 mg/L	14.0 ± 9.5	8.3 ± 4.0	100
Angiotensin			
0.0 mM	14.8 ± 9.3	15.2 ± 7.2	40–72
3.95×10^{-3} mM	5.7 ± 5.4	15.9 ± 6.7	54–63
3.95×10^{-2} mM	5.0 ± 5.3	8.9 ± 0.1	57–78
3.95×10^{-1} mM	13.2 ± 0.6	11.7 ± 1.6	39–66
Bradykinin			
0.0 M	8.7 ± 4.6	6.0 ± 1.0	100
5×10^{-8} M	8.7 ± 4.2	13.0 ± 3.6	10
5×10^{-7} M	7.3 ± 4.0	14.3 ± 4.9	100
5×10^{-6} M	8.3 ± 2.5	11.7 ± 1.5	100
5×10^{-5} M	8.3 ± 4.0	10.0 ± 1.7	100
5×10^{-4} M	8.3 ± 6.1	10.7 ± 2.3	100

cantly higher than controls at pH 1.1 using HCl. Gurley (1894) observed extrusion of *Henneguya creplini* myxospores induced by glacial acetic acid (concentration or pH not given). Gurley (1894) also noted myxospores of a wide variety of myxozoans extruded polar filaments when exposed to sulfuric acid. Sulfuric acid did not give similar results with *Henneguya rupestris*, nor did hydro-chloric acid (Herrick 1941). Similar failure of acids to extrude filaments has been noted for *M. cerebralis* (Uspenskaya 1957) and *M. cartilaginis* (Hoffman et al. 1965). However, Herrick (1936) reported that some filaments of *M. kostiri* extruded after using concentrated sulfuric acid. Stomach acids have also generally failed to induce extrusion of myxospore polar filaments in tests with

M. *cerebralis* (Uspenskaya 1957), M. *muelleri* (Lom 1964), and M. *cartilaginis* (Hoffman et al. 1965). However, Iversen (1954) used the digestive juices of a freshly killed rainbow trout and was successful in extruding polar filaments of *Myxobolus squamalis*.

Bases were more effective discharge agents than acids in this study. This has also been noted for myxospores of other myxozoans (Hoffman et al. 1965). Strong bases such as potassium and sodium hydroxide, at concentrations of greater than 1–2%, have been used to extrude polar filaments in *Myxidium minteri*, *Chloromyxum majori*, M. *cerebralis*, M. *kisutchi*, M. *notemigoni*, and M. *cartilaginis* (Yasutake and Wood 1957; Uspenskaya 1957; Lewis and Summerfelt 1964; Hoffman et al. 1965). The effect of KOH on discharge of the polar filaments was temperature-dependent; using 1.5% KOH, Hoffman et al. (1965) noted that no filaments of M. *cartilaginis* myxospores fired at 6°C, but at 23°C and 40°C, 57% and 96% of polar filaments were extruded. Herrick (1936) used ammonium hydroxide (no concentration or duration noted) but was unsuccessful in extruding polar filaments of *Myxobolus kostiri*. Herrick (1941) was similarly unsuccessful using potassium hydroxide on *Henneguya rupestris*. Durations, pH, temperatures, and concentrations were not reported in most of these studies; therefore, it was difficult to ascertain whether or not there are true genetically-based differences among Myxozoans in susceptibility to acids and bases.

Among Cnidaria, the effects of pH have depended on whether cnidae are isolated from the cnidocyte. In-situ cnidae discharged at pH values below 4 and above 11, whereas isolated cnidae discharged only at pH values of 2 or below (Blanquet 1970). Sulfuric acid solutions in seawater (10%) caused discharge (Parker and Van Alstyne 1932). Exposure of isolated cnidae to sulfuric, citric, hydrochloric, and acetic acids, with mean pH values ranging from 2.9 to 3.3, caused 50% of the cnidae to discharge (Yanagita and Wada 1953). This is higher than the discharge rates observed in this study at similar pH levels (22–28%). Yanagita and Wada (1953) used potassium hydroxide, sodium hydroxide, and ammonium hydroxide, at pH values ranging from 10.8 to 11.0, and caused 50% of the cnidae to discharge. These results are also higher than the rates of discharge observed in this study at a similar pH.

Salts

Anions have varied in their ability to discharge cnidae. Salleo et al. (1983b) noted that the discharge potency of sodium salts isosmotic with seawater followed the lyotropic series: SO_4^{2-} < Ch_3COO^- < F^- < Cl^- < Br^- < NO_3^- < I^- < ClO_4^- < SCN^-. In the present study, Cl^- was superior to I^- and PO_4^- for discharge of triactinomyxon polar filaments. Chloride ions, especially with divalent cations ($CaCl_2$ and $MgCl_2$), were effective discharge agents for the isolated cnidae of the anemone *Diadumene*, at concentrations above 6.9‰ (Yanagita 1959). In the present study, Cl^- concentrations of 6.2‰ or higher were required for significant increases in discharge.

The overall effect of high salt concentrations on discharge of triactinomyxon polar filaments indicated that the influx of ions can play a role in polar filament discharge. In this study, K^+ was the most effective cation for triactinomyxon discharge. Gurley (1894) also noted the ability of KI to induce discharge for a variety of myxozoan myxospores. The efficacy of K^+ for discharge has also been observed for many cnidarian species (Parker 1905; Yanagita 1960; Santoro and Salleo 1991; Kawaii et al. 1997). This effect was muted in isolated cnidae (Yanagita 1959; Blanquet 1970; McKay and Anderson 1988). The effect of potassium was also dependent on the presence of calcium (McKay and Anderson 1988). Sodium ion depletion has also blocked discharge (Brinkmann et al. 1996).

Among other cations tested, NH_4^+ also stimulated in-situ cnidae discharge (Yanagita 1960). Blanquet (1970) noted that Ca^{2+} and Mg^{2+} did not induce discharge. Similarly, Na^+ and Li^+ as chloride salts were not effective (Yanagita 1960). In this study, NH_4^+ was not as effective for inducing discharge as K^+, Mg^{2+}, or Ca^{2+}, yet still was capable of low levels of polar filament discharge.

The discharge induced by the calcium chelators sodium citrate and EGTA, in this study, did not differ from the salt or pH controls, but discharge increased relative to the triactinomyxon stock controls. This indicated that high osmolality and pH were more conducive to discharge than the loss of Ca^{2+}. Results also indicated that a drop in Ca^{2+} is not needed for polar filament discharge.

In isolated cnidae, contrary to in-situ cnidae, removal of calcium has induced discharge (Lubbock and Amos 1981; Hidaka and Mariscal 1988). Isolated cnidae of *Pelagia noctiluca*, pretreated with $CaCl_2$, were inhibited from discharging when exposed to trypsin, an effect not observed in controls or for cnidae treated with $MgCl_2$ (Salleo et al. 1983a).

The effect of calcium ion does not apply to all cnidae. Salleo et al. (1990) noted that very little calcium was discharged from the cnidae of *Calliactis parasitica*. These authors also found that Ca^{2+} and Mg^{2+} free artificial seawater did not induce discharge of isolated cnidae. Mariscal (1984) has similarly observed holotrichous isorhiza cnidae of *Haliplanella luciae* that appear to lack calcium, suggesting that other ions and chemical messengers aid in triggering these.

Mucus and Tricaine Methanesulfonate

Mucus would seem to be a logical candidate for a discharge agent, but results appear to vary among species of Myxozoa. Actinospore forms, such as aurantiactinomyxon, raabeia, neoactinomyxum, and echinactinomyxon, have discharged polar filaments in response to mucus (Yokoyama et al. 1995; Xiao and Desser 2000). The percentage of extruded polar filaments varied with species and actinospore, but no actinospore was species specific (Xiao and Desser 2000). However, Xiao and Desser (2000) noted that some triactinomyxon forms were discharged by mucus from a narrower range of species. For example, for the 'triactinomyxon C' form (Xiao and Desser 1998), discharge was induced in more than 90% of the actinospores, by common shiner *Luxilus cornutus*, fathead minnow *Pimephales promelas*, and golden shiner *Notemigonus crysoleucas* mucus, but discharge was 12% or less for the other species, including the other cyprinid *Semotilus atromaculatus*. Triactinomyxon F had a narrower range, discharging 77% in mucus of common shiner, 12% in golden shiner, and 8% or less in the remaining species. Recent work with mucus has shown that it contains compounds that can alter membrane permeability (Ebran et al. 2000). This may be a part of the signaling mechanism for those Myxozoa that respond to mucus.

Yokoyama et al. (1995) noted that mucin from bovine submaxillary gland (0.1% w/v) was also a discharge agent. In this study, bovine mucin failed to induce extrusion at concentrations from 5 to 1000 mg/L. Similarly, El-Matbouli et al. (1999) failed to extrude polar filaments of M. *cerebralis* triactinomyxons, using rainbow trout mucus. This study also failed to find a link between triactinomyxon discharge and contact with mucus. This was true whether there was a mechanical stimulus or not. It is possible that a larger force may be required than was applied by the stirring rod used in this study. Propulsive forces of 0.0–0.6 N have been recorded for swimming cod *Gadus morhua* (Videler 1993), but none were measured in this study. However, triactinomyxon attachment to the respiratory epithelium and buccal cavity (El-Matbouli et al. 1999), where presumably these forces are small, indicates that other mechanisms may be responsible for discharge.

Tricaine induced significant increases in polar filament discharge, but the concentrations required for discharge were much higher than those used for routine fisheries studies (Eisler and Backiel 1960). At concentrations of 50–300 mg/L no significant increase in discharge was observed. This observation would indicate that the polar filament extrusion for triactinomyxons exposed to mucus were not influenced by anesthetic use for mucus collection. Further study is needed to examine the combined effect of mucus and MS-222.

Electrical potentials are an integral part of living organisms (Crane 1950; McCaig and Robinson 1980). These result from the high electrical resistance of the lipid membrane of cells and charged ions in the fluids on either side of the membrane (Zhelev and Needham 1994). The intercellular electrical currents are generally in the mV range (McCaig and Robinson 1980), so the associated magnetic field produced by a living fish would be much smaller than that provided by the magnet used in this study. The magnetic field had no effect on discharge in this study, either with or without mucus present. The hypothesis that a magnetic field might be required for discharge was therefore rejected.

Electricity

Pulses of direct current in this study were capable of inducing polar filament extrusion. The effect was influenced by pulse length. The longest pulse duration tested (99 μsec) produced the greatest discharge. The number of pulses was also a factor, with maximum mortality and discharge observed after 25 pulses of 3 kV. This is the first reported attempt to use electricity for polar filament discharge in Myxozoa.

Electricity has stimulated cnidae discharge, primarily in-situ. Glaser and Sparrow (1909) found that in-situ cnidae of *Metridium* responded to electricity, but isolated cnidae did not. This effect was later corroborated by Parker and Van Alstyne

(1932), for *Metridium* and *Physalia*, and by Yanagita (1960), for *Diadumene*. The threshold direct current Yanagita (1960) needed to induce extrusion varied from 8 to 20 μA (delivered by a 67.5 V battery through a 100 kΩ potentiometer). Pantin (1942) experimented with *Anemonia sulcata* and found that electrical shocks (240 V from a 2-μF condenser and passed through a 400 Ω potentiometer) stimulated cnidae discharge. Anderson and McKay (1987) found that isolated cnidae of *Cladonema* and *Chrysaora* failed to discharge after electrical stimulation, despite detection of a variety of voltage-dependent ionic currents. In-situ cnidae of *Physalia* similarly failed to discharge (Anderson and McKay 1987). In *Hydra* tentacles, stenotele and desmoneme nematocysts readily discharged, using 24-V DC, but the isorhizas did not (Tardent 1988).

Electrical pulses in this study were deadly. Despite the positive correlation between discharge and triactinomyxon mortality, it is not clear if discharge was the cause of death. Nonetheless, the results do indicate a potential for using electricity as a means of disinfection. The recommended lethal direct current would be 25 99-μsec pulses of 3 kV.

Other Discharge Agents

In sea anemones, two distinct classes of chemoreceptors, located on the supporting cells adjacent to cnidocytes, predispose these to discharge cnidae in the event of prey contact (Thorington and Hessinger 1988; Watson and Hessinger 1987). One class of chemoreceptor binds N-acetylated sugars, and the other binds certain amino compounds, including proline (Watson and Roberts 1994). N-acetylated sugars, which occur as conjugates of the surface mucins or external chitin of prey, sensitize cnidocytes to discharge (Thorington and Hessinger 1988). For example, glycine and N-acetylneuraminic acid (NANA) have chemosensitized cnidae discharge (Thorington and Hessinger 1990; Thorington and Hessinger 1998). Chemoreceptors for N-acetylated sugars may stimulate adenylate cyclase to sensitize and tune mechanoreceptors involved in initiating cnidae discharge (Watson and Hessinger 1992). Other chemosensitizing agents include glutathione (Loomis and Lenhoff 1956), glucose-6-phosphate, fructose-6-phosphate, fructose-1,6-diphosphate, acetylcholine (1/100,000), and acetylcholine with physostigmine (10^{-5} M, a cholinesterase inhibitor; Lentz 1966).

In this study, mixing proline, NANA, or glutathione with triactinomyxons did not induce polar filament extrusion at concentrations similar to those used for cnidarians. Also, the neurochemicals acetylcholine chloride, angiotensin, and bradykinin were ineffective discharge agents. For hydra, the presence of a primitive nerve network may be the reason that discharge is affected by neurochemicals such as acetylcholine (Lentz 1966). The absence of any nerves in myxozoans may explain the negative results observed in this study. The inability of chemosensitizing agents such as NANA to induce polar filament discharge tends to indicate that the supporting cell complex that is present among cnidarians may not be present in Myxozoa. Further work is needed to elucidate the exact biochemical and physical mechanisms of polar filament discharge.

In summary, there are some similarities and differences between cnidae and polar capsules in their response to discharge agents, including the sensitivity to K^+, electricity, and extreme pH values. However, the discharge rate in response to extreme pH was lower than for cnidarians. Major differences between the two structures include the lack of response by triactinomyxons to neurochemicals and chemosensitizing agents. The loss of Ca^{2+} induced by the chelators tested did not induce discharge of triactinomyxons, whereas this can cause discharge in some cnidarians. Bovine mucin and rainbow trout mucus were not effective discharge agents for triactinomyxons, despite mechanical contact and magnetic fields. Electrical pulses were effective in discharging and inactivating triactinomyxons. Further research is needed to explore the potential for premature discharge of polar capsules as a means of preventing infection of fish by myxozoan parasites.

ACKNOWLEDGMENTS

This work has been funded by the Federal Aid in Sport Fish Restoration program and the Utah Division of Wildlife Resources.

REFERENCES

Anderson, P. A. V., and M. C. McKay. 1987. The electrophysiology of cnidocytes. Journal of Experimental Biology 133:215–230.

Blanquet, R. S. 1970. Ionic effects on discharge of the isolated and *in situ* nematocysts of the sea anemone, *Aiptasia pallida*: a possible role of calcium. Comparative Biochemistry and Physiology 35:451–461.

Brinkmann, M., D. Oliver, and U. Thurm. 1996. Mechanoelectric transduction in nematocytes of a hydropolyp (Corynidae). Journal of Comparative Physiology 178:125–138.

Crane, E. E. 1950. Bioelectric potentials, their mainte-
nance and function. Progress in Biophysics and Bio-
physical Chemistry 1:85–136.

Ebran, N., S. Julien, N. Orange, B. Auperin, and G.
Molle. 2000. Isolation and characterization of novel
glycoproteins from fish epidermal mucus: correlation
between their pore-forming properties and their
antibacterial activities. Biochimica et Biophysica
Acta 1467:271–280.

Eisler, R., and T. Backiel. 1960. Narcotization of chinook
salmon fingerlings with tricaine methanesulfonate
(MS-222). Transactions of the American Fisheries
Society 89:164–167.

El-Matbouli, M. R. W. Hoffman, H. Schoel, T. S. McDowell,
and R. P. Hedrick. 1999. Whirling disease: host specificity and interaction between the actinospore stage of
Myxobolus cerebralis and rainbow trout Oncorhynchus
mykiss. Diseases of Aquatic Organisms 35:1–12.

Glaser, O. C., and C. M. Sparrow. 1909. The physiology
of nematocysts. Journal of Experimental Biology
6:361–382.

Guilford, H. G. 1963. New species of Myxosporidia found
in percid fishes from Green Bay (Lake Michigan).
Journal of Parasitology 49:474–478.

Gurley, R. R. 1894. The Myxosporidia, or psorosperms of
fishes, and the epidemics produced by them. Report
of the U.S. Fish Commission 26:65–304.

Herrick, J. A. 1936. Two new species of Myxobolus from
fishes of Lake Erie. Transactions of the American
Microscopical Society 55:194–198.

Herrick, J. A. 1941. Some myxosporidian parasites of
Lake Erie fishes. Transactions of the American
Microscopical Society 60:163–170.

Hidaka, M., and R. N. Mariscal. 1988. Effects of ions on
nematocysts isolated from acontia of the sea
anemone Calliactis tricolor by different methods.
Journal of Experimental Biology 136:23–34.

Hoffman, G. L., R. E. Putz, and C. E. Dunbar. 1965. Studies on Myxosoma cartilaginis n. sp. (Protozoa: Myxosporidea) of Centrarchid fish and a synopsis of the
Myxosoma of North American freshwater fishes.
Journal of Protozoology 12:319–332.

Iversen, E. S. 1954. A new Myxosporidian, Myxosoma
squamalis, parasite of some salmonoid fishes. Journal
of Parasitology 40:397–404.

Jones, K. H., and J. A. Senft. 1985. An improved method
to determine cell viability by simultaneous staining
with fluorescein diacetate-propidium iodide. Journal
of Histochemistry and Cytochemistry 33:77–79.

Kawaii, S., K. Yamashita, N. Nakai, and N. Fusetani.
1997. Intracellular calcium transients during nematocyst discharge in Actinulae of the hydroid, Tubu-
laria mesembryanthemum. Journal of Experimental
Zoology 278:299–307.

Kent, M. L., L. Margolis, and J. O. Corliss. 1994. The
demise of a class of protists: taxonomic and nomenclatural revisions proposed for the protist phylum
Myxozoa Grassé, 1970. Canadian Journal of Zoology
72:932–937.

Kudo, R. 1918. Experiments on the extrusion of polar filaments of cnidosporidian spores. Journal of Parasitology 4:143–147.

Lentz, T. L. 1966. The cell biology of hydra. North-Hol-
land Publishing Company, Amsterdam.

Lewis, W. M., and R. C. Summerfelt. 1964. A Myxosporidian, Myxobolus notemigoni sp. n., parasite of the gold-
en shiner. Journal of Parasitology 50:386–389.

Lom, J. 1964. Notes on the extrusion and some other features of myxosporidian spores. Acta Protozoologica
2:321–327.

Lom, J., and I. Dyková. 1992. Protozoan parasites of fishes.
Elsevier, Amsterdam.

Lom, J., and P. de Puytorac. 1965. Studies on the myxosporidian ultrastructure and polar capsule development. Protistologica 1:53–65.

Loomis, W. F., and H. M. Lenhoff. 1956. Growth and sexual differentiation of Hydra in mass culture. Journal
of Experimental Zoology 132:555–574.

Lubbock, R., and W. B. Amos. 1981. Removal of bound
calcium from nematocyst contents causes discharge.
Nature (London) 290:500–501.

Mariscal, R. N. 1974. Nematocysts. Pages 129–178 in L.
Muscatine and H. M. Lenhoff, editors. Coelenterate
biology. Academic Press, New York.

Mariscal, R. N. 1984. Cnidaria: cnidae. Pages 57–68 in J.
Bereiter-Hahn, A. G. Maltowsky, and K. S.
Richards, editors. Biology of the integument. I.
Invertebrates. Springer-Verlag, Berlin.

Markiw, M. E. 1992a. Salmonid whirling disease. U.S. Fish
and Wildlife Service Leaflet 17, Washington, D.C.

Markiw, M. E. 1992b. Experimentally induced whirling
disease II. Determination of longevity of the infective
triactinomyxon stage of Myxobolus cerebralis by vital
staining. Journal of Aquatic Animal Health 4:44–47.

McCaig, C. D., and K. R. Robinson. 1980. The ontogeny
of the transepidermal potential difference in frog
embryos. Journal of General Physiology 76:14a.

McKay, M. C., and P. A. V. Anderson. 1988. Preparation and
properties of cnidocytes from the sea anemone Antho-
pleura elegantissima. Biological Bulletin 174:47–53.

Palaniappan, S., S. K. Sastry, and E. R. Richter. 1990.
Effects of electricity on microorganisms: a review.
Journal of Food Processing and Preservation
14:393–414.

Pantin, C. F. A. 1942. The excitation of nematocysts.
Journal of Experimental Biology 19:294–310.

Parker, G. H. 1905. The reversal of ciliary movement in
metazoans. American Journal of Physiology 13:1–16.

Parker, G. H., and M. A. Van Alstyne. 1932. The control
and discharge of nematocysts, especially in Metridium and Physalia. Journal of Experimental Zoology
63:329–344.

Picken, L. E. R., and R. J. Skaer. 1966. A review of
researches on nematocysts. Symposia of the Zoological Society of London 16:19–50.

Salleo, A., G. LaSpada, and M. Alfa. 1983a. Blockage of
trypsin-induced discharge of nematocysts of Pelagia
noctiluca by Ca^{2+}. Molecular Physiology 3:89–97.

Salleo, A., G. LaSpada, G. Falzea, and M. G. Denaro. 1983b. Discharging effectiveness of lyotropic anions on nematocysts of *Pelagia noctiluca*. Molecular Physiology 6:19–26.

Salleo, A., G. La Spada, and E. A. Robson. 1990. Discharge characteristics of nematocysts isolated from acontia of *Calliactis parasitica*. Marine Biology 104:459–464.

Santoro, G., and A. Salleo. 1991. The discharge of in situ nematocysts of the acontia of *Aiptasia mutabilis* is a Ca^{2+}- induced response. Journal of Experimental Biology 156:173–185.

Tardent, P. 1988. History and current state of knowledge concerning discharge of cnidae. Pages 309–333 *in* D. A. Hessinger and H. M. Lenhoff, editors. The biology of nematocysts. Academic Press, New York.

Thorington, G. U., and D. A. Hessinger. 1988. Control of discharge: factors affecting discharge of cnidae. Pages 233–253 *in* D. A. Hessinger and H. M. Lenhoff, editors. The biology of nematocysts. Academic Press, New York.

Thorington, G. U., and D. A. Hessinger. 1990. Control of cnidae discharge: III. Spirocysts are regulated by three classes of chemoreceptors. Biological Bulletin 178:74–83.

Thorington, G. U., and D. A. Hessinger. 1998. Efferent mechanisms of discharging cnidae: II. A nematocyst release response in the sea anemone tentacle. Biological Bulletin 195:145–155.

Uspenskaya, A. V. 1957. The ecology and spreading of the pathogen of *Myxosoma cerebralis* (Hofer, 1903; Plehn, 1905) of trout in the fish ponds of the Soviet Union. Pages 47–55 *in* G. K. Petrusheveski, editor. Parasites and diseases of fish. Bulletin USSR Scientific Institute of Freshwater Fisheries 42. English translation, Office of Technical Services, U.S. Department of Commerce, Washington, No. 60-51169.

Uspenskaya, A. V. 1995. Alternation of actinosporean and myxosporean phases in the life cycle of *Zschokkella nova* (Myxozoa). Journal of Eukaryotic Microbiology 42:665–668.

Videler, J. J. 1993. Fish swimming. Chapman and Hall, London.

Watson, G. M., and D. A. Hessinger. 1987. Receptor-mediated endocytosis of a chemoreceptor involved in triggering the discharge of cnidae in a sea anemone tentacle. Tissue Cell 19:747–755.

Watson, G. M., and D. A. Hessinger. 1992. Receptors for n-acetylated sugars may stimulate adenylate cyclase to sensitize and tune mechanoreceptors involved in triggering nematocyst discharge. Experimental Cell Research 198:8–16.

Watson, G. M., and J. Roberts. 1994. Localization of proline receptors involved in regulating nematocyst discharge Journal of Experimental Zoology 270:527–537.

Xiao, C., and S. S. Desser. 1998. Actinosporean stages of myxozoan parasites of oligochaetes from Lake Sasajewun, Algonquin Park, Ontario: new forms of triactinomyxon and raabeia. Journal of Parasitology 84:998–1009.

Xiao, C., and S. S. Desser. 2000. The longevity of actinosporean spores from Oligochaetes of Lake Sasajewun, Algonquin Park, Ontario and their reaction to fish mucus. Journal of Parasitology 86:193–195.

Yanagita, T. M. 1959. Physiological mechanism of nematocyst responses in sea anemone II. Effects of electrolyte ions upon the isolated cnidae. Journal of the Faculty of Science University of Tokyo, Section IV. Zoology 8:381–400.

Yanagita, T. M. 1960. Physiological mechanism of nematocyst responses in sea-anemone III. Excitation and anaesthetization of the nettling response system. Comparative Biochemistry and Physiology 1:123–139.

Yanagita, T. M., and T. Wada. 1953. Discharge-inducing concentrations of acids and bases for the nematocysts of sea-anemone. Natural Science Report, Ochanomizu University, Volume 4, No. 1.

Yasutake, W. T., and E. M. Wood. 1957. Some Myxosporidia found in Pacific Northwest salmonids. Journal of Parasitology 43:633–642.

Yokoyama, H., K. Ogawa, and H. Wakabayashi. 1995. Chemoresponse of actinosporean spores of *Myxobolus cultus* to skin mucus of goldfish *Carassius auratus*. Diseases of Aquatic Organisms 21:7–11.

Zhelev, D. V., and D. Needham. 1994. The influence of electric fields on biological and model membranes. Pages 105–142 *in* D. O. Carpenter and S. Ayrapetyan, editors. Biological effects of electric and magnetic fields, volume 1. Academic Press, New York.

Section 3

Oligochaete Research

American Fisheries Society Symposium 29:79–85, 2002
© 2002 by the American Fisheries Society

The Role of *Tubifex tubifex* (Annelida: Oligochaeta: Tubificidae) in the Transmission of *Myxobolus cerebralis* (Myxozoa: Myxosporea: Myxobolidae)

WILLARD O. GRANATH, JR.* AND MICHAEL A. GILBERT

Division of Biological Sciences, University of Montana, Missoula, Montana 59812, USA

ABSTRACT. *Myxobolus cerebralis*, the causative agent of salmonid whirling disease, was described nearly a century ago as a one-host parasite. It was not until 1984 that the true two-host life cycle of the parasite was discovered. Consequently, numerous studies have examined interactions between the parasite and its fish hosts but, in comparison, there has been relatively little work conducted on interactions between *M. cerebralis* and its obligate, oligochaete host *Tubifex tubifex*. Recently, though, several research groups have begun examining the interplay between the parasite and the oligochaete in hopes of determining the parameters necessary for the successful transmission of *M. cerebralis* by *T. tubifex*. These studies can be broken into three broad categories: 1) individual host factors that affect the transmission of *M. cerebralis*, such as species-specific host susceptibility and an individuals' capacity to support and release varying numbers of the parasite; 2) *T. tubifex* population factors that affect the transmission of the parasite, including genetic and geographic variations in susceptibility; and 3) ecological and epidemiological parameters that influence the transmission of *M. cerebralis* by *T. tubifex*, such as habitat conditions, time of season, water temperature, and distribution of infected oligochaetes within a watershed. Therefore, the goals of this review are to discuss what is known for each category and to evaluate these results in context of their role in the transmission of *M. cerebralis*. By doing so, continued and future research, necessary to completely understand the epidemiology of whirling disease, should become apparent.

Myxobolus cerebralis, the causative agent of salmonid whirling disease, was first described in Germany (Hofer 1903) after signs of the disease were observed in rainbow trout (*Oncorhynchus mykiss*) that had been imported from North America. At this time, it was thought that *M. cerebralis* had a direct life cycle and was thus transmitted from fish to fish. It was not until much later, when Wolf and Markiw (1984) proposed a two-host life cycle for *M. cerebralis*, that the parasite was shown to have an indirect life cycle, involving an aquatic oligochaete as an obligatory host (which was subsequently shown to be *Tubifex tubifex* [Wolf et al. 1986]). Consequently, there have been numerous studies describing various interactions and effects of the parasite on its fish hosts, but, in comparison, relatively few studies have examined the interactions and parameters involved in the transmission of *M. cerebralis* by *T. tubifex*. Recently, there has been a growing interest in the oligochaete host,

and this review will discuss the current state of knowledge and place this information in relation to their role in the transmission of *M. cerebralis*. Further, variables remaining unknown, but which are important to understanding the transmission dynamics of *M. cerebralis*, as well as future research needs, will be discussed.

TUBIFEX TUBIFEX LIFE HISTORY

To fully appreciate the role of *T. tubifex* in the transmission of *M. cerebralis*, it is important to understand the basic biology and life history traits of the oligochaete. Although *T. tubifex* was first described in 1774, there continues to be a debate over the identification and taxonomy of this species and the possibility of subspecies (reviewed by Brinkhurst 1996). Much of this debate stems from the fact that the positive identification of *T. tubifex* is based on the morphology of reproductive structures (e.g., penis sheath, atrium, vas deferens). Complicating identification of these worms is the fact that after breeding they resorb their reproductive organs (Poddubnaya 1984) and the chaetae of *T. tubifex* can change forms depending on different

*Address correspondence to: Willard O. Granath, Jr.,
University of Montana, Division of Biological Sciences, 32
Campus Drive #4824, Missoula, Montana 59812-4824, USA.

environmental conditions (Chapman and Brinkhurst 1987). Fortunately, there are current efforts underway to develop molecular markers that will clarify these taxonomic issues (Beauchamp et al. 2001; C. Rasmussen, Western Fisheries Research Center, personal communication).

Although no worldwide survey for *T. tubifex* has been conducted it is widely believed that these worms are ubiquitous, with different subspecies and races spread throughout the continents. The distribution of *M. cerebralis* over several continents (Hoffman 1970) tends to support this view.

Tubifex tubifex inhabits sediments at the bottom of lakes and streams where it feeds by ingesting whole sediments and selectively digesting some of the bacterial species present (Wavre and Brinkhurst 1971). The worm plays an important role in the ecosystem by reworking sediments and providing a valuable food source for leeches, crustaceans, fish, and numerous species of insects. Unfortunately, *T. tubifex* is probably best known for its ability to survive eutrophic conditions and has thus earned the nickname "sludge worms." Although this species is quite capable of surviving in pristine conditions, in most such cases the numbers of *T. tubifex* are kept low by competition and/or predation and are therefore difficult to detect without extensive sampling efforts. Distribution within a watershed is influenced by the composition and organic content of the substratum, although this relationship is indirect. That is, the microflora associated with different substrata is the primary factor in substratum selection by the worms, not the substratum itself (McMurtry et al. 1983; Lazim and Learner 1987; Lestochova 1994). The abundance of *T. tubifex* has also been found to positively correlate with both sedimentation rates and organic carbon flux (Robbins et al. 1989). *Tubifex tubifex* is often found in close association with other oligochaete species such as *Limnodrilus hoffmeisteri*. These characteristic species associations are believed to be due to mutualistic relationships in which one species feeds on the bacteria associated with the fecal pellets of the other species and vice versa (Brinkhurst 1971, 1974; Milbrink 1993).

In addition to feeding on bacteria, *T. tubifex* is able to absorb small, dissolved, organic molecules directly through its body wall. In areas with large concentrations of dissolved organic materials it is estimated that the worms can obtain up to 40% of their oxidative requirements by absorbing molecules directly from the water (Hoffman et al. 1987). *Tubifex tubifex* is capable of anaerobic respiration and has survived 16 weeks in the laboratory under anoxic conditions (Reynoldson 1987). The end products of glycogen breakdown in the absence of oxygen are acetate and propionate and, as mentioned above, these products can be reabsorbed through the body wall upon a return to aerobic conditions (Hoffman et al. 1987). This adaptation is one of the reasons why *T. tubifex* is able to survive in eutrophic environments that often result from large amounts of organic pollution. In extreme cases, *T. tubifex* and *L. hoffmeisteri* may be the only benthic macroinvertebrates present in such an environment (Brinkhurst 1996). *Tubifex tubifex* also has the unique ability to survive drought and food shortages by forming a protective cyst and lowering its metabolic rate. Cysts containing live *T. tubifex* have been recovered from a cattle pond after five months of drought, and worms in the laboratory survived a six-month starvation experiment in cysts (Anlauf 1990). Food shortage, which often occurs as a result of drought, is believed to be the main factor in cyst formation, as the addition of food is the only treatment that results in worms exiting the cysts (Anlauf 1990). In fact, some researchers have speculated that *T. tubifex* can be dispersed to new locations in these cysts. If worms infected with *M. cerebralis* could use this type of dispersal mechanism it could have significant impacts on efforts to control the spread of whirling disease.

Like many other oligochaetes, *T. tubifex* is hermaphroditic. They usually breed sexually but have the ability to reproduce asexually by parthenogenesis (Poddubnaya 1984). Fertilized eggs are deposited in a cocoon and young worms hatch directly from the cocoon 7–26 d later, depending on temperature. The newly hatched worms will become sexually mature within one to three months, and after breeding, the worms can undergo several cycles of resorption and regeneration of their sexual organs (Poddubnaya 1984). Reports concerning the lifespan of *T. tubifex* are highly variable. Poddubnaya (1984) reported that the lifespan of reproducing parthenogenetic individuals varies from 70 to 530 d. Matsumoto and Jammoto (1966) indicated that *T. hattai* live for 4–6 years. *Tubifex tubifex* in our laboratory have been kept for more than three years and their age at the time of acquisition was unknown. Although the exact numbers vary, it is clear that these worms live long enough to span several seasons. This may be significant in the epidemiology of whirling disease in that it would allow for a seasonal periodicity in triactinomyxon (TAM) release by *T. tubifex*, provided that they remain infected with *M. cerebralis* for the duration of their lives.

INDIVIDUAL HOST FACTORS
AFFECTING TRANSMISSION

Before discussing specific factors of individual *T. tubifex* that affect the transmission of *M. cerebralis*, it should be noted that this species of oligochaete is the only known suitable host for the parasite. For example, *Limnodrilus hoffmeisteri, Ilyodrilus templetoni, Quistadrilus multisetosus* (Wolf et al. 1986), *Dero* spp., *Stylaria* spp., *Aeolomsoma* spp. (Markiw and Wolf 1983), and *Tubifex ignotus* (El-Matbouli and Hoffman 1989) have been found to be unsuitable hosts for *M. cerebralis*. In addition, our laboratory, as well as several others, has examined numerous species of aquatic oligochaetes and found only *T. tubifex* to support development and transmission of *M. cerebralis*. Further, the so-called "Great Lakes" *T. tubifex* has been shown by us and numerous other investigators to be an unsuitable host for the parasite. Thus, not only is *T. tubifex* the only suitable species that is capable of transmitting *M. cerebralis*, there may be discrete subsets of susceptible species or, perhaps, subspecies.

An important aspect of *T. tubifex* is that individuals can remain persistently infected with *M. cerebralis* throughout their lifespan. Gilbert and Granath (2001) conducted a series of experiments to quantify and determine the periodicity and length of the release of the TAM stage of *M. cerebralis* by *T. tubifex*. In this study, individual *T. tubifex* (initially infected as a group) were examined daily for the release of *M. cerebralis* TAMs, and the number of waterborne TAMs released by each worm was quantified. The duration of the infection in these worms was also monitored using a PCR diagnostic test (Andree et al. 1998). Triactinomyxons were first released 74 d post exposure (PE) and continued to be released until 132 d PE. During this period, each worm released, on average, 1.5×10^3 waterborne TAMs 12 times; however, no pattern or periodicity was noted. The results of the PCR diagnostic tests conducted at 5, 7, 9, and 15 months PE were positive, and the persistent infection was confirmed at 606 d PE when the remaining worms began releasing TAMs again. Similar results were observed in naturally infected

T. tubifex, indicating that these worms remain infected for the duration of their natural lifespan and are capable of shedding viable TAMs, in temporally separate periods. The finding that *M. cerebralis* infections persist in *T. tubifex* for the life of the worm, and that the worms release TAMs more than once during their lives, are significant in that they open the possibility of a seasonal periodicity in TAM release. For example, a seasonal event such as this would help explain data from live box studies done by the Montana Department of Fish, Wildlife, and Parks (MFWP). These live box studies involved maintaining sentinel fish in infected watersheds throughout the state of Montana, where two distinct peaks in the infection of sentinel fish were observed during the late Spring and early Fall seasons (R. Vincent, MFWP, personal communication). Thus, information concerning a seasonal periodicity in TAM release would be extremely valuable to agencies responsible for stocking salmonid fish. If young fish were stocked between periods of TAM release it would delay (possibly for months) the exposure of the fish to the parasite, and it has been shown that when older fish are exposed they suffer fewer signs of disease (Markiw 1992).

Overall, the data available to date indicate that *T. tubifex* is the only species of oligochaete that can transmit *M. cerebralis*, that individuals vary in their ability to produce TAMs, and that individual worms show no periodicity in TAM release (although as a population, there may be a seasonal periodicity). Further, *T. tubifex* can remain infected with the parasite for its entire lifespan. All of these factors are important considerations when evaluating the transmission dynamics of *M. cerebralis* by *T. tubifex*.

HOST POPULATION FACTORS
AFFECTING TRANSMISSION

Recent data indicate that, at the population level, *T. tubifex* varies in its ability to transmit *M. cerebralis* and, further, that the parasite may affect worm populations. Stevens et al. (2001) examined the effects of various doses of *M. cerebralis* myxospores on TAM production and the biology of *T. tubifex* from two geographic regions where whirling disease is enzootic (California and Montana). They exposed these worms to 50, 500, or 1,000 myxospores and determined TAM production and the effect of the parasite on biomass, abundance and individual weight of the oligochaetes. Results indicated that the California worms produced significantly more TAMs than the Montana *T. tubifex*, but within each population, TAM production did not vary with spore dose. Further, when compared with uninfected controls, total worm biomass, abundance and individual weight was decreased in both geographic populations. Thus, this study shows that *T. tubifex* populations vary in their ability to produce TAMs and that the parasite appears to have a deleterious effect on worm populations. Since the ability of *T. tubifex* to produce TAMs varies by geographic region, this could significantly affect the severity of whirling disease in these locations.

Current studies also are revealing that genetic differences occur within and between *T. tubifex* populations. For example, Stevens et al. (2001), using a conserved genetic marker (the internal transcribed spacer region 1 of the ribosomal RNA gene locus) showed that the California and Montana worms used in the above study were genetically distinct. Our laboratory has been conducting a long-term study on the epidemiology of whirling disease throughout an entire drainage (Rock Creek, Montana). As part of this study, we are examining the genetic heterogeneity of *T. tubifex* throughout Rock Creek using randomly amplified polymorphic DNA techniques. Preliminary analysis indicates that there are at least three genotypes in the creek, and only individuals possessing the dominant genotype were found to be infected with *M. cerebralis* (C. Rasmussen, Western Fisheries Research Center, personal communication). However, laboratory studies will be needed to determine if any of these genetic differences are related to the ability of *T. tubifex* to produce TAMs and/or resist infection.

EPIDEMIOLOGICAL FACTORS
AFFECTING TRANSMISSION

One of the most important abiotic factors in the transmission of *M. cerebralis* by *T. tubifex* is temperature. El-Matbouli et al. (1999) examined the effect of water temperature on the development, release and survival of TAMs in *T. tubifex*. They examined infected worms held at 5, 10, 15, 20, 25, and 30°C and found that the highest level of TAM production occurred at 10°C and 15°C; at higher or lower temperatures, TAM release was minimal. This finding probably explains why researchers detect the most severe cases of whirling disease in trout in the

10–15°C temperature range. For example, Baldwin et al. (2000), using sentinel cage methods, observed a direct correlation between both the prevalence and severity of infection of *M. cerebralis* in rainbow trout, with temperature. The highest prevalence of infection and the most severe lesions were found in trout that were held in an enzootic area for the parasite when water temperatures were between 10°C and 12°C. Thus, if young trout, which are the most susceptible to the disease, are emerging from their redds at these temperatures, it could severely affect the fish population. Many additional studies have been conducted by MFWP that show a correlation between water temperature and infection rates in sentinel fish. Typically, peak infection rates occurred in these fish during the spring warming and fall cooling periods when average daily water temperatures were between 11°C and 14°C (R. Vincent, MFWP, personal communication). This information, coupled with the fact that *T. tubifex* can remain persistently infected with the parasite, release TAMs more than once during its life, and live for several years (Gilbert and Granath 2001, discussed above), leaves open the possibility of a seasonal periodicity in TAM release. A periodicity such as this would explain the peaks in infection rates of sentinel fish seen by MFWP and likely account for the correlation between water temperature and the severity of *M. cerebralis* infections in rainbow trout (Baldwin et al. 2000).

The high densities of T. tubifex often seen in eutrophic environments has lead many people to believe that habitat degradation is absolutely necessary for the establishment of T. tubifex populations large enough to effectively transmit the parasite. There are numerous examples of heavily degraded streams with a high incidence of whirling disease. However, in our Rock Creek studies mentioned above, we have been able to collect M. cerebralis-infected T. tubifex from a variety of habitats, including high impact/high sedimentation areas as well as low impact/low sedimentation regions of the creek. There also are tributaries of the stream that have both high and low habitat degradation from which M. cerebralis has not been detected. Overall, this stream contains pristine areas and highly degraded regions, but there is no obvious correlation between habitat and the presence/absence of M. cerebralis-infected T. tubifex. In fact, the ability of T. tubifex to occupy a variety of habitat types has led some researchers to suggest that this species may consist of different ecological races

(Milbrink 1973; Poddubnaya 1980). To begin addressing such ecological questions, Anlauf (1994, 1997) used horizontal starch gel electrophoresis to examine polymorphisms of two different enzymes from *T. tubifex*: isocitrate-dehydrogenase (IDH) and phosphoglucose-isomerase (PGI). That study concluded that *T. tubifex* populations from Germany consisted primarily of three dominant genotypes, designated A, B, and C, and that these genotypes differed in their adult weights, reproduction and relative mortality in response to temperature shifts. That study also found that these genotypes were stable in the laboratory for eight weeks over temperature shifts from 5°C to 28°C. Anlauf and Neumann (1997) used the same techniques to address the question of possible genetic variability of *T. tubifex* in relation to habitat type. For this study, they examined the polymorphisms of six different enzymes from 20 different populations of worms that were collected from four habitat types. These habitats were classified as I—lakes without an anoxic period in the hypolimnion; II—lakes with an extended anoxic period in the hypolimnion; III—creeks with a permanent flow of water; and IV—pools without a permanent flow of water and seasonal drying or freezing. Their results indicated a good correlation between habitat type and the frequency of some alleles. That is, some alleles were observed mainly from shallow-water habitats (types III and IV) while others were observed mainly from deepwater habitats (types I and II). In addition, individuals classified as genotype C (based on criteria designated by Anlauf 1997) were found at almost every shallow-water site tested, but were only found at two deepwater sites. Most of the alleles that were restricted to shallow-water habitats were seen in these individuals as well.

Although the results of the above studies support the concept of different ecological races as postulated by Milbrink (1973) and Poddubnaya (1980), they must be interpreted cautiously. For example, Kinne (1962) looked at the effects of osmotic shock and rearing temperature on developmental stages of fish. This study found that the variation seen in some physiological characters become fixed during the course of ontogeny under different environmental conditions. Therefore, if only individuals that are adults when collected are tested using protein-based molecular markers, irreversible "nongenetic" adaptations might be interpreted as ecological or geographical variation (that

remain stable in the laboratory), but are, in reality, manifestations of the environment during ontogeny. Although the development of *T. tubifex* most likely is not similar to that of fish, until more is known about the developmental biology of tubificid worms, it should be noted that Anlauf and Neumann (1997) only used mature *T. tubifex*.

Despite the above concerns, if it turns out that there are different ecological races of *T. tubifex*, and that they differ in their susceptibility to M. *cerebralis* infection, the impacts of different worm populations on the transmission of M. *cerebralis* could be very important. For example, Anlauf and Neumann (1997) found that shallow-water populations exhibited a lower level of heterozygosis than expected, based on Hardy-Weinberg calculations, which would result from a high degree of inbreeding or parthenogenic reproduction. Thus, if a catastrophic event were to significantly lower a population and select for an ecological race of *T. tubifex* that was susceptible to infection, or if a susceptible race were to colonize a new watershed, rapid reproduction via parthenogenesis and subsequent inbreeding may create a large population of susceptible worms. This, in turn, could exacerbate an existing whirling disease problem or put a watershed at greater risk of a whirling disease epizootic. Conversely, if a nonsusceptible race were selected for or colonized a new area, then *T. tubifex* populations could increase with little or no effect on the transmission of M. *cerebralis*. Clearly more work is needed on *T. tubifex* genetics and its relationship to susceptibility to infection with M. *cerebralis*.

FUTURE STUDIES AND UNKNOWN VARIABLES

The completion of numerous current studies, as well as those planned for the near future, should greatly increase our understanding of the role of *T. tubifex* in the transmission M. *cerebralis*. For example, continued oligochaete surveys in combination with sentinel fish studies should determine the range of influence an infected *T. tubifex* population has within a river system. The development of molecular markers, to identify *T. tubifex* as a species, as well as to differentiate worms with varying susceptibilities to M. *cerebralis* infection, will be a significant advancement. Not only would this knowledge make new experiments possible, it may also allow for the identification of watersheds at risk of whirling disease epizootics. Further studies will be necessary before an accurate assessment of the true impact of habitat conditions on the ability of *T. tubifex* to establish themselves and transmit the parasite can be made.

There are many other questions that need to be answered to truly understand the epidemiology of whirling disease within an ecosystem. For example, more must be learned about the dispersal mechanism(s) of *T. tubifex*, the dispersal of myxospores and the general range of movement of the worms (e.g., how far can infected *T. tubifex* move? or, do they move at all after becoming infected?).

The lack of information concerning pathology in *T. tubifex* caused by *M. cerebralis* is more likely due to a lack of knowledge in the area of tubificid physiology rather than the absence of pathology itself. Alterations in the coloration as well as distortions of the intestine can be observed in infected *T. tubifex* that are actively shedding TAMs (personal observation). Also, observations in our laboratory and by Stevens et al. (2001) indicate that infected *T. tubifex* are at least partially inhibited from breeding or do not breed at all. Although no developmental stages of *M. cerebralis* were seen in the gonads of worms during histological studies of infected *T. tubifex* (El-Matbouli and Hoffman 1998), the parasite may have an indirect effect on worm reproduction. Whether or not this is a result of parasitic castration remains to be determined, but one common form of parasitic castration can occur when a parasite secretes molecules that destroy gametes or prevent gametogenesis in a reversible or nonreversible manner (Cheng 1986).

CONCLUSIONS

Due to the relatively recent discovery of the two-host life cycle of *M. cerebralis*, it is not surprising that there are gaps in information pertaining to the oligochaete host. Although it is apparent that *T. tubifex* is the only suitable worm host, there is solid evidence that its ability to transmit *M. cerebralis* varies considerably, and depends on individual, population and epidemiological characteristics. Persistently infected *T. tubifex*, seasonal release of TAMs, and geographic, genetic and habitat differences all contribute to determining the severity of whirling disease within an ecosystem. Although much remains to be learned, if the results of recent studies are any indication, a much clearer picture of the role of *T. tubifex* in the transmission *M. cerebralis* will be emerging in the near future.

ACKNOWLEDGMENTS

The authors thank the organizers of the 7th Annual Whirling Disease Symposium held 8–9 February 2001, for which we were asked to present the subject of this manuscript. We also are grateful to The National Partnership on the Management of Wild and Native Coldwater Fisheries, who supported some of the work presented here.

REFERENCES

Andree, K. B., E. M. MacConnell, and R. P. Hedrick. 1998. A nested polymerase chain reaction for the detection of genomic DNA of *Myxobolus cerebralis* in rainbow trout *Oncorhynchus mykiss*. Diseases of Aquatic Organisms 34:145–154.

Anlauf, A. 1990. Cyst formation in *Tubifex tubifex* (Müller)—an adaptation to survive food deficiency and drought. Hydrobiologia 190:79–82.

Anlauf, A. 1994. Some characteristics of genetic variants of *Tubifex tubifex* (Müller 1774) (Oligochaeta: Tubificidae) in laboratory cultures. Hydrobiologia 278:1–6.

Anlauf, A. 1997. Enzyme variability of *Tubifex tubifex* (Müller)(Oligochaeta, Tubificidae) and seven other tubificid species. Archiv für Hydrobiologie 139:83–100.

Anlauf, A., and D. Neumann. 1997. The genetic variability of *Tubifex tubifex* (Müller) in 20 populations and its relation to habitat type. Archiv für Hydrobiologie 139:145–162.

Baldwin, T. J., E. R. Vincent, R. M. Silflow, and D. Stanek. 2000. *Myxobolus cerebralis* infection in rainbow trout (*Oncorhynchus mykiss*) and brown trout (*Salmo trutta*) exposed under natural stream conditions. Journal of Veterinary Diagnostic Investigation 12:312–321.

Beauchamp K. A., R. D. Kathman, T. S. McDowell, and R. P. Hedrick. 2001. Molecular phylogeny of tubificid oligochaetes with special emphasis on *Tubifex tubifex* (Tubificidae). Molecular Phylogenetics and Evolution 19:216–224.

Brinkhurst, R. O. 1971. Interspecific interactions and selective feeding by tubificid oligochaetes. Limnology and Oceanography 17:122–133.

Brinkhurst, R. O. 1974. Factors mediating interspecific aggregation of tubificid oligochaetes. Journal of the Fisheries Research Board of Canada 31:460–462.

Brinkhurst, R. O. 1996. On the role of tubificid oligochaetes in relation to fish disease with special reference to the Myxozoa. Annual Review of Fish Diseases 6:29–40.

Chapman, P. M., and R. O. Brinkhurst. 1987. Hair today, gone tomorrow: induced chaetal changes in tubificid ogliochaetes. Hydrobiologia 155:45–55.

Cheng, T. C. 1986. General parasitology, 2nd edition. Academic Press, Orlando, Florida.

El-Matbouli, M., and R. W. Hoffman. 1989. Experimental transmission of two *Myxobolus* spp. developing bisporogeny via tubificid worms. Parasitology Research 75:461–464.

El-Matbouli, M., and R. W. Hoffman. 1998. Light and electron microscopic studies on the chronological development of *Myxobolus cerebralis* to the actinosporean stage in *Tubifex tubifex*. International Journal for Parasitology 28:195–217.

El-Matbouli, M., T. S. McDowell, D. B. Antonio, K. B. Andree, and R. P. Hedrick. 1999. Effect of water temperature on the development, release and survival of the triactinomyxon stage of *Myxobolus cerebralis* in its oligochaete host. International Journal for Parasitology 29:627–641.

Gilbert, M. A., and W. O. Granath, Jr. 2001. Persistent infection of *Myxobolus cerebralis*, the causative agent of salmonid whirling disease, in *Tubifex tubifex*. Journal of Parasitology 87:101–107.

Hofer, B. 1903. Ueber die drehkrankheit der regenbogenforelle. Allgemeine Fischerei-Zeitung 28:7–8.

Hoffman, G. L. 1970. Intercontinental and transcontinental dissemination and transfaunation of fish parasites with emphasis on whirling disease (*Myxosoma cerebralis*). Pages 69–81 in S. F. Snieszko, editor. A symposium on diseases of fish and shellfish. American Fisheries Society, Special Publication No. 5, Washington, D.C.

Hoffmann, K. H., E. Hipp, and U. A. Sedlmeier. 1987. Aerobic and anaerobic metabolism of the freshwater oligochaete *Tubifex* sp. Hydrobiologia 155:157–158.

Kinne, O. 1962. Irreversible non-genetic adaptation. Comparative Biochemistry and Physiology 5:265–282.

Lazim, M. N., and M. A. Learner. 1987. The influence of sediment composition and leaf litter on the distribution of tubificid worms (Oligochaeta): a field and laboratory study. Oecologia 72:131–136.

Lestochova, E. I. 1994. Influence of small river conditions on the abundance of Tubificidae. Hydrobiologia 278:129–131.

Markiw, M. E. 1992. Experimentally induced whirling disease. I. Dose response of fry and adults of rainbow trout exposed to the triactinomyxon stage of *Myxobolus cerebralis*. Journal of Aquatic Animal Health 4: 40–43.

Markiw, M. E., and K. Wolf. 1983. *Myxosoma cerebralis* (Myxozoa: Myxosporea) etiologic agent of salmonid whirling disease requires tubificid worm (Annelida: Oligochaeta) in its life cycle. Journal of Protozoology 30:561–564.

Matsumoto, M., and G. Jammoto. 1966. On the seasonal rhythmicity of oviposition in the aquatic oligochaete *Tubifex hattai*. Japanese Journal of Ecology 16:134–139.

McMurtry, M. J., D. J. Rapport, and K. E. Chua. 1983. Substrate selection by tubificid oligochaetes. Canadian Journal of Fisheries and Aquatic Sciences 40:1639–1646.

Milbrink, G. 1973. On the use of indicator communities of Tubificidae and some Lumbriculidae in the assessment of water pollution in Swedish lakes. Zoon 1:125–139.

Milbrink, G. 1993. Evidence for mutualistic interactions in freshwater oligochaete communities. Oikos 68:317–322.

Poddubnaya, T. L. 1980. Life cycles of mass species of Tubificidae. Pages 175–184 in R. O. Brinkhurst and D. G. Cook, editors. Aquatic Oligochaete Biology. Plenum, New York.

Poddubnaya, T. L. 1984. Parthenogenesis in Tubificidae. Hydrobiologia 115:97–99.

Reynoldson, T. B. 1987. The role of environmental factors in the ecology of tubificid oligochaetes—an experimental study. Holarctic Ecology 10:241–248.

Robbins, J. A., T. Keilty, D. S. White, and D. N. Edgington. 1989. Relationships among tubificid abundances, sediment composition, and accumulation rates in Lake Erie. Canadian Journal of Fisheries and Aquatic Sciences 46:223–231.

Stevens, R., B. L. Kerans, J. C. Lemmon, and C. Rasmussen. 2001. The effects of *Myxobolus cerebralis* myxospore dose on triactinomyxon production, and biology of *Tubifex tubifex* from two geographic regions. Journal of Parasitology 87:315–321.

Wavre, M., and R. O. Brinkhurst. 1971. Interactions between some tubificid oligochaetes and the bacteria found in the sediments of Toronto Harbour, Ontario. Journal of the Fisheries Research Board of Canada 28:335–341.

Wolf, K., and M. E. Markiw. 1984. Biology contravenes taxonomy in the Myxozoa: new discoveries show alternation of invertebrate and vertebrate hosts. Science 225:1449–1452.

Wolf, K., M. E. Markiw, and J. K. Hiltunen. 1986. Salmonid whirling disease: *Tubifex tubifex* (Müller) identified as the essential oligochaete in the protozoan life cycle. Journal of Fish Diseases 9:83–85.

American Fisheries Society Symposium 29:87–91, 2002

Triactinomyxon Production as Related to Rearing Substrate and Diel Light Cycle

RONNEY E. ARNDT

Fisheries Experiment Station, Utah Division of Wildlife Resources, Logan, Utah 84321, USA
Phone: (435) 752-1066, Fax: (435) 752-6977, E-mail: nrdwr.rarndt@state.ut.us

ERIC J. WAGNER, QUINN CANNON, MARK SMITH

Fisheries Experiment Station, Utah Division of Wildlife Resources, Logan, Utah 84321, USA

ABSTRACT. The culture of the aquatic worm *Tubifex tubifex*, the alternate host of whirling disease, is necessary to conduct research regarding triactinomyxon (TAM) viability, controlled infection studies, and methods of disease control. Presumed infected worms collected from the field may produce TAMs for several months, but production generally decreases after a few months. To ensure a stable supply of TAMs, we investigated the effects of rearing substrate on TAM production and worm survival. We also analyzed the time of TAM release during 24-h periods divided into 12 h light:12 h dark.

Mixed species of field-collected worms were placed into culture with the following substrates: organic debris, silt, and sand. The controls were reared without a substrate. After nine weeks, worms in the silt substrate produced significantly more TAMs than worms in either the control or sand substrate. The worms in silt also produced significantly more TAMs than the worms in organic debris in four of seven bi-weekly samples where significant treatment differences were found. The estimated average daily TAM production during weeks 9–27 averaged 149 for control worms, 754 for the organic, 231 for the sand, and 2,573 for worms in the silt substrate. By the end of the study (week 27), average total worm weight for the controls was 0.8 g, 3.3 g for the organic, 2.6 g for the sand, and 7.5 g for the silt. The silt treatment had significantly higher worm mass than any of the other treatments, and was the only treatment to increase in mass from the starting weight of 4.5 g.

During two separate periods, worms were also observed for TAM release within 24-h cycles composed of 12 h light:12 h dark cycles. During the first period, there were slightly more TAMs released during the light period (59%) compared with 41% for the dark. For the second period, slightly more TAMs were released during the light period (56%) compared with 44% for the dark. For both tests however, these differences were not significant. These results indicate that, at least in the laboratory, there is no relationship between daylight hours and TAM production.

The culture of the aquatic worm *Tubifex tubifex*, the alternate host of *Myxobolus cerebralis*, has been conducted for several years at the Fisheries Experiment Station (FES, Logan, Utah) to support research on triactinomyxon (TAM) viability, controlled infection, and disease control. Traditionally, presumed infected worms were collected from a field site, brought to the FES laboratory, cleaned of organic debris and sediment, and placed in culture. The worms typically produced TAMs for several months (~20,000 TAMs/culture per d) after which production decreased and the worms were either combined with other cultures or discarded. Using this approach, TAM production could not be predicted, thus compromising research projects. Because of these deficiencies, we undertook several projects to determine if

TAM production could be increased, and the survival of worms could be enhanced.

The first study tested the hypothesis that different types of rearing substrates (sand, silt, or organic debris) could enhance worm survival and TAM production. In a stream or river environment, we can generalize that *T. tubifex* is found in higher numbers in shallow regions of low current velocity in substrates rich in fine sediment and organic matter compared with other habitat types (Lazim and Learner 1987). Sauter and Güde (1996) found that in Lake Constance (Germany), sediments composed of 70–100% clay and silt sustained higher densities of *T. tubifex* compared with lower levels of silt and clay. In a more detailed analysis, McMurtry et al. (1983) studied the preferences of *T. tubifex* and *Limnodrilus hoffmeisteri* for two types of substrate that were similar in physical structure but differed in elemental and organic composition. Mixed species groups and individual *T. tubifex* worms preferred the sediment that was lower in organic matter and elemental content but that had a higher level of bacterial colonization.

For the second part of our research we analyzed the time of TAM release during 24 h periods divided into light and dark cycles. Yokoyama et al. (1993) reported a circadian rhythm in the release of *Echinactinomyxon* spp. spores by the oligochaete *Branchiura sowerbyi*. When kept under a 12 h light and 12 h dark regiment, the highest numbers of spores were released between 2200 and 0200 hours. When the photoperiod was reversed, the release of actinospores was also reversed, indicating that the timing of actinospore release was cued by the daily light:dark cycle.

METHODS

Substrate Test

Worms were collected from a private aquaculture facility that has been infected with *M. cerebralis* since the mid 1990s. The worms were separated from the sediment by rotation on an oscillation table (35 rpm) until a clean collection of worms was obtained. The worms were then divided into 12 separate containers with 4.5 g per container. Worms were not separated by species, but past collections have been composed chiefly of *T. tubifex* and *L. hoffmeisteri*. The containers were plastic 400 mL beakers that contained control and experimental substrate types in triplicate. The substrates included organic debris, silt, and sand. The organic substrate was the debris collected when the worms were separated after field collection. It consisted of pieces of plants, seed husks, decomposed wood, pine needles, and empty clam and snail shells. The average size of the particles was 2.8 × 1.2 mm. The silt was collected from the lower Logan River and was run through a series of sieves so that 92% of particles were less than 125 μm. The sand was purchased from a local landscape company, with the following size composition: 27% less than 250 μm, 32% 250–500 μm, 20% 500 μm-1 mm, 15% 1–2 mm, and 6% greater than 2 mm. The controls had no substrate. All three types of substrate were autoclaved at 100°C for 60 min prior to use, thereby ensuring that bacteria and other living organisms were killed. Each beaker contained 2 cm of substrate and 6 cm of well water.

Beakers were placed into a refrigerator, aerated by an air stone, and maintained at 15°C throughout the study. Lighting was provided by a full spectrum fluorescent bulb on a 12 h light:12 h dark basis. Individual cultures were fed 0.5 g of spirulina pellets on a weekly basis.

For TAM enumeration counts were made three times per week during weeks 1–7, on a weekly basis during weeks 8–12, and on a biweekly basis during weeks 13–27. Regardless of sampling regimen, water exchanges were made on Monday, Wednesday, and Friday of every week. TAM collection and enumeration were accomplished by the following methodology. For TAM collection, water was individually drawn from each replicate container via a siphon tube onto a 20 μm Nitex mesh filter (Aquaculture Research/Environmental Associates, Homestead, Florida). The material retained on the filter was then washed into a 50 mL plastic test tube and the total volume recorded. From each tube, three 50 μL subsamples were withdrawn and placed into individual glass test tubes. This quantity was then stained with crystal violet, prepared as a wet mount, TAM counts made, and total TAM production calculated.

For tracking TAM production over the course of the study, data from weeks 1–12 were pooled into biweekly units and average TAM production was calculated based on the number of sampling events that occurred during that specific two-week period. These discrete biweekly values were then used to test for significance by a one-way analysis of variance (ANOVA), which was also used to test for differences in the final worm weights. For each replicate, all worms were removed from their respective sediments at the conclusion of the test, cleaned of debris, and their gross weight was measured. Thirty worms from each replicate were also preserved in 10% buffered formalin. After seven days they were transferred to 70% alcohol and sent to Aquatic Resources Center (College Grove, Tennessee) for species identification.

Diel Light Cycle Test

For this work, two different sets of experimental cultures were checked for TAM release during two separate time frames. The cultures used during the first period were part of a study on infectivity and TAM recovery on different mesh sizes. For the second period, the cultures from the substrate test were used. For both tests, cultures were maintained at 15°C with aeration provided to the individual cultures. Lighting was provided by a full spectrum fluorescent bulb on a 12 h light:12 h dark basis. Light intensity was not measured for the first test, but was 73 lux at the air/water interface for the second test. For the first test, water from six cultures was exchanged 12 h before the first TAM harvest beginning at 1930 hours on 13 March 2000. Water was siphoned onto a 100 μm Nitex filter placed over a 20 μm Nitex filter. The 100 μm prefilter was used to remove debris and organisms found in the worm cultures. The retentate was placed into a vial and fixed with 5% formalin, and TAM counts were made the following day according to the methods previously discussed. This procedure was repeated every 12 h at 0730 hours and 1930 hours for each culture through 15 March 2000. The experiment was extended an additional four days, during which time a 10 μm Nitex filter was used for comparison with the 20 μm Nitex mesh.

Concerns about the accuracy of lighting manipulation and worm acclimation time from the first study lead us to repeat the test. For the second test the protocol listed above was repeated during the week of 4 December 2000, using the three silt treatment worm cultures from the substrate study discussed above, as well as one of our main TAM producing cultures. This second test was run for a total of 4 light cycle and 4 dark cycle samples over a period of four days. Samples were examined for TAMs immediately after collection without fixation. For both tests, data from a given 24-h cycle were expressed as a percentage of TAM production for a 12 h light or 12 h dark cycle based on the total production from that 24-h period. Values from a given 24-h cycle were then analyzed for significance by the Wilcoxon Signed Rank Test.

RESULTS AND DISCUSSION

Substrate Test

This test indicated a significant impact of culture substrate type on TAM production and worm survival. Worms in the silt substrate produced significantly more TAMs than either worms in no substrate (controls) or worms in the sand substrate from weeks 9–15, and then again from weeks 19–23 (Figure 1). The worms cultured in the silt substrate also produced significantly more TAMs than the worms in the organic substrate in four of seven bi-weekly samples where differences were

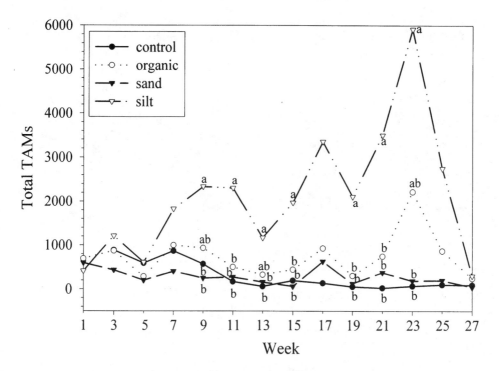

Figure 1. Average (*N* = 3) biweekly TAM production of tubifex worms reared in organic, sand, silt, or no substrate. A letter next to a treatment symbol indicates a significant difference (*P* ≤ 0.05) within a given biweekly period.

found (weeks 11, 15, 19, 21). No significant differences were found in TAM production between treatments during weeks 17, 25, and 27 due to wide variation in TAM production within replicates. The estimated daily TAM ($N = 3$) production during weeks 9–27 averaged 149 TAMs for the controls, 754 for the organic, 231 for the sand, and 2,573 for the silt substrate.

The ability to assess worm survival by gross worm weight was compromised by the apparent reproduction within the silt treatments, although by the end of the study (week 27), total average worm weight for the controls was 0.8 g, 3.3 g for the organic, 2.6 g for the sand, and 7.5 g for the silt. The silt treatment had significantly higher worm mass than any other treatments ($P < 0.001$), and was the only treatment to increase in mass from the starting weight of 4.5 g. There were no significant differences between treatments in the final oligochaete species composition. The largest proportion of worms, 47%, were immature tubificids with hair and pectinate chaetae. Immature tubificids with bifid chaetae comprised 17% of the total. The identifiable adults were broken down by the following percentages: *T. tubifex*, 16%; *Quistadrilus multisetosus*, 9%; *L. hoffmeisteri*, 8%; *L. profundicola*, less than 1%; *Potamothrix bavaricus*, less than 1%.

The ability to explain differences in the treatment responses is compromised by our limited evaluation of the physical composition and size description of the various substrate types. It is clear that worms in the silt substrate produced more TAMs and survived better than the other treatments. Whether this difference can be attributed to the physical makeup of the substrates or to the elemental and bacterial composition is unknown. Lazim and Learner (1987) demonstrated that *T. tubifex* shows a clear preference for silt-clay substrates compared with sand. They also showed that worms demonstrated a preference for substrates that contained leaf litter over substrates that did not. It has been shown by McMurtry et al. (1983) that when tubificids are given a choice of two substrates, one of which contains a richer concentration of elements and organic compounds and the other a higher level of bacteria, the worms preferred the substrate containing the higher concentration of bacteria. In our test all substrates were autoclaved prior to use and all cultures were fed the same amount of spirulina. It is possible that the silt substrate contributed more to microbial colonization and therefore better worm survival.

Diel light cycle

For both tests conducted there was not a strong relationship between the light or dark cycle and TAM production. For the first test there was wide variation in TAM production (Figure 2). Overall, a slightly higher percentage of TAMs was released during the light cycle, 59%, compared with 41% for the dark cycle, although this difference was not significant ($P = 0.219$). Only one of the seven cycles showed a significant difference in TAM production between the two periods. During the fifth cycle, significantly ($P = 0.031$) more TAMs were released during the light cycle (69%), compared with 31% during the dark cycle. The results from the second test were similar to the first (Figure 3). Overall, for the four cycles, an array of 56% of TAMs were released during the light cycles compared with 44% for the dark. For the second 24-h cycle, significantly ($P = 0.022$) more TAMs were released during the light cycle, 72%, compared with 28% for the dark. No other significant relationships were found. Our results indicate that, at least in the laboratory, there does not appear to be a relationship between light or dark hours and TAM production. These results differ from those of Yokoyama et al. (1993) who revealed a circadian rhythm of actinospore release by the oligochaete *B. sowerbyi*. They observed a peak in actinospore release during 2200–2400 hours. It is difficult to determine whether such laboratory results are transferable to the natural environment. In the field, the seasonality of TAM production may be related to a changing diel light cycle, water flows, water temperatures, or timing of worm infection.

REFERENCES

Lazim, M. N., and M. A. Learner. 1987. The influence of sediment composition and leaf litter on the distribution of tubificid worms (Oligochaeta). Oecologia 72:131–136.

McMurtry, M. J., D. J. Rapport, and K. E. Chua. 1983. Substrate selection by tubificid oligochaetes. Canadian Journal of Fisheries and Aquatic Sciences 40:1639–1646.

Sauter, G., and H. Güde. 1996. Influence of grain size on the distribution of tubificid oligochaete species. Hydrobiologia 334:97–101.

Yokoyama, H., K. Ogawa, and H. Wakabayashi. 1993. Some biological characteristics of actinosporeans from the oligochaete *Branchiura sowerbyi*. Diseases of Aquatic Organisms 17:223–228.

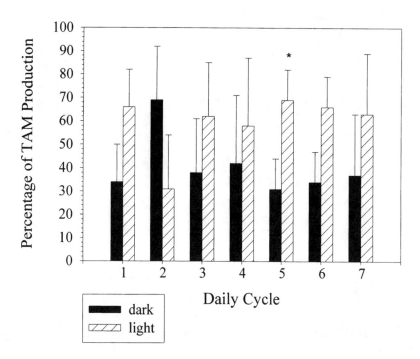

Figure 2. Diel TAM release: Test 1. Average (N = 3, ± SD) percentage of daily TAM release during either the 12 h light or 12 h dark period within a 24-h cycle, sampled over seven cycles. An asterisk above a bar indicates a significant difference ($P \leq 0.05$) within a given 24 h cycle.

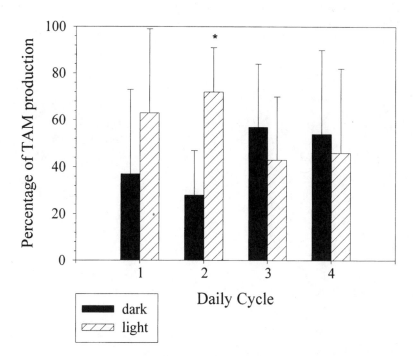

Figure 3. Diel TAM release: Test 2. Average (N = 3, ± SD) percentage of daily TAM release during either the 12 h light or 12 h dark period within a 24 h cycle, sampled over four cycles. An asterisk above a bar indicates a significant difference ($P \leq 0.05$) within a given 24 h cycle.

Section 4

Salmonid Research

American Fisheries Society Symposium 29:95–107, 2002

The Effects of *Myxobolus cerebralis* on the Salmonid Host

ELIZABETH MACCONNELL* AND E. RICHARD VINCENT

Montana Fish, Wildlife and Parks, 1400 S. 19th Avenue, Bozeman, Montana 59718-5496, USA

ABSTRACT. Whirling disease is a chronic inflammatory disease in salmonid fish caused by the myxosporean parasite *Myxobolus cerebralis*. The disease, first recognized in cultured rainbow trout *Oncorhynchus mykiss*, is characterized by the radical tail-chasing behavior of infected fish. The parasite is enzootic in free ranging and cultured populations of trout in areas of salmonid aquaculture, except South America. Recent reports of severe population declines among wild trout associated with whirling disease has caused great concern and renewed interest in this pathogen. The parasite has a sequential affinity for the skin, nerves, and, finally, skeletal cartilage, in the fish host. Abundant cartilage in the skeleton of young trout renders them extremely susceptible to the effects of the disease. Most salmonid species are susceptible to infection with *M. cerebralis*, but susceptibility varies among species and may also vary among strains and individual fish within a population or a similarly exposed group. Rainbow trout and anadromous steelhead are highly susceptible to whirling disease. The severity of the disease and mortality are related to age of the fish, when first exposed, and parasite dose. Trophozoite stages of the parasite lyse cartilage, cause an inflammatory response, and may interfere with normal bone deposition in the salmonid host. In severely infected fish, growth rates are depressed during active infection, and behavioral effects and severe skeletal deformities compromise functions such as swimming and feeding. The effects of *M. cerebralis* on the salmonid host are determined by factors such as species, age, size, parasite dose, immune response, and water temperature. Understanding the interaction of these factors in waters where *M. cerebralis* is present is critical to the future management of healthy wild trout populations.

Whirling disease is a chronic inflammatory disease in salmonid fish caused by the myxosporean parasite *Myxobolus cerebralis* (Hoffman 1990). The parasite has a complex two-host life cycle involving the salmonid fish and oligochaete worm *Tubifex tubifex* (Wolf and Markiw 1984). *Myxobolus cerebralis* has selective tropism for cartilage present in the salmonid skeleton throughout their lives. Abundant cartilage in the skeleton of young trout renders them extremely susceptible to the effects of the parasite. The disease, characterized by the radical tail-chasing behavior of infected fish, was first recognized in cultured rainbow trout *Oncorhynchus mykiss* in Europe (Hofer 1903). Whirling disease has been studied primarily in the rainbow trout host.

The parasite is enzootic in free-ranging and cultured populations of trout in areas of salmonid aquaculture worldwide, except South America (Hoffman 1990; Margolis et al. 1996). Recent reports of severe population declines among wild trout associated with whirling disease in several intermountain states has caused great concern and renewed interest in this pathogen and its effects on the salmonid host (Vincent 1996; Nehring and Walker 1996). The actual impacts of *M. cerebralis* on wild fish depend greatly on numerous variables encountered in natural habitats. Experimental infections in the laboratory can isolate some of the variables experienced in the field. Therefore, comparisons of field and laboratory studies are essential to understanding the effects of whirling disease on wild trout populations.

PATHOGENESIS

The parasite has a sequential affinity for the skin, nerves, and skeletal cartilage in the salmonid host. Waterborne triactinomyxons attach and penetrate the epithelium in less than 10 min after contact with the fish (Markiw 1989; El-Matbouli et al. 1992). Portals of entry in rainbow trout include the epidermis, respiratory epithelium, and buccal cavity, with the greatest numbers of triactinomyxons found in caudal fin one hour after exposure (Markiw 1989; El-Matbouli et al. 1995; El-Matbouli et al. 1999). The process by which triactinomyxon stages enter the fish, develop, and migrate to the cartilage has been described in detail for the rainbow trout host by El-Matbouli et al. (1995). The

*Corresponding author: Phone: (406) 582-8656;
Fax: (406) 587-3998; e-mail: beth_macconnell@fws.gov

parasite begins to multiply shortly after penetrating the epidermis and reaches the peripheral nerves within a few days. Developmental stages of the parasite have been detected interaxonally in most areas of the central nervous system and some peripheral nerves 4–24 d post exposure. Nervous tissue containing the parasite appears normal, and there is no tissue reaction. The parasite continues to replicate and remains sequestered in peripheral nerves for several weeks before reaching cartilage. Presumably, the parasite is sheltered from the host's humoral and cellular immune response during this time.

During the active phase of infection, multicellular trophozoite stages of the parasite continue to replicate and phagocytize chondrocytes (El-Matbouli et al. 1995). Lysis of cartilage stimulates a cellular response in the salmonid host. In most susceptible species the parasite invokes an intense inflam-

matory response. Large numbers of macrophages migrate to areas of cartilage necrosis. Lesions typically contain remnant cartilage, developmental and sporogonic stages of the parasite, and focal to diffuse granulomatous inflammation. At this stage, clinical signs, such as whirling and black tail, may appear. The onset of disease and severity of infection is determined by the number of tissue sites infected, damage to cartilage, and degree of inflammation. Infection intensity or severity of infection is often measured by scoring lesions in histological sections of parasite-exposed fish (Hedrick et al. 1999a; Baldwin et al. 2000; Andree et al. 2002, this volume).

In rainbow trout, lesions progress from small foci of parasites and cartilage degeneration to extensive cartilage necrosis, numerous parasites, granulomatous inflammation, and involvement of surrounding tissues (Figures 1a–d). Initially (<1 month post

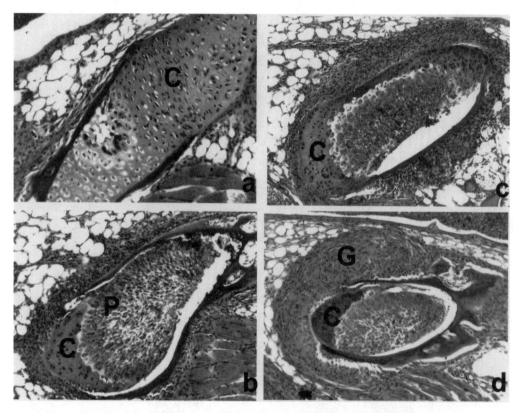

Figure 1. Tissue sections of rainbow trout cartilage in the lower pharyngeal region showing sequential pathology of *Myxobolus cerebralis* infection less than 1 to 3 months post exposure at 13–15°C: a) discrete focal area of trophozoite stages of *M. cerebralis*, distinct zone of cartilage degeneration, no inflammation; b) numerous parasites, widespread cartilage necrosis and infiltration of mononuclear leukocytes; c) increase in inflammatory cells, primarily macrophages, closely associated with infected cartilage; and d) granulomatous inflammation surrounding infected cartilage. Normal cartilage (C); parasites (P); granulomatous inflammation (G) (H&E).

exposure at 13–15 C), trophozoite stages are found in small, discrete foci in cartilage. There is minimal tissue damage and no inflammation. As the infection progresses (2 months post exposure), parasites replicate and cause widespread cartilage degeneration and necrosis. Mononuclear leukocytes, primarily macrophages, increase in numbers and are found closely associated with infected cartilage in adjacent soft tissues. Large granulomatous lesions may have necrotic centers containing spores. Coalescing areas of granulomatous inflammation may become so extensive that normal architecture is lost (Figure 2). The inflammatory response to the trophozoite stage of the parasite can disrupt osteogenesis, resulting in irregular bone formation and permanent skeletal deformities. Eventually, sporogenesis gives rise to clusters of spores that remain trapped in bone for the life of the fish (Figure 3). Opercular cysts containing *M. cerebralis* spores have been reported in cutthroat trout *O. clarki henshawii* and may provide for spore dissemination prior to death of the fish host (Taylor and Haber 1974).

In adult fish, spores of the parasite are found in isolated pockets in bone and are rarely associated with inflammatory lesions. The greatest numbers of spores are found in fish five months after parasite exposure; spore numbers then continue to decline throughout the life of the fish (Markiw 1992a).

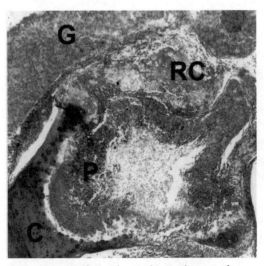

Figure 2. Large lesion located near the eye of a wild rainbow trout naturally exposed to *Myxobolus cerebralis* with a necrotic center containing spores, coalescing areas of granulomatous inflammation (G), numerous trophozoite stages of the parasite (P), and remnant cartilage (RC). Normal cartilage (C) (H&E).

CLINICAL SIGNS

Clinical signs of whirling disease may be seen 3–8 weeks post exposure to the parasite and last for a year. The signs include whirling or radical tail-chasing behavior, blacktail, and skeletal deformities that range from subtle to striking (Figure 4). Cartilage and bones in the head and spine are commonly impacted by the parasite in infected rainbow trout. Blacktail in the fish is produced when the parasite infects cartilage in the posterior spinal column, causing pressure on the caudal nerves, which control pigmentation in the caudal area (Halliday 1976). Damage to the auditory-vestibular organ was believed to be the cause of the whirling behavior in infected fish, but a recent study demonstrated that whirling is a neuropathological consequence of lower brain stem and spinal cord constriction (Rose et al. 2000). Survivors of long-term infections rarely show whirling behavior or blacktail but may have permanent skeletal deformities. Blacktails and skeletal deformities in salmonids can also be due to causes other than M. cerebralis infection (Wolf et al. 1981; Margolis et al. 1996).

GROWTH AND SURVIVAL

Reduced growth rates may also result from M. cerebralis infections. When compared with unexposed controls, severely infected steelhead and chinook salmon have depressed growth rates during active infection (Hedrick et al. 2001). Coho salmon with moderate infections showed no difference in growth rate, compared with unexposed control fish. Normal growth rate appears to resume after clinical disease subsides, with the exception of severely crippled fish. Mortality in young salmonids is highly variable and may be direct, as a result of exposure to the parasite, or indirect, as a result of the infection. Survival increases with increasing age of fish when first exposed to the parasite. Only 9% of sac fry survived exposure to 100 triactinomyxons per fish, but more than 50% of the 2-month-old fry survived the same parasite exposure (Markiw 1991, 1992a). Stressful environmental conditions or secondary pathogens in the natural environment may significantly increase mortality. Schisler et al. (2000) evaluated the effects of M. cerebralis and multiple stressors on morbidity and mortality in rainbow trout and found that elevated water temperatures significantly increased mortality in fish exposed to M. cerebralis.

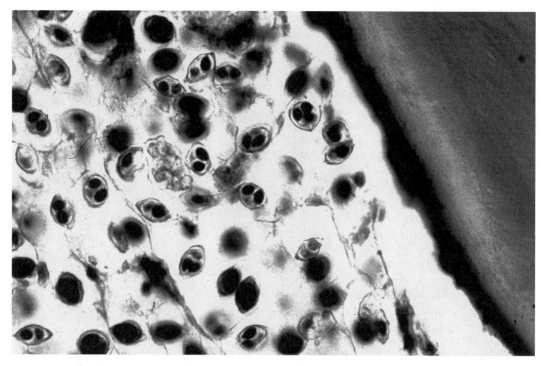

Figure 3. *Myxobolus cerebralis* spores surrounded by bone in rainbow trout (Giemsa stained). Mature spores contain two darkly stained polar capsules.

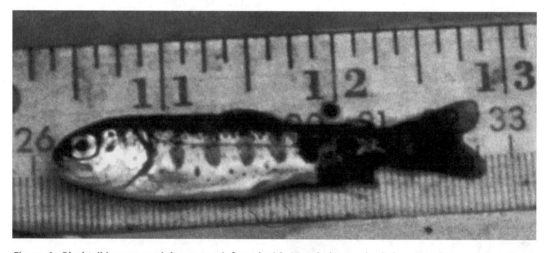

Figure 4. Blacktail in young rainbow trout infected with *Myxobolus cerebralis* by natural exposure.

Whirling behavior and severe skeletal deformities compromise functions such as swimming and feeding. It is unlikely that severely infected fish survive in the wild, as clinical and subclinical infections may negatively impact swimming stamina. Generally, as severity of infection increases, swimming stamina decreases (Ryce et al. 2001).

SPECIES SUSCEPTIBILITY

Rainbow trout is the principal salmonid host for the parasite and, therefore, the fish host studied in most detail. The high susceptibility of rainbow trout to whirling disease was documented in Europe shortly after the species was introduced

(Hofer 1903). Prior to the discovery of the parasite life cycle, patterns of susceptibility in several species of salmonids were examined (Halliday 1976; O'Grodnick 1979; Schaperclaus 1986). More recently, the relative susceptibility based on possible genetic variation between strains or species has been investigated. Experimental exposures of different ages of fish to graded doses of the infectious stage have been conducted in controlled laboratory environments or under variable natural conditions in the field. Susceptibility to M. *cerebralis* infection was evaluated by one or more of the following criteria: prevalence, severity of infection, numbers of spores, mortality, or clinical signs. The comparative testing of species or strains has demonstrated quantitative and qualitative differences in development of the parasite and subsequent disease in various salmonid species.

Most salmonid species examined in these field and laboratory trials were susceptible to infection with M. *cerebralis*. Susceptibility, however, varies among species and may also vary among strains and individual fish within a population or a similarly exposed group. Rainbow trout and anadromous steelhead are highly susceptible to whirling disease. Variability between strains of rainbow trout has been investigated, and some strains showed increased resistance at lower parasite doses, but few overall differences in susceptibility among strains were identified. To date, no naturally resistant strains of rainbow trout have been found. *Ceratomyxa shasta*-resistant, Deschutes and Pit River strains of rainbow trout were tested and found to be susceptible to whirling disease (Hedrick 2001b). This study demonstrated that mechanisms of resistance that have developed among salmonid populations to one myxosporean, C. *shasta*, do not provide protection against another myxosporean, M. *cerebralis*. Field and laboratory studies of the susceptibility of the DeSmet strain of rainbow trout showed reduced infection intensities, compared with other rainbow trout strains, which were more similar to that of cutthroat trout (Vincent, this volume). Comparison of spore concentrations, following natural exposures of Colorado River rainbow trout, suggested that progeny of fish recruited before M. *cerebralis* (pre-Whirling Disease) was present in the system may have an increased resistance to whirling disease, compared with progeny of fish that recruited during the early years of M. *cerebralis* infection (post-whirling disease; Thompson et al. 1999). Subsequent testing of this strain in

laboratory trials showed no significant difference in spore counts, microscopic lesions, or swimming performance between the pre- and post-Whirling Disease fish (Ryce et al. 2001). Experimental exposures of strains of steelhead trout, including one from an enzootic site, showed high susceptibility to infection and disease (Hedrick et al. 2001a; Densmore et al. 2001).

Laboratory exposures of westslope cutthroat trout O. *clarki lewisii* and Yellowstone (Snake River) cutthroat trout O. *clarki bouveri* showed similar infection prevalence, but severity of infection was less in cutthroat trout than rainbow trout when exposed to equal parasite doses at the same age (Hedrick et al. 1999a). This study provided preliminary evidence that subspecies of cutthroat trout are not equally susceptible to whirling disease. Field trials showed poorer survival, but lower spore numbers, in Colorado River O. *c. pleuriticus*, Greenback O. *c. stomias*, and Rio Grande O. *c. virginalis* subspecies of cutthroat trout, compared with rainbow trout (Thompson et al. 1999). Yellowstone cutthroat, added to the trial at a later date, showed better survival and fewer spores than other cutthroat subspecies or rainbow trout tested. Westslope cutthroat trout exposed in the laboratory to 100 and 1,000 triactinomyxons per fish consistently showed lower infection rate and intensity, compared with the rainbow trout tested (Hedrick et al. 1999a). When these test results were compared with similarly exposed Yellowstone cutthroat trout, westslope cutthroat trout again showed a lower infection rate and intensity (Vincent 2002, this volume). Markiw (1992b) reported that after laboratory exposure greenback cutthroat trout were 7.5-fold less susceptible to whirling disease than rainbow trout.

Prevalence of infection, appearance of clinical signs, severity of infection, and spore counts were similar in chinook salmon O. *tshawytscha* and age-matched rainbow trout (Hedrick et al. 2001b). Studies of sockeye salmon O. *nerka* showed that this species is highly susceptible to whirling disease (O'Grodnick 1979). More recently, deformed wild kokanee O. *nerka* with significant cranial cartilage pathology have been reported (C. Wilson, Utah Department of Natural Resources, personal communication). Brook trout *Salvelinus fontinalis* showed high prevalence of clinical signs of whirling disease and mortality when exposed to M. *cerebralis* under natural conditions (Thompson et al. 1999). Mountain whitefish *Prosopium williamsoni* suffered direct mortality when exposed to the parasite, and survivors developed clinical whirling disease (author's unpublished data).

Partial resistance describes species that develop clinical whirling disease only when exposed to high parasite doses. Rainbow trout have significantly higher infection rates and more severe lesions than bull trout *Salvelinus confluentus*, brown trout *Salmo trutta*, coho salmon *O. kisutch*, or Atlantic salmon *Salmo salar*. Both rainbow trout and bull trout showed the same prevalence of infection after exposure to doses of 100 (0%) or 1,000 (100%) triactinomyxons per fish, but only rainbow trout developed clinical signs of whirling disease (Hedrick et al. 1999a). When bull trout were exposed to doses of 2,000 triactinomyxons per fish, 75% developed bent tails (Figure 5), but this condition was rare when fish were exposed to doses of 1,000 triactinomyxons (author's unpublished data). In laboratory studies, brown trout developed clinical whirling disease only when exposed to a very high dose of 10,000 triactinomyxons per fish (Hedrick et al. 1999b). Clinical signs of whirling disease were seen in brown trout after prolonged exposure under natural conditions in field trials (Thompson et al. 1999). Prevalence and severity of infection were significantly higher in rainbow trout than brown trout sequentially exposed under natural conditions (Baldwin et al. 2000). Coho salmon demonstrated no clinical signs of infection or reduced growth, lower infection prevalence and number of spores than similarly exposed rainbow trout (Hedrick 2001b). El-Matbouli et al. (1992) listed Atlantic salmon as intermediate in susceptibility to whirling disease. Controlled experimental exposures of Atlantic salmon suggest that their susceptibility is similar to that of brown trout (V. Blazer, U.S. Geological Survey, personal communication).

In resistant species, no clinical disease or spores develop, regardless of dose. It is unclear, however, at what stage (attachment, penetration, migration, replication, or sporogenesis) the parasite is unsuccessful. Lake trout *Salvelinus namaycush* has long been considered a resistant species, based on field and laboratory exposures (O'Grodnick 1979; Hoffman 1990; V. Blazer, U.S. Geological Survey, personal communication). However, in a recent laboratory study, *M. cerebralis* spores were extracted from lake trout exposed at 10.5 weeks posthatch (E. Wagner, Utah Department of Natural Resources, personal communication). Laboratory studies of Arctic grayling *Thymallus arcticus* showed that the species is resistant to whirling disease (Hedrick et al. 1999a). In one field exposure of Arctic grayling, trophozoites and necrotic cartilage were seen in histological sections of a few exposed fish (<5%), but parasites were few in number and appeared to be undergoing degeneration, and no spores were recovered (author's unpublished data).

Salmonid species are not equally susceptible to infection with *M. cerebralis* or development of whirling disease (Table 1). Among the *Oncorhynchus* spp., inland trout are highly susceptible to whirling disease, but susceptibility is variable among anadromous salmonids. Coho salmon are much more resistant to whirling disease than steelhead, sockeye, or chinook salmon (Hedrick et al. 2001a). The widest variation in susceptibility is seen among the chars or *Salvelinus* spp. Brook trout are susceptible, bull trout are partially resistant, and lake trout are resistant (O'Grodnick 1979; Hedrick et al. 1999a). Contrary to previous reports, splake,

Figure 5. Bull trout with spinal deformity (lordosis) 3 months post exposure to 2,000 triactinomyxons.

Table 1. Susceptibility to whirling disease among species of salmonids by laboratory or natural exposure to *M. cerebralis* at vulnerable life stages. Scale of 0—3 or S: 0 = resistant, no spores develop; 1 = partial resistance, clinical disease rare and develops only when exposed to very high parasite doses; 2 = susceptible, clinical disease common at high parasite doses, but greater resistance to disease at low doses; 3 = highly susceptible, clinical disease common; S = susceptibility is unclear (conflicting reports, insufficient data, lack of *M. cerebralis* confirmation).

Genus	Species	Common name	Susceptibility	Reference*
Oncorhynchus	mykiss	Rainbow trout	3	Baldwin et al. 2000; Hedrick et al. 2001b
	mykiss	Steelhead trout	3	Hedrick et al. 2001a
	clarki	Cutthroat trout		
	c. bouveri	Yellowstone cutthroat	2	Hedrick et al. 1999a; Thompson et al. 1999
	c. lewisi	Westslope cutthoat	2	Hedrick et al. 1999a
	c. pleuriticus	Colorado River cutthroat	2	Thompson et al. 1999
	c. viriginalis	Rio Grande cutthroat	2	Thompson et al. 1999
	c. stomias	Greenback cutthroat	2	Thompson et al. 1999
	tshawytscha	Chinook salmon	2	Hedrick et al. 2001a
	nerka	Sockeye salmon	3	O'Grodnick 1979
	keta	Chum salmon	1 S	Halliday 1976
	gorbuscha	Pink salmon	1 S	Halliday 1976
	masu	Cherry salmon	1 S	Halliday 1976
	kisutch	Coho salmon	1	Hedrick et al. 2001a
Salvelinus	fontinalis	Brook trout	2	O'Grodnick 1979
	malma	Dolly Varden	1 S	Halliday 1976
	confluentus	Bull trout	1	Hedrick et al. 1999a
	namaycush	Lake trout	0 S	O'Grodnick 1979
	Salmo salar	Atlantic salmon	2 S	El-Matbouli et al. 1992
	trutta	Brown trout	1	Baldwin et al 2000; Hedrick et al. 1999b
Prosopium	williamsoni	Mountain whitefish	2 S	Taylor 1974
Thymallus	thymallus	European grayling	2 S	Volf 1957
	arcticus	Arctic grayling	0	Hedrick et al. 1999a
Hucho	hucho	Danube salmon	3	El-Matbouli et al. 1992

*most recent report

a hybrid-cross between lake and brook trout, are not refractory to M. cerebralis (Markiw 1992a, 1992b, E. Wagner, Utah Department of Natural Resources, personal communication). The susceptibility of *Salmo* species Atlantic salmon, and brown trout, is similar. Whirling disease was reported in European grayling *T. thymallus* (Volf 1957), based on the observation of clinical signs, whirling, and blacktail. However, this report is suspect because rainbow trout and brown trout reared simultaneously at the same facility showed no signs of whirling disease. In addition, recent laboratory studies demonstrate that Arctic grayling are resistant (Hedrick et al.1999a). Danube salmon *Hucho hucho*, native to Europe, are highly susceptible to whirling disease (El-Matbouli et al. 1992).

FACTORS THAT AFFECT RESISTANCE

Susceptibility, onset of disease, and severity of signs are determined by a variety of factors. In addition to genetics, the degree of susceptibility or resistance to whirling disease changes with fish age, size, parasite dose, or environmental factors at initial exposure. Details of these variables on the severity of whirling disease have been recently, or are currently, under investigation.

Age

Age is a key factor, as development and severity of whirling disease are directly dependent on the age of fish when first exposed to the parasite. Resistance to whirling disease increases with fish age. In a laboratory experiment to determine the earliest susceptible age eggs and 1 d post hatch fry were found to be refractory to infection (Markiw 1991). Severity of infection decreases with age but cartilage is a key structural component of the salmonid skeleton throughout life so larger older fish can become infected but are less likely to suffer disease or produce many spores. Rainbow trout exposed at 12 months of age for 2.5 months developed spores of M. cerebralis but no clinical signs of whirling disease (Hoffman 1974). As ossification of the skeleton progresses in older fish, greater doses of parasite are required to induce clinical signs, and by the age of 2–3 years, rainbow trout show no external signs of whirling disease following infection (Markiw 1992a).

Size

Size at first exposure may also affect susceptibility to whirling disease. The skeleton progressively ossifies as fish age. It is unclear if rate of ossification is determined by growth rate, age, or a combination of both. Thompson et al. (1999) suggest that size at first exposure influences susceptibility to whirling disease. The smallest sizes of same-age fish introduced into sentinel cages suffered the highest mortality. Exposure to doses of 10,000 triactinomyxons per fish caused direct mortality in mountain whitefish at 3 weeks of age, but no mortality occurred during exposure when larger whitefish 11 weeks of age were exposed (author's unpublished data). Age-matched rainbow trout, three times larger than 3-week whitefish suffered no direct mortality as a result of exposure to the same dose. A relationship between infection severity and size was observed in field trials when fish of different sizes, matched in degree-days, were exposed at the same time under natural conditions. The severity of infection was greater in the smaller fish (author's unpublished data). Laboratory exposure of different sizes and ages, matched in degree-days, will determine if age is more important than size in susceptibility to whirling disease.

Dose

In general, disease severity increases with increasing parasite dose. Clinical whirling disease and mortality are likely in small, newly-hatched rainbow trout exposed to low doses of triactinomyxons. Rainbow trout sac fry, 2 d post hatch, exposed to a dose of 10 triactinomyxons per fish, became severely infected, but no spores were recovered from fish

exposed at two months of age to the same dose (Markiw 1991, 1992a).

Markiw (1992a) demonstrated a linear relationship in rainbow trout between parasite dose and spores produced five months post exposure, up to a dose of 10,000 triactinomyxons per fish. In fish exposed at two months of age, significantly more spores were produced at each higher dose. Some strains or species show increased resistance to whirling disease when exposed to low doses of triactinomyxons. Exposure to high doses of triactinomyxons, however, can mask the resistance seen at lower doses. Severity of infection was lower in cutthroat than in rainbow trout, when exposed at the same age to doses of 1,000 or fewer triactinomyxons per fish, but similar when exposed to higher doses (Hedrick et al. 1999a). As discussed previously, brown and bull trout developed clinical signs of whirling disease only when exposed to very high doses of triactinomyxons.

Immune Response

The host immune response to M. cerebralis varies widely among salmonid so far examined. Certain salmonid species may more effectively eliminate infections, presumably due to a more effective immune recognition and destruction of M. cerebralis. Our current knowledge of the immune response to M. cerebralis in the salmonid host, however, is limited. The high degree of variability seen in fish with light infections suggests that the host immune response can be effective in eliminating the parasite. Clearly, the cellular immune response is stimulated during the active feeding phase of developmental stages of the parasite, during cartilage destruction (Halliday 1974). In contrast, once formed, spores of M. cerebralis do not elicit a cellular immune response. Strong, naturally acquired resistance to reinfection has been documented in rainbow trout (Hedrick et al. 1998). Antibodies to M. cerebralis have been detected in rainbow trout with whirling disease (Griffin and Davis 1978). The antibody response is highly variable, and a repeated pattern of antigen recognition has not been demonstrated (M. Adkison, University California Davis, personal communication). Passive immunization of young trout with serum containing anti-M. cerebralis spore antibodies has shown only marginal efficacy (Hedrick et al. 1998). Active acquired immunity has been found only among exposed fish that developed cartilage lesions.

Similar to other chronic infectious diseases in fish, the macrophage is the key host cell involved in the immune response. Early in infection, macrophages are found in the subcutis attacking the residual epithelial stages of the parasite (El–Matbouli et al. 1999). Rainbow trout mount an extensive cell-mediated response that contributes to disease. The cellular immune response varies in species from an extensive, diffuse granulomatous response to a well-defined, encapsulated granuloma containing few intact parasites (Figure 6). Lesions in brown trout contain more multinucleated giant cells than rainbow trout and rarely progress beyond small, discrete foci (Baldwin et al. 2000). The level of infection, as measured by spore numbers, decreases over time, suggesting that the host immune response effectively eliminates some parasites.

By utilizing the nervous system to migrate to cartilage, the parasite is presumed to escape the host immune response early in infection. A marked eosinophilic granular leukocyte response (Figure 7) has been observed in the nerves of M. *cerebralis*-exposed brown trout, coho salmon, and Atlantic salmon, suggesting that these cells play a role in eliminating the parasite or preventing the parasite from reaching cartilage (Hedrick et al. 1999b, 2001; V. Blazer, U.S. Geological Survey, personal communication). Specific major histocompatibility (MHC) class I or II alleles that confer resistance or protection have yet to be identified in trout but are under investigation.

The principal location of parasite lesions varies among species infected with M. *cerebralis*. Cartilage throughout the body of the fish, including cranium, spine, fins, vertebrae, ribs, and operculum, can be infected (Antonio et al. 1998). Typically, head cartilage is examined for parasite detection. In rainbow trout and other highly susceptible salmonids, parasites and lesions are found in cartilage throughout the body but are consistently located in the cranial regions, primarily the ventral calvarium. In contrast, in brown trout infection, most commonly occurs in gill arches, rarely the ventral calvarium (Hedrick et al.1999b; Baldwin et al. 2000). Lesions may be found in the cranium but are often limited to the axial skeleton in bull trout and mountain whitefish. Wild mountain whitefish collected from a positive river showed no clinical signs of whirling disease or spores in the cranium, but abundant M. *cerebralis* spores were found in the caudal peduncle area (author's unpublished data).

Water Temperature

Water temperature greatly influences initiation, onset, and progression of whirling disease. The developmental cycle of the parasite is greatly extended, and the onset of clinical signs in the fish host is delayed in colder water temperatures. Trophozoite and spore stages of M. *cerebralis* are first found in cartilage 35 d and 52 d, respectively, at 17°C, but temperatures of 7°C delay sporogenesis until 120 d post exposure (Halliday 1973). Parasite attachment and the subsequent ability to

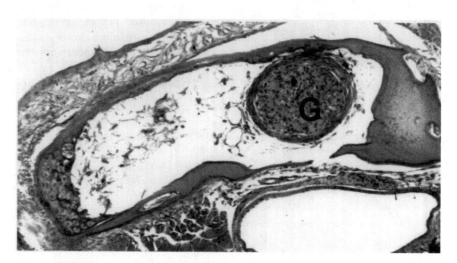

Figure 6. Encapsulated granuloma (G) containing few intact parasites in gill arch of *Myxobolus cerebralis* infected brown trout (H&E).

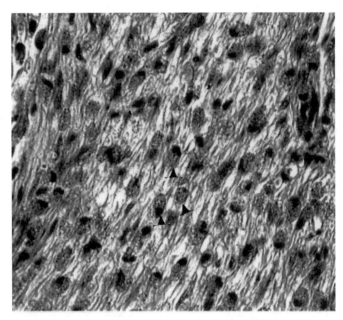

Figure 7. Tissue section of cranial nerve in *Myxobolus cerebralis* exposed brown trout, showing marked infiltration of eosinophilic granular cells (arrowheads) (H&E).

successfully infect the fish may be influenced by water temperature. Severity of infection varied when fish were exposed to triactinomyxons at different temperatures and then shifted to the same temperature for 90 d. Rainbow trout exposed at 4°C showed significantly less average infection intensity when compared with fish exposed to the same doses at 13°C (author's unpublished data).

FACTORS THAT AFFECT DISEASE SEVERITY IN THE NATURAL ENVIRONMENT

Environmental factors play a significant role in determining the result of host and parasite interactions, but many are poorly understood. The impacts of M. *cerebralis* on wild fish populations may be highly variable within and between river systems. Spring creek environments appear to be excellent habitats for both the salmonid and oligochaete hosts. Determining if whirling disease is enhanced in this environment is currently under investigation. Seasonal variation in infectivity has been described for other salmonid diseases caused by myxosporean parasites (Grupcheva et al. 1985; Foott and Hedrick 1987). Critical to understanding the influences of seasonal variation, environmental variables need to be measured over time.

Life History

Life histories of salmonids in North America vary significantly among species and strains. The most critical period for salmonids that live in M. *cerebralis* positive waters is during hatching and emergence, when fish are at the most susceptible age and size. There are two main spawning seasons used by many salmonids: fall (brown trout, brook trout, and bull trout) and spring–early summer (rainbow trout, steelhead, and cutthroat trout). Fall spawning generally occurs October–December, with eggs incubating during winter and hatching emergence occurring February–April. Spring spawning usually occurs during March–June, followed by a shorter incubation period and hatching emergence occurring May–July. Studies of wild trout streams have shown that the triactinomyxon stage of M. *cerebralis* has a seasonal pattern, with the highest density of parasites occurring during June through September (Thompson and Nehring 2000). Salmonid species with life histories where very young susceptible fish are exposed to high concentrations of triactinomyxons suffer the highest young-of-the-year loses.

Species at risk for whirling disease can be determined by comparing the distribution and seasonal occurrence of M. *cerebralis* with resident and anadromous salmonid life histories in a particular river system. Resident rainbow trout, steelhead, and chinook salmon that emerge during times of peak infection are the salmonid species most likely to become infected with M. *cerebralis* in the Lostine River, Oregon (Sandell et al. 2001). Species of fall spawning salmonids generally hatch and emerge during periods of low concentrations of triactinomyxon and, therefore, are often at low risk for whirling disease. Recent studies utilizing sentinel rainbow trout, in Madison River, Montana, found that in years when infection intensity is low (<2.7 average histological score) during the critical hatching-emergence period, the survival of young rainbow trout is high, but, in years when infection intensity is high (>2.7 average histological score), survival is very poor (Figure 8). Population declines in wild rainbow trout follow years of high infection prevalence and intensity (author's unpublished data).

Water Temperature

The entire life cycle of the parasite in both the fish and worm hosts is temperature dependent. Natural outbreaks of whirling disease occur during periods of temperatures optimal for the parasite. A positive correlation between water temperature, infection rates, and severity of infection was demonstrated for rainbow trout in Montana (Baldwin et al. 2000). Only conductivity, not water temperature, pH, or total dissolved solids, had a significant correlation with prevalence of infection among sentinel rainbow trout fry in Oregon (Sandall et al. 2001), although infections were most prevalent in fish exposed in the lower river, where temperatures were higher. Studies in three different watersheds in Montana have shown that infection intensity was most severe when fish were naturally exposed between 12°C and 15°C, declined rapidly when water temperatures were above or below this range and prevalence of infection was near zero at temperatures below 5°C or above 17°C (author's unpublished data).

Water Flow

Flow rate is another environmental factor that may have a significant influence on infection intensities in wild salmonid populations but has yet to be thoroughly investigated. While no statistical correlation among water temperature, flow rate, and density of triactinomyxons was established, data

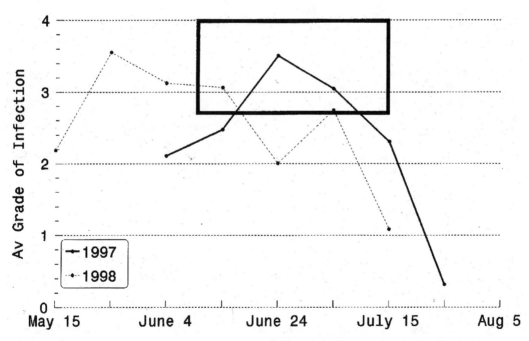

Figure 8. Comparison of average infection intensity measured in sentinel rainbow trout, 1997 and 1998, Madison River, Montana. Black box represents the hatch-emergence period for wild rainbow trout in the Madison River.

suggested that both flow rate and temperature influence the density of triactinomyxons in the Colorado River, Colorado (Thompson and Nehring 2000). Higher water flows may dilute the concentration of triactinomyxons, in effect reducing the actual parasite dose for individual fish and subsequent infection intensities, even during periods of optimal water temperatures. Conversely, low flow rates could result in higher than normal infection intensities. Studies to determine the effect of increasing flow rate at a given water temperature on infection intensity are currently underway.

CONCLUSION

Pathogens are inherent to aquatic ecosystems and may be of major importance in controlling population abundance. Determining which variables are important, as well as how they are measured and analyzed are critical to understanding diseases and their impacts on fish populations (Hedrick 1998). The parasite M. cerebralis has successfully adapted to life in the aquatic environment, and the disease it causes presents a serious threat to the future of wild salmonid populations. Recent studies have yielded considerable new knowledge of the fish host and complex interactions with the parasite and environment. The effects of M. cerebralis on the salmonid host are determined by numerous factors, including species, age, size, parasite dose, immune response, and water temperature. Despite our greatly increased knowledge of these variables in the last 8 years, a more thorough understanding of the interactions with M. cerebralis will be critical to the future management of healthy wild trout populations.

REFERENCES

Andree, K. B., R. P. Hedrick, and E. MacConnell. 2002. A review of the approaches to detect Myxobolus cerebralis, the cause of salmonid whirling disease. Pages 197–211 in J. Bartholomew and C. Wilson, editors. Whirling Disease: reviews and current topics. American Fisheries Society, Symposium 29, Bethesda, Maryland.

Antonio, D. B., K. B. Andree, T. S. McDowell, and R. P. Hedrick. 1998. Detection of Myxobolus cerebralis in rainbow trout and oligochaete tissues by using a non-radioactive in situ hybridization (ISH) protocol. Journal of Aquatic Animal Health 10:361–371.

Baldwin, T. J., E. R. Vincent, R. M. Siflow, and D. Stanek. 2000. Myxobolus cerebralis infection in rainbow trout (Oncorhynchus mykiss) and brown trout (Salmo trutta) exposed under natural stream conditions. Journal of Veterinary Diagnostic Investigations 12:312–321.

Densmore, C. L., V. Blazer, D. Cartwright, W. B. Schill, P. Pooler, J. Schachte, C. Petrie, M. Batur, T. Waldrop, and A. Mack. 2001. A comparison of susceptibility to Myxobolus cerebralis among strains of rainbow and steelhead trout in field and laboratory trials. Journal of Aquatic Animal Health 13:220–227.

El-Matbouli, M., T. Fischer-Scherl, and R. W. Hoffman. 1992. Present knowledge of the life cycle, taxonomy, pathology, and therapy of some Myxosporea spp. important for freshwater fish. Annual Review Fish Disease 3:367–402.

El-Matbouli M., Hoffmann R. W., and Mandok C. 1995. Light and electron microscopic observations on the route of the triactinomyxon-sporoplasm of Myxobolus cerebralis from epidermis into rainbow trout (Oncorhynchus mykiss) cartilage. Journal of Fish Biology 46:919–935.

El-Matbouli M., Hoffmann R. W., H. Schoel, T. S. McDowell, and R. P. Hedrick. 1999. Whirling disease: host specificity and interaction between the actinosporean stage of Myxobolus cerebralis and rainbow trout Oncorhynchus mykiss. Diseases of Aquatic Organisms 35:1–12.

Foott, J. S., and R. P. Hedrick. 1987. Seasonal occurrence of the infectious stage of proliferative kidney disease (PKD) and resistance of rainbow trout, Salmo gairdneri Richardson, to reinfection. Journal of Fish Biology 30:477–483.

Griffin, B. R., and E. M. Davis. 1978. Myxosoma cerebralis; detection of circulating antibodies in infected rainbow trout (Salmo gairdneri). Journal of Fisheries Research Board of Canada 35:1186–1190.

Grupcheva, G., I. Dykova, and J. Lom. 1985. Seasonal fluctuation in the prevalence of Shaerospora renicola and myxosporean bloodstream stages in carp fingerlings in Bulgaria. Folia Parasitologica 32:193–203.

Halliday, M. M. 1973. Studies of Myxosoma cerebralis, a parasite of salmonids. II. The development and pathology of Myxosoma cerebralis in experimentally infected rainbow trout (Salmo gairdneri) fry reared at different water temperatures. Nordisk Veterinaermedicin 25:349–358.

Halliday, M. M. 1976. The biology of Myxosoma cerebralis: the causative organism of whirling disease of salmonids. Journal of Fish Biology 9:339–357.

Hedrick, R. P. 1998. Relationships of the host, pathogen, and environment: implication for disease of cultured and wild fish populations. Journal of Aquatic Animal Health 10:107–111.

Hedrick, R. P., M. El-Matbouli, M. A. Adkison, and E. MacConnell. 1998. Whirling disease: re-emergence among wild trout. Immunological Reviews 166:365–376.

Hedrick, R. P., T. S. McDowell, K. Mukkatira, M. P. Georgiadis, and E. MacConnell. 1999a. Susceptibility of selected inland salmonids to experimentally induced infections with Myxobolus cerebralis, the causative agent of whirling disease. Journal of Aquatic Animal Health 11:330–339.

Hedrick, R. P., T. S. McDowell, M. Gay, G. D. Marty, M. P. Georgiadis, and E. MacConnell. 1999b. Comparative susceptibility of rainbow trout *Oncorhynchus mykiss* and brown trout *Salmo trutta* to *Myxobolus cerebralis*, the cause of salmonid whirling disease. Diseases of Aquatic Organisms 37:173–183.

Hedrick, R P., T. S. McDowell, K. Mukkatira, M. P. Georgiadis, and E. MacConnell. 2001a. Susceptibility of three species of anadromous salmonids to experimentally induced infections with *Myxobolus cerebralis*, the causative agent of whirling disease. Journal of Aquatic Animal Health 13:43–50.

Hedrick, R. P., 2001b. Salmonids resistant to *Ceratomyxa shasta* are susceptible to experimentally induced infections with *Myxobolus cerebralis*. Journal of Aquatic Animal Health 13:35–42.

Hofer, B. 1903. Ueber die Drehkrankheit der Regenbogenforelle. Allgemeine Fischerei-Zietung 28:7–8.

Hoffman, G. L. 1974. Fish age as related to susceptibility to *Myxobolus cerebralis*, cause of whirling disease. Progressive Fish-Culturist 36:151.

Hoffman, G. L. 1990. *Myxobolus cerebralis*, a worldwide cause of salmonid whirling disease. Journal of Aquatic Animal Health 2:30–37.

Margolis, M. L., M. L. Kent, P. Bustos. 1996. Diseases of salmonids resembling myxosporean whirling disease, and the absence of *Myxosoma cerebralis*, in South America. Diseases of Aquatic Organisms 25:33–37.

Markiw, M. E. 1989. Portals of entry for salmonid whirling disease in rainbow trout. Diseases of Aquatic Organisms 6:7–10.

Markiw, M. E. 1991. Whirling disease: earliest susceptible age of rainbow trout to the triactinomxid of *Myxobolus cerebralis*. Aquaculture 92:1–6.

Markiw, M. E. 1992a. Experimentally induced whirling disease I. Dose response of fry and adults of rainbow trout exposed to the triactinomyxon stage of *Myxobolus cerebralis*. Journal of Aquatic Animal Health 4:44–47.

Markiw, M. E. 1992b. Salmonid whirling disease. U.S. Fish and Wildlife Service, Fish and Wildlife Leaflet 17.

Nehring R B., and Walker PG. 1996. Whirling disease in the wild: the new reality in the Intermountain West. Fisheries 21:28–32.

O'Grodnick, J. J. 1979. Susceptibility of various salmonids to whirling disease (*Myxobolus cerebralis*). Transactions of the American Fisheries Society 108:187–190.

Rose, J. D., G. S. Marrs, C. Lewis, and G. Schisler. 2000. Whirling behavior and its relation to pathology of brain stem and spinal cord in rainbow trout. Journal of Aquatic Animal Health 12:107–118.

Ryce, E. M., A. Zale, and R. B. Nehring. 2001. Lack of selection for whirling disease among progeny of Colorado River rainbow trout. Journal of Aquatic Animal Health 13:63–68.

Sandell, T. A., H. Lorz, D. G. Stevens, and J. L. Bartholomew. 2001. Dynamics of *Myxobolus cerebralis* in the Lostine River, Oregon: implications for resident, and anadromous salmonids. Journal of Aquatic Animal Health 13:142–150.

Schaperclaus W. 1986. Fish diseases. Volume 2. Aka107 demie-Verlag: Berlin. 665–666.

Schisler, G. J., E. P. Bergersen, and P. G. Walker. 2000. Effects of multiple stressors on morbidity, and mortality of fingerling rainbow trout infected with *Myxobolus cerebralis*. Transactions of the American Fisheries Society 129:859–865.

Taylor, R. E, and M. H. Haber. 1974. Opercular cyst formation in trout infected with *Myxosoma cerebralis*. Journal of Wildlife Diseases 10:347–351.

Thompson, K. G., R. B. Nehring, D. C. Bowden, and T. Wygant. 1999. Field exposures of seven species or subspecies of salmonids to *Myxobolus cerebralis* in the Colorado River, Middle Park, Colorado. Journal of Aquatic Animal Health 11:312–329.

Thompson, K.B, and R. B. Nehring. 2000. A simple technique used to filter, and quantify the actinospore of *Myxobolus cerebralis* and determine its seasonal abundance in the Colorado River. Journal of Aquatic Animal Health 12:316–323.

Vincent E. R. 1996. Whirling disease and wild trout: the Montana experience. Fisheries 21(6):32–34.

Vincent, E. R. 2002. Relative susceptibility of various salmonids to whirling disease with emphasis on rainbow and cutthroat trout. Pages 109–115 *in* J. Bartholomew and C. Wilson, editors. Whirling Disease: reviews and current topics. American Fisheries Society, Symposium 29, Bethesda, Maryland.

Volf, F. 1957. The first cases of whirling disease among our salmonid fish in our hatcheries. Shornick CSAV, *Zivocisna Vyroba* 30:425–428.

Wolf, K., M. E. Markiw, J. M. Cruz, M. H. Galhano, J. Eiras, and R. L. Herman. 1981. Non-myxosporidan blacktail of salmonids. Journal of Fish Diseases 4:355–357.

Wolf, K., and M. E. Markiw. 1984. Biology contravenes taxonomy in the Myxozoa. Science 255:1449–1452.

American Fisheries Society Symposium 29:109–115, 2002
© 2002 by the American Fisheries Society

Relative Susceptibility of Various Salmonids to Whirling Disease with Emphasis on Rainbow and Cutthroat Trout

E. Richard Vincent*

Montana Fish, Wildlife, and Parks, Whirling Disease Research Coordinator
1400 South 19th Avenue, Bozeman, Montana 59718-5496, USA

ABSTRACT.   There are many factors that govern the susceptibility of salmonids to the parasite *Myxobolus cerebralis*, such as fish size and age, the water temperature in which the fish is exposed to the infective spore stage (triactinomyxon), and the dosage (concentration level) of the infective spore. This study examined changes in infection intensity (degree of cartilage loss and cell inflammation in cranial cartilage) following exposure to 100–4,000 triactinomyxons per fish in 7.6 L of water for 2 h. The infection intensity in each fish was determined by histological examination of the cranial cartilage of each exposed fish, and then, a numerical score of zero to five was assigned to each fish based on this cartilage tissue damage. A total of ten rainbow trout *Oncorhynchus mykiss* strains, three strains of cutthroat trout *O. clarki*, and five other salmonid species that included brown trout *Salmo trutta*, bull trout *Salvelinus confluentus*, chinook salmon *O. tshawytscha*, eastern brook trout *Salvelinus fontinalis*, and kokanee salmon *O. nerka* were examined. Rainbow trout was the most susceptible species tested, having the most severe cranial cartilage damage and cell inflammation at all dosage levels. The only exception within the rainbow trout strains tested was the DeSmet strain, which showed a much lower infection severity, especially at the lower triactinomyxon dosages (100 triactinomyxons/fish and 1,000 triactinomyxons/fish), but even this strain showed severe infection intensities when exposed to higher triactinomyxon dosages (>2,000 triactinomyxons/fish). Of the nine other salmonids tested, only eastern brook trout showed an infection intensity similar to rainbow trout.

The relative susceptibility of various species or strains of salmonids to infection by the parasite *Myxobolus cerebralis* is governed by a number of factors, including genetics, age and size of the fish at the time of exposure, water temperature, and dosage. Most of Montana's major wild trout streams contain a mixture of rainbow trout *Oncorhynchus mykiss* and brown trout *Salmo trutta* with smaller populations of native westslope cutthroat *O. clarki lewisi* or Yellowstone cutthroat trout *O. c. bouvieri*. Comparisons between *M. cerebralis*-infected brown trout and rainbow trout, where they coexist in a Montana wild trout stream, show that rainbow trout not only exhibited a much higher lesion severity and rate of infection than brown trout, but also that the parasite tended to concentrate in the cranial cartilage in rainbow trout versus gill arches in the brown trout (Baldwin et al. 2000). In controlled laboratory exposures, Hedrick et al. (1999) found rainbow trout to be much more susceptible to whirling disease infections when compared with either westslope cut-

throat trout or Yellowstone cutthroat trout. In contrast, lake trout *Salvelinus namaycush* and Arctic grayling *Thymallus arcticus* were found to experience a very low disease severity.

The age of fish when exposed to various dosages of triactinomyxon appears to regulate the rate and intensity of the whirling disease infection in rainbow trout (E. Ryce, Montana State University, personal communication). Water temperatures have also been found to regulate the infection intensity in rainbow trout during field exposures in the Madison River, Missouri River, Willow Creek, and Little Prickley Pear Creek (author's unpublished data). In these studies, field exposures of rainbow trout showed the highest infection intensities at water temperatures between 12°C and 16°C, with infection intensities declining rapidly as mean daily water temperatures decreased or increased from these optimum water temperatures

METHODS

The objective of this study was to determine the relative susceptibility of various salmonids to infection by *M. cerebralis* following controlled exposures

*Phone: (406) 994-3551; Fax: (406) 994-4090;
e-mail: rvincent@montana.edu

to triactinomyxons under laboratory conditions and following natural exposure at an enzootic site.

Fish Species and Strains

A total of 20 salmonid strains or species was tested in field and laboratory exposures. The species and strains of rainbow trout were obtained from eggs taken from spawning wild rainbow trout in various Montana streams and from state and federal fish hatcheries (Table 1).

Natural Exposures

To compare susceptibility following natural exposure, each species/strain was exposed in a stream known to have wild trout infected with M. cerebralis (Willow Creek, a small tributary to the Jefferson River in southwestern Montana). These exposures used sentinel cages consisting of a cylindrical screened container 46 cm in diameter and 51 cm in depth that was placed along the stream bank. All exposures used a single sentinel cage containing at least 50 fish and kept in the stream for 123 d during June through November. The start of the exposure periods varied because of the inability to match the starting date with age (60 d old). Most field exposures occurred between 13 July and 13 November 1996. The field exposures included Deschutes rainbow trout, DeSmet rainbow trout, Eagle Lake rainbow trout, westslope

cutthroat trout, Yellowstone cutthroat trout, kokanee salmon, and brown trout (Table 2).

Laboratory Exposures

The second approach was to test the various strains or species using known dosages of triactinomyxons under laboratory conditions. Each strain or species was exposed to triactinomyxon dosages ranging from 100 to 3,000 triactinomyxons per fish in 7.6 L of water for 2 h. All test lots consisted of 50 fish that were approximately 60 d old. Triactinomyxons were obtained from a colony of Tubifex tubifex worms that had been previously infected by the parasite M. cerebralis. Triactinomyxons were extracted from the water in the worm colony, concentrated using a 20 mm filter, and then placed in a known volume of water and counted in a petri dish with a marked grid. Triactinomyxons were then placed on ice packs for transport to the Pony Whirling Disease Laboratory, where the test fish were held. Prior to exposure, the triactinomyxons were acclimated to 10°C. After the 2 h exposure, fish were transferred to 114 L aquaria and held at 10°C for 90 d (900 degree-days C).

Assessment of Infection

At the end of this period, all of the surviving fish were sacrificed, and the heads placed in a buffered solution of formaldehyde (v/v n-phosphate AFIP

Table 1. Source of salmonids used in the susceptibility to *M. cerebralis* testing.

Species or strain of a species	Source of test fish
Arlee rainbow trout	Montana State Hatchery System
Deschutes rainbow trout	Oregon State Hatchery System
Erwin rainbow trout	USFWS Federal Hatchery, Ennis, MT
Randolph rainbow trout	New York State Hatchery System
DeSmet rainbow trout	Willow Creek, MT spawning trap
Eagle Lake rainbow trout	Clark Canyon Reservoir, MT spawning trap
Finger Lake rainbow trout	Finger Lake, NY
Firehole rainbow trout	Firehole River, YNP
Madison rainbow trout	Madison River, MT
Missouri rainbow trout	Little Prickley Pear Creek, MT
Snake River cutthroat trout	Wyoming State Hatchery System
Westslope cutthroat trout	Montana State Hatchery System
McBride Yellowstone cutthroat trout	Montana State Hatchery System
Eastern brook trout	Private Hatchery in Montana
Kokanee salmon	Montana State Hatchery System
Chinook salmon	Fort Peck Reservoir, MT spawning trap
Bull trout	USFWS Federal Hatchery, Creston, MT
Brown trout	Willow Creek, MT
Arctic grayling	USFWS Federal Hatchery, Bozeman, MT

Table 2. Comparison of relative susceptibility to *M. cerebralis* infections using average infection intensities from histological grading of cranial cartilage in various rainbow trout strains. Fish were ranked in the three categories based on lesions from an exposure level of 1000 triactinomyxons/fish. All fish tested were approximately 60 d of age and exposed in 10 C °water for 2 h. Histological scoring scale: 0–5.

| Rainbow trout strains | # of Triactinomyxons | | |
	100/ fish (n)	1,000/fish (n)	2,000/fish (n)
Very high susceptibility			
Eagle Lake	3.08 (70)	3.94 (64)	
Finger Lake	2.14 (76)	3.73 (78)	
Firehole River	3.96 (93)	4.29 (47)	
Madison River	2.58 (67)	3.78 (67)	
Missouri River	2.70 (69)	4.00 (64)	
Randolph	1.33 (97)	3.70 (95)	
High susceptibility			
Arlee rainbow trout	No data	3.63 (100)	4.23 (129)
Erwin rainbow trout	2.80 (50)	3.04 (50)	
Moderate susceptibility			
DeSmet rainbow trout	0.86 (147)	2.24 (138)	

fixative—pH 7.0) for 72 h. Each individual head was then placed in a whirl-pack bag with 2 mL of isopropyl alcohol and sent to the Washington State University Animal Diagnostic Laboratory in Pullman, Washington. With exception of bull trout, the cranial area was chosen for the histological evaluation. In bull trout, a large number of severe caudal deformities were observed, and additional histological examination was preformed on this tissue. The histological scoring system (Table 3) for each fish was based on the degree of cartilage damage and was developed by Dr. Tom Baldwin (Washington State University at Pulllman, Washington) and Elizabeth MacConnell (U.S. Fish and Wildlife Service at Bozeman, Montana). All of the results for each exposure group are stated as mean infection intensity, which is the average of all of the histological scores from each individual fish within each test lot. These mean histological scores were placed in four categories as follows: Low 0–2.00; Medium 2.01–2.74; High 2.75–3.70; and Very High 3.71–5.00. The categories were arbitrary, except that the lower boundary of the high category fit with field data in Montana, which indicates that average infection intensities greater than 2.75 usually cause wild rainbow trout population declines. Statistical differences between the average infection intensities for each species or strain tested was computed using the two-tailed t-test.

Table 3. Grading scheme used to quantify *M. cerebralis* lesion severity.

Grade	Description
0	No abnormalities noted
1	Discrete, rare (usually single), small foci of cartilage degeneration with intralesional *M. cerebralis* myxospores /generative stages present. Any associated leukocyte infiltrates are small.
2	Single, locally extensive focus or several smaller foci (usually 2) of cartilage degeneration/necrosis with intralesional M. cerebralis myxospores /generative stages. Lytic foci typically surrounded and/or infiltrated by few to moderate numbers of leukocytes.
3	Multiple foci of cartilage degeneration/necrosis (usually 3 or 4) with intralesional *M. cerebralis* myxospores / generative stages. Moderate numbers of leukocytes typically associated with lytic cartilage.
4	Multifocal (usually > 4 sites) to coalescing, often locally extensive areas of cartilage degeneration /necrosis with intralesional *M. cerebralis* myxospores/ generative stages. Moderate to large numbers of leukocytes typically border and/or infiltrate degenerate cartilage.
5	Multifocal (6 or more) to coalescing areas of cartilage necrosis, with locally extensive destruction in at least one focus (preferably more) are present and have intralesional *M. cerebralis* myxospores/generative stages. Moderate to large numbers of leukocytes typically border and/or infiltrate necrotic cartilage.

RESULTS

Laboratory Exposures

Rainbow Trout

With the exception of the DeSmet strain, only small differences in average infection intensity were found among the strains of rainbow trout exposed to either the 100 or the 1,000 triactinomyxons per fish (Table 2). The DeSmet rainbow trout strain showed a significantly lower average infection intensity than all of the other rainbow trout strains tested at both the low (l00) and moderate (1,000) triactino-myxon dosages (P-values ranging from 0.10 to <0.001). Based on the 1,000-triactinomyxon/fish dosages, average infection intensity differences were small but significant among the remaining eight strains. These differences allowed the strains to be divided into two additional categories. The first group (high) included two domesticated strains (the Arlee and Erwin), and the second group (very high) contained the remaining six strains (Eagle Lake, Finger Lake, Firehole River, Madison River, Missouri River, and Randolph). The high group had

average infection intensities ranging from 3.00 to 3.65, and the very high group had average infection intensities greater than 3.65. The differences between the average infection intensities for the high and very high groups were significant with P-values ranging from 0.05 to 0.001. At the low tri-actinomyxon dosage (100), the average infection intensities for the eight rainbow trout strains were more variable. At the low dosage, the Firehole River and Eagle Lake strains showed the highest average infection intensities.

Cutthroat Trout

Of the three subspecies of cutthroat trout tested, westslope cutthroat trout showed the lowest average infection intensities at the 1,000 triactino-myxon/fish dosage (Table 4). The average infection intensity was significantly lower than among Yel-lowstone cutthroat trout at both 100 and 1,000 tri-actinomyxons per fish (P < 0.001). The average infection intensities were also lower than in the Snake River cutthroat trout for both dosages, but differences were not significant. Differences between the Yellowstone and Snake River subspecies were small and not significant at either dosage.

Table 4. Comparison of relative susceptibility to M. cerebralis infections using average infection intensities from histological grading of cranial cartilage in all salmonids tested. Species, subspecie or strains were ranked on the basis of average severity of infection from exposure to 1,000 triactinomyxons/fish. All fish tested were approximately 60 d of age and exposed in 10 C° water for 2 h. Histological scoring scale: 0–5.

Species, strain, or subspecie	Number of triactinomyxons /fish	
	100 (n)	1,000 (n)
Very high susceptibility		
Average of six rainbow trout strains (Table 2)	2.68 (472)	3.90 (415)
Brook trout	2.75 (98)	3.76 (93)
High susceptibility		
Arlee rainbow trout	No data	3.63 (100)
Erwin rainbow trout	2.80 (50)	3.04 (50)
Moderate Susceptibility		
Kokanee salmon	0.50 (65)	2.53 (93)
DeSmet rainbow trout	0.86 (147)	2.24 (186)
Yellowstone cutthroat trout	0.86 (146)	2.21 (148)
Snake River cutthroat trout	1.39 (138)	1.93 (94)
Westslope cutthroat trout	0.59 (131)	1.41 (139)
Low susceptibility		
Bull trout	0.31 (66)	1.12 (33)
Chinook salmon[1]	0.00 (90)	0.22 (95)
Brown trout		0.06[2]
Arctic grayling		0.002[2]

[1]Triactinomyxon viability was suspect during this experiment.
[2]Based on 123 day Willow Creek exposure.

Other Salmonids

Of the four other salmonids tested for their relative susceptibility to M. *cerebralis* infections, only brook trout showed a high susceptibility to the parasite with average infection intensities not significantly different from any rainbow trout strain, except the DeSmet rainbow trout (P = 0.001). The kokanee salmon were placed with the DeSmet rainbow trout in the moderate category showing significantly lower average infection intensities than all other rainbow trout (P = 0.001). The remaining two species (bull trout and chinook salmon) were placed in the low susceptibility category showing significantly lower average infection intensities than all other salmonids tested under laboratory experiments. As the dosage level of triactinomyxons per fish increased, the prevalence of caudal deformities in bull trout increased from 0% at 100 to 75% at 1,000 to 100% at 4,000. Histological examination of this caudal area found that visible deformities were associated with severe cartilage damage and cell inflammation.

Field Exposures

Of the three rainbow trout strains and two cutthroat trout subspecies exposed to Willow Creek water for 123 d, all showed higher average infection intensities than experienced under the highest dosage tested in the laboratory experiments. The DeSmet rainbow and westslope cutthroat trout average infection intensity increased from the moderate to high infection intensity category (3.12 and 3.19, respectively), and the Yellowstone cutthroat trout increased to the very high category (3.91). Kokanee salmon increased from an average infection intensity of 2.25 at the 1,000 triactinomyxon/fish dosage under laboratory conditions to 2.86 in the field exposures. Brown trout exposed to Willow Creek water had a very low infection intensity (0.06), and arctic grayling showed no signs of infection.

DISCUSSION

Of the species tested during this study, rainbow and brook trout were found to be the most susceptible to severe M. *cerebralis* infections. This result is similar to other reported findings. Hedrick et al. (1999) tested five salmonid species at two different triactinomyxon dosages (100–200 and 1,000–2,000). Using clinical signs, spore concentrations in the head, and severity of microscopic lesions to assess infection severity, they concluded that rain-

bow trout were the most susceptible. Baldwin et al. (2000) examined rainbow and brown trout sampled from Montana coldwater salmonid streams and found that rainbow trout showed significantly higher infection rates and more severe lesions than did brown trout in age-matched fish.

Currently, there are four methodologies used to describe M. *cerebralis* infections in salmonid fish: 1) polymerase chain reaction (PCR), 2) external clinical signs, 3) spore counts, and 4) histology. Of these four methodologies, presently, only histology and spore counts are capable of quantification of disease severity. The histological lesion scoring method was used in this study because it appears to best reflect a measurement of disease severity. This method evaluates cell cartilage damage and associated inflammation. The cranial cartilage was chosen as the primary target to examine because it is an area of parasite concentration and because damage in this area may compromise long-term survival. Another critical site, which was not examined in this study, was the vertebral column. Intense infections in less critical sites (such as fin rays, opercula, or jaws) may have less impact on the fish's ability to survive in wild populations. External clinical signs often reflect damage to less critical cartilage sites and may not accurately reflect the true level of disease in the fish.

In a related study, a total of 4,339 rainbow trout exposed to various levels of M. *cerebralis* spores were examined for five external clinical signs of whirling disease (shortened snout, protruding eyes, deformed opercula, lower jaw deformities, and deformed vertebral column) prior to cranial histology (author's unpublished data). All of these rainbow trout were examined externally and sacrificed approximately 900 degree-days C, post exposure. Of the 4,339 fish examined, 36.5% (1,588) showed one or more of these external deformities, but later examination by histology showed no M. *cerebralis* infection. Of the five external deformity categories, shortened snout appeared to be the most reliable, although 23.2% of those characterized for this trait were not shown to be positive by histology. Many of these deformities could have genetic or environmental causes not related to M. *cerebralis* infections. Two other external features (whirling behavior and black tail) were much more reliable in determining M. *cerebralis* infections, with only 3.7% and 15.6% of the black tail and whirling fish, respectively, negative by histology. However, only 5% of fish positive by histology showed either black tail or whirling behavior.

The spore count method was not used in this experiment because it requires holding exposed fish for a minimum of five months post exposure. Salmonids scored histologically are generally sacrificed at 900 degree-days C, post exposure (90 d at 10 C at the Pony Whirling Disease Laboratory). Histological examination at that point in time shows few myxospores are present. This is the optimum period for histological scoring, since maximum inflammation is occurring. However, by sampling at this time, we were unable to compare spore counts with histological evaluation for assessing severity of whirling disease.

In this study, there were two distinct categories of infection intensity (high and moderate) among the nine strains of rainbow trout compared. However, because two strains within the high category have significantly lower lesion scores (Arlee and Erwin), this category was further divided into high and very high infection intensity. Although these differences were statistically significant, average infection intensities in both groups were so high that survival was considered to be poor in all of these rainbow trout strains. While average infection intensities between the rainbow trout strains were consistent at the 1,000 triactinomyxon/fish dosage, there was significantly more variability at the 100 triactinomyxon/fish dosage. This variability could have been the result of minor size/age differences that was masked at the higher dosages. One strain of rainbow trout (DeSmet) showed significantly lower average infection intensities than the other eight strains and was placed in the moderate category. This strain's lower susceptibility may allow its survival in streams where moderate M. cerebralis infections are found, even though more highly susceptible strains could not survive.

In a previous study (author's unpublished data), we were able to show a relationship between average histological scores and its impact on the survival of young-of-the-year rainbow trout in a section of wild trout stream. Sentinel cage exposures with DeSmet rainbow trout in Willow creek showed an average infection intensity of 3.12. DeSmet rainbow trout populations have been self-sustaining in this stream since 1983, and with the advent of whirling disease in 1995, spawning age fish numbers have declined by more than 80%. The reason for this decreased susceptibility to M. cerebralis infections is unknown. One explanation is that it might be a result of cross breeding with Yellowstone and Westslope cutthroat trout and natural development of resistance.

All of the cutthroat trout subspecies tested in this study showed a significantly lower average infection intensity than all of the rainbow trout strains, except DeSmet rainbow trout, at both the low (100 triactinomyxon) and moderate (1,000 triactinomyxon) dosage. In streams that are known to have moderate M. cerebralis infection rates, these species should experience lower infection intensities and better survival. Although the cutthroat trout species tested showed a moderate average infection intensity when exposed to moderate dosages of triactinomyxons (1,000/fish) in laboratory tests, they are not necessarily resistant to higher doses of triactinomyxons, as shown by field exposures in Willow Creek. Sentinel cage exposures in this stream resulted in much higher average infection intensities than observed in laboratory exposures.

Of the other salmonids tested, only brook trout showed a high relative susceptibility to severe M. cerebralis infections. In this study, brook trout were placed in the very high susceptibility category with no significant difference from rainbow trout strains in this category. Thompson et al. (1999) found a high prevalence of clinical signs and mortality in brook trout when exposed to natural infections of M. cerebralis in the Colorado River. Since they have such a high susceptibility to M. cerebralis infections, it would be reasonable to expect population declines due to whirling disease in some wild populations.

Because of the relatively low average infection intensity in the cranial area at all dosages, bull trout were placed in the low susceptibility category. Even at triactinomyxon dosages of 4,000 per fish, bull trout showed an average cranial infection score of only 1.59. Using the cranial infection scores, one would not expect to see population level declines, even in streams where high M. cerebralis infections are prevalent, but severe tail deformities were noted before fish were sacrificed for histological scoring. There appeared to be a relationship between triactinomyxon dosage and the prevalence of young bull trout experiencing caudal deformities. This severely deformed caudal area may reduce the ability of these young bull trout to survive. Thus, even though the average cranial infection intensity scores placed this species in the low susceptibility category, the concentration of infection in the caudal area may still significantly compromise the young fry's ability to survive under wild conditions.

Although this study demonstrates clear differences in species and even strain susceptibility in controlled laboratory exposures, it cannot be concluded that salmonid species that show either a low or moderate susceptibility to the parasite M. *cerebralis* are immune to impacts or effects in the wild. The relative susceptibility to M. *cerebralis* infections is only a part of the formula in determining whether this parasite can cause severe populations declines. Other factors that can influence damage by this parasite in wild salmonid populations are high triactinomyxon concentrations, water temperatures, flow rates, matching of triactinomyxon production periods with ages and sizes of young salmonids most vulnerable to the parasite, location of primary fry habitat in relation to triactinomyxon production areas, and movement of fry in relation to these high infection sites. This relative susceptibility data suggests that if all other conditions are equal, the species, subspecies, or strains that have the highest severe average infection intensities are the most likely to suffer severe wild population declines.

ACKNOWLEDGMENTS

These studies were supported by the National Partnership on the Management of Wild and Native Cold Water Fisheries and the Montana Fish, Wildlife and Parks. Elizabeth MacConnell provided all of the triactinomyxon used in the laboratory exposures. Jeff Bagdanov, Jody Hupka, and Jennie Miles conducted the actual exposure of triactinomyxon with the fish at the Pony Whirling Disease Laboratory, maintaining the fish in aquaria and sending the appropriate cranial tissue to the Washington State Animal Diagnostic Laboratory in Pullman, Washington. Bob McFarland provided assistance with statistical analysis of the exposure data. The salmonids species/ strains that were used in these tests were provided by the states of Wyoming, Oregon, Montana, and New York, plus Erwin rainbow trout from the Ennis National Fish Hatchery.

REFERENCES

Baldwin, T. J., E. R. Vincent, R. M. Siflow, and D. Stanek. 2000. *Myxobolus cerebralis* infection in rainbow trout (*Oncorhynchus mykiss*) and brown trout (*Salmo trutta*) exposed under natural stream conditions. Journal of Veterinary Diagnostic Investigations 12:312–321.

Hedrick, R. P., T. S. MacDowell, K. Mukkataira, M. P. Georgiadis, and E. MacConnell. 1999. Susceptibility of selected inland salmonids to experimentally induced infections with *Myxobolus cerebralis*, the causative agent of whirling disease. Journal of Aquatic Animal Health 11:330–339.

Thompson, K. G., R. B. Nehring, D. C. Bowden, and T. Wygant. 1999. Field exposures of seven species or subspecies of salmonids to Myxobolus cerebralis in the Colorado River, Middle Park, Colorado. Journal Of Aquatic Animal Health 11:312–329.

American Fisheries Society Symposium 29:117–124, 2002
© 2002 by the American Fisheries Society

Relative Susceptibility of Selected Deschutes River, Oregon, Salmonid Species to Experimentally Induced Infection by *Myxobolus cerebralis*

SARAH A. SOLLID, HARRIET V. LORZ, DONALD G. STEVENS, AND JERRI L. BARTHOLOMEW*

Department of Microbiology and the Center for Fish Disease Research
Oregon State University, Corvallis, Oregon 97331-3804, USA

ABSTRACT. Laboratory challenges to determine the susceptibility of indigenous Deschutes River, Oregon, salmonids to *Myxobolus cerebralis* were conducted as part of a study to assess the risk of reintroducing anadromous salmon above a migration barrier on that river. Replicate groups of progeny from wild rainbow trout *Oncorhynchus mykiss*, steelhead (anadromous rainbow trout), kokanee *O. nerka*, and chinook salmon *O. tshawytscha*, were exposed to doses of 0, 200, or 2,000 triactinomyxons per fish. Fish were evaluated at 5 months postchallenge for spore concentration in the cranial cartilage, severity of microscopic lesions in the cartilage, and clinical signs of disease. The wild rainbow trout (0.7 g at exposure) were most susceptible to infection, with infection prevalence and spore concentrations similar to those of a susceptible (Mt. Lassen) control rainbow trout strain (0.7 g at exposure), although clinical disease signs were less common in the wild strain. Two year classes of steelhead, exposed at different sizes (0.3 g and 1.0 g), both showed fewer clinical disease signs, a lower prevalence of infection, a lower spore concentration, and a decreased mean lesion score, compared with the control rainbow trout (0.6 g and 1.2 g). Kokanee (1.5 g at exposure) became infected but less severely than the control rainbow trout (1.8 g at exposure). Clinical signs were not evident in the kokanee or the susceptible rainbow trout, possibly because of the large size at exposure. No signs of infection were detected in the chinook salmon (1.0 g at exposure) at either dose, despite high infection prevalence in the control rainbow trout (0.6 g at exposure). These results demonstrate that the indigenous salmonids present in the Deschutes River, both above and below the barrier, are susceptible to infection, but the rainbow trout would be most at risk should introduction of the parasite occur in this system.

The myxozoan parasite *Myxobolus cerebralis* is the causative agent of whirling disease, a condition first recognized for its effects in fish culture but more recently known for its impacts on wild trout populations (Höfer 1903; Nehring and Walker 1996; Vincent 1996; Hedrick et al. 1998). Spores of the parasite can develop in many species of salmonids, but the expressed severity of the disease may vary between species (Hoffman and Putz 1969; O'Grodnick 1979; Markiw 1992). Clinical signs of the disease manifest as cranial or spinal deformities, a darkened tail region (black tail), and the erratic swimming behavior that gave the disease its name. After penetration of the epidermis and migration through the nervous system, *M. cerebralis* preferentially attacks the cartilage, digesting it as the parasite progresses in its development (El-Matbouli et al. 1999). Young fish are more susceptible to infection by *M. cerebralis* and the subsequent effects of the disease, as they have proportionally more abundant cartilaginous tissue than adult fish (Halliday 1976; Markiw 1992).

A number of studies have examined the susceptibility of various salmonid species (O'Grodnick 1979; Hedrick et al. 1998, 1999b; Thompson et al. 1999). These reports are conflicting, as some conclusions were derived from outcomes of natural epizootics and exposures in live cages rather than under the controlled conditions of laboratory studies. However, there is a generally accepted ranking of relative susceptibilities that places rainbow trout *Oncorhynchus mykiss* as highly susceptible; sockeye salmon *O. nerka*, chinook salmon *O. tshawytscha*, Atlantic salmon *Salmo salar*, cutthroat trout *O. clarki*, and brook trout *Salvelinus fontinalis* as intermediate in their susceptibility; and brown trout *S. trutta* and coho salmon *O. kisutch* as having low

---

*Corresponding author.

susceptibility to the disease. Most resistant are lake trout S. *namaycush*, which are considered refractory to infection (O' Grodnick 1979). The ability to conduct controlled parasite challenges is beginning to result in a more clear understanding of species susceptibility. However, differences in age and size at challenge, fish strain, challenge dose, and exposure conditions continue to complicate interpretations. Nevertheless, these data are critical to understanding how this parasite might spread within a system and allow for the prediction of potential impacts on resident salmonid populations.

Since the late 1980s, M. *cerebralis* has been established in the upriver tributaries of the Columbia River (Lorz et al. 1989). Anadromous fish from wild or naturally-reproducing parents may become infected in rivers, where they hatch and complete their early rearing, or, as juveniles, when they migrate through enzootic waters on their way to the ocean. However, hatchery-reared salmon often first encounter the parasite as juveniles, when released into their native rivers before commencing their seaward migration. Evidence that salmonids can become infected as juveniles is demonstrated by detection of M. *cerebralis* spores in adult hatchery fish returning to these rivers (Engelking 2001, this volume). Adult salmon are known to stray during their return migration (Groot and Margolis 1991), and stray adult steelhead (anadromous rainbow trout) and chinook salmon infected with M. *cerebralis* have been documented in the lower Deschutes River, a major tributary in the mid-Columbia River basin. Some of these fish have differential fin clips and coded wire tags that identify their origin from upper Columbia and Snake River basin hatcheries (H. Mark Engelking, Oregon Department of Fish and Wildlife, personal communication).

On the Deschutes River, Oregon, the Pelton Round Butte (PRB) hydroelectric project serves as a barrier to natural migration, isolating upriver resident fish populations. Sampling of resident fish (H. Mark Engelking, Oregon Department of Fish and Wildlife, personal communication) and examination of *Tubifex tubifex* populations (authors' unpublished data) indicate that M. *cerebralis* is not established above the PRB project. However, transporting returning adult salmonids above the PRB project has been proposed in an effort to reestablish migration runs (Ratliff et al. 1998). This reintroduction brings with it the danger of stray adult salmon inadvertently introducing M. *cerebralis* (and other pathogens) into this isolated portion of the river.

Salmonid species considered for passage above the PRB project are spring-run chinook salmon, summer-run steelhead, sockeye, and bull trout S. *confluentus*. Prior to construction of the project, wild runs of these species spawned in the Metolius, Crooked, and upper Deschutes River tributaries, which are now inaccessible to returning adults. Wild sockeye runs were eliminated early in this century, but naturally-reproducing populations of kokanee O. *nerka* exist in the lakes and reservoirs above the project. Bull trout populations, that were once common throughout the Deschutes River, became extinct above Steelhead Falls, by the 1960s (D. Ratliff, Biologist, Portland General Electric Company, personal communication). However, passage is considered important, as it would provide the opportunity for genetic exchange that will permit the long-term conservation of this species.

As plans for fish passage were developed, critical uncertainties were identified (Ratliff et al. 1998). Along with likelihood for success of the effort, these uncertainties also included the risk of upstream pathogen transfer and the susceptibility to pathogens that are considered at risk for introduction. The information needed to assess these risks include habitat evaluation to determine if conditions exist for establishing the parasite life cycle, monitoring stray anadromous salmon to estimate the frequency and potential degree of introduction, and determining species susceptibility. This study focused on the latter objective and examined the susceptibility of indigenous rainbow trout, steelhead, chinook salmon, and kokanee, by controlled laboratory exposure, to the infectious stage of M. *cerebralis*.

METHODS

Fish Source

Gametes from Metolius River kokanee (DR-K), Crooked River rainbow trout (DR-R), Deschutes River summer steelhead (DR-S), and Deschutes River spring chinook salmon (DR-C) were collected from wild, spawning adults. Rainbow trout (RB-C), obtained as eyed eggs from Mt. Lassen Trout Farm (Red Bluff, California), served as a susceptible exposure control for each test. Concurrent exposure of age-matched Mt. Lassen rainbow trout allowed for verification of parasite viability and for comparison of susceptibility between the test species. Eggs from each species were incubated and fry-reared at the Salmon Disease Laboratory (SDL; Oregon State University, Corvallis, Oregon), in 12.8°C specific pathogen-free (SPF) well water. After absorption of the egg yolk, fry were fed a commercial trout food ad libitum. The source, age, and size at exposure for each species are presented in Table 1.

Parasite Source

Susceptible *T. tubifex* cultures were seeded with M. *cerebralis* spores harvested from the head cartilage of infected fish (Andree et al. 1998). Cultures were maintained in SPF water in a 15°C incubator, fed every 3 d, and monitored for triactinomyxon production. At approximately 12 weeks, triactinomyxons were harvested, as described by Hedrick et al. (1999b). For fish challenges, triactinomyxons were stained for viability, enumerated, and held briefly at 4°C prior to exposure.

Challenge

For each species, challenges were conducted as described by Hedrick et al. (1999b). Duplicate groups of 25 fish were used for each of three treatments: a high dose of 2,000 triactinomyxons per fish, a low dose of 200 triactinomyxons per fish, and no dose (unexposed control group). The weight and age in degree-days of the RB-C were matched as closely as possible to that of each test species (Table 2). In 1999, the kokanee (1.5 g), wild rainbow trout (0.7 g), and steelhead (1.0 g) had separate RB-C groups. Because of space limitations in the isolation facility the following year, the steelhead (0.3 g) and chinook salmon (1.0 g) shared a RB-C (0.6 g). Fish were challenged in 5-L tanks, and triactinomyxons were introduced with

500 mL of static aerated water. After a 2-h exposure, the fish and parasites were transferred to 25-L tanks and maintained for 5 months in 12.8°C SPF water to allow for myxospore development. During this period, the fish were fed a semimoist commercial trout food at a rate of approximately 2% body weight per day.

Evaluation

Fish were observed for clinical signs of the disease (black tail, skeletal deformities) and behavioral changes ("whirling"). Surviving fish were euthanatized 5 months postexposure by an overdose (500 mg/L) of tricaine methanesulfonate (Finquel, Argent Laboratories, Redmond, Washington) and examined for clinical signs. Heads were removed behind the operculum and divided into equal halves along a midsaggital plane. One half of each head was stored at 4°C, for determination of spore concentration, and the remaining half was secured in a histological cassette and stored in 10% buffered formalin.

Spore Enumeration

For exposures conducted in 1999, cranial cartilage was disrupted for 30 s in 2 mL of filter-sterilized water, using a Stomacher tissue homogenizer (VWR Scientific Products, Seattle, Washington). Spores were then enumerated following the homogenization of the sample. This procedure was used to enable subsequent use of the live spores to infect *T. tubifex* populations. Due to the necessity for protocol standardization, samples collected from exposures conducted in 2000 were processed by pepsin-trypsin digestion (Thoesen 1994). Following both extraction processes, samples were

Table 1. Origin, age, and weight of species experimentally infected with *Myxobolus cerebralis*.

Species	Age[a] (weight) at exposure	Broodstock origin
Experimental species		
Rainbow trout	63 d (0.7g)	Crooked River
Kokanee	129 d (1.5g)	Metolius River
Steelhead-1999	60 d (1.0g)	Deschutes River
Steelhead-2000	15 d (0.3g)	
Chinook salmon	104 d (1.0g)	Deschutes River
Control species		
Mt. Lassen rainbow trout	[b]	Mt. Lassen Trout Farm (Red Bluff, California)

[a]Fish age expressed as time based on a constant 12.8°C water temperature
[b]Control fish matched as closely as possible with the test species compared

Table 2. Infection intensity as determined by percent of surviving fish with spores and clinical signs, mean lesion score, and mean number of spores per half head of infected fish at 5 months post exposure to 200 or 2,000 triactinomyxons (TAMs) per fish. Unexposed fish were not infected with *Myxobolus cerebralis*.

Species (exposure weight)	% surviving with spores (% surviving with signs)		Mean lesion score		Mean spore count (method)	
	200 TAMs/fish	2,000 TAMs/fish	200 TAMs/fish	2,000 TAMs/fish	200 TAMs/fish	2,000 TAMs/fish
Rainbow trout	100 (0)	100 (21)	2.0	4.5	3.7×10^5 (H)	8.9×10^5 (H)
(0.7g)	100 (17)	100 (12)	2.0	5.0	1.8×10^5 (H)	8.2×10^5 (H)
Control rainbow trout	100 (0)	100 (88)	4.7	4.9	2.5×10^5 (H)	1.2×10^6 (H)
(0.7g)	93 (0)	100 (95)	3.7	5.0	2.7×10^5 (H)	1.6×10^6 (H)
Kokanee	29 (0)	13 (0)	0.8	0.1	1.5×10^4 (H)	2.0×10^4 (H)
(1.5g)	9 (0)	33 (0)	0.5	0.8	1.5×10^4 (H)	2.0×10^4 (H)
Control rainbow trout	75 (0)	96 (0)	2.6	4.2	3.0×10^4 (H)	1.2×10^5 (H)
(1.8g)	75 (0)	100 (0)	2.4	3.0	3.7×10^4 (H)	5.7×10^4 (H)
Steelhead 1999[a]	33 (0)	72 (0)	2.4	1.5	4.5×10^4 (H)	4.5×10^4 (H)
(1.0 g)	53 (0)		0.6		5.5×10^4 (H)	
Control rainbow trout[a]	100 (5)	100 (44)	4.8	4.8	1.4×10^5 (H)	4.7×10^5 (H)
(1.2 g)	100 (10)		3.9		1.1×10^5 (H)	
Steelhead 2000	4 (0)	70 (13)	0.9	2.4	3.3×10^4 (P)	1.0×10^4 (P)
(0.3g)	16 (0)	33 (0)	0.6	2.3	1.0×10^4 (P)	9.8×10^3 (P)
Control rainbow trout	92 (13)	92 (42)	3.9	4.2	1.1×10^5 (P)	3.9×10^5 (P)
(0.6g)	100 (10)	30 (9)	4.1	0.8	1.0×10^5 (P)	1.9×10^4 (P)
Chinook salmon	0 (0)	0 (0)	0.0	0.0	ND (P)	ND (P)
(1.0g)	0 (0)	0 (0)	0.0	0.0	ND (P)	ND (P)
Control rainbow trout	92 (13)	92 (42)	3.9	4.2	1.1×10^5 (P)	3.9×10^5 (P)
(0.6g)	100 (10)	30 (9)	4.1	0.8	1.0×10^5 (P)	1.9×10^4 (P)

[a]Only one group of this species was exposed to 2,000 TAMs/fish
(H): Processed by homogenization
(P): Processed by pepsin-trypsin digest
ND: None detected

examined and duplicate spore counts were made using a hemacytometer. Calculations were made for each half head as follows: Number of spores/half head = (number of spores counted × dilution factor × mL sample)/number of grids counted.

Histological Assessment

A random selection of 10 half heads from each exposure dose was processed using standard histological methods. Two sections from each fish were stained with either hematoxylin and eosin or May-Grünwald Giemsa (Yasutake and Wales 1983) and examined microscopically. Cartilaginous lesions were evaluated as described previously by Hedrick et al. (1999b), based on parasite abundance, severity of lesions, and presence of host immune response.

Statistical Analysis

The data collected following spore enumeration was log10 transformed, in an attempt to bring the distribution closer to normal. Mean spore concentrations for each tank were then calculated, using only those fish in which spores were observed (as described by Hedrick et al. 1999a). Using the S-PLUS 2000 statistical software package (Statsci 2000), a two-way analysis of variance (ANOVA) was performed to determine if fish species and/or parasite dose affected mean spore concentrations. Because the digest protocol was modified in the second year of this study, direct comparisons between spore concentrations were not made. Statistical comparison of the presence of clinical signs, prevalence of infection, and mean lesion score between fish species was not performed. For all tests, tanks were used as the experimental unit, and significance was defined as $P < 0.05$.

RESULTS

Rainbow Trout

At both the high and low dose, all of the Deschutes River rainbow trout (DR-R) examined at 5 months postchallenge were infected with *M. cerebralis* (Table 2). Infection prevalence in the RB-C was similar, with spores detected in 93% of those challenged at the low dose and 100% at the high dose. There were few clinical signs of disease in either replicate group of DR-R exposed to a low dose (0%, 17%), and none of the RB-C showed clinical signs at that dose. In the high dose groups, however, clinical disease signs (whirling and black tail) were more evident in the RB-C strain (88% and 95%) compared with the DR-R strain (21% and 12%). Differences in lesion severity between the strains were evident only at the low dose, where the mean lesion score in the DR-R was 2.0, compared with mean lesion scores of 4.7 and 3.7 in the replicate RB-C groups. At the high dose, the mean lesion severity in the DR-R was not different from the RB-C groups, with scores between 4.5 and 5.0. Mean spore concentrations were similar at the low dose for both strains of rainbow trout. However, at the high parasite dose, the RB-C groups had slightly higher mean spore concentrations. None of the unexposed control groups showed infection by *M. cerebralis*.

The two-way ANOVA showed that the spore concentrations per half head were not affected by the difference in rainbow strains ($P = 0.2990$). For both strains, there was a significant effect, due to differences in parasite dose, upon the spore concentration ($P < 0.0010$). The interaction between strain and parasite dose had a nonsignificant effect on spore concentrations ($P = 0.4029$).

Steelhead

Two year classes of Deschutes River steelhead (DR-S) were challenged in this study. In the 1999 exposure, all RB-C fish at both exposure doses were infected with *M. cerebralis*, as determined by the presence of spores (Table 2). The DR-S, which were 1.0 g at exposure, had a lower prevalence of infection at the low dose (33%, 53%) than at the high dose (72%). The RB-C developed black tail and/or cranial deformities at both the low dose (5%, 10%) and the high dose (44%). None of the DR-S, at either exposure dose, displayed clinical disease signs. Mean lesion scores for the DR-S were

lower (2.4, 0.6) in comparison to the RB-C (4.8, 3.9) at the low dose exposure. The mean lesion score in high dose exposure groups also showed lower levels of pathology in the cranial cartilage of the DR-S (1.5) than in the RB-C (4.8). The mean spore concentrations in the DR-S groups exposed to a low parasite dose were lower than the RB-C groups at the same exposure dose. At the high dose, the difference in cranial spore concentration between the species was 10-fold higher in the RB-C groups. None of the unexposed control groups showed infection by *M. cerebralis*.

The ANOVA showed a significant effect on the concentration of spores due to the difference in fish species ($P = 0.0240$). The parasite dose ($P = 0.0879$) and interaction between species and dose ($P = 0.0764$) both showed nonsignificant effects on spore concentrations.

In the 2000 exposure, the steelhead trout were smaller (0.3 g), yet the data were similar to the steelhead exposures from the previous year. At the low exposure dose, relatively few of DR-S became infected (4%, 16%), compared with the high exposure dose (33%, 70%; Table 2). Infection prevalence in the RB-C was high, except for one exposure group at the high dose where only 30% of the fish became infected. This exposure group also had a low mean histology score. Signs of clinical disease were similar to those in the previous exposure, with none of the DR-S showing disease signs at the low dose exposure, compared with 10–13% of the exposed RB-C. At the high dose, 42% of the RB-C showed clinical signs. Clinical signs were apparent only in 13% of one group of DR-S exposed at the high dose. Infection severity in the RB-C groups was again higher at the low dose (mean lesion scores of 3.9 and 4.1) compared with the DR-S (mean lesion scores of 0.9 and 0.6). Mean lesion scores, among DR-S exposed at the high dose, were 2.4 and 2.3, which was lower than the one RB-C group (4.2). Spore concentrations were lower for each exposure level in the DR-S than in the corresponding RB-C groups, even for the high dose group in which the infection prevalence was decreased. None of the unexposed control groups showed infection by *M. cerebralis*.

The two-way ANOVA showed nonsignificant effects on spore concentrations, due to the difference in fish species ($P = 0.0928$) and parasite dose ($P = 0.5834$). The interaction between species and dose was also nonsignificant ($P = 0.6366$).

Kokanee

Kokanee (DR-K) from the Metolius River tributary showed a lower prevalence of infection compared with the RB-C at both parasite exposure doses (Table 2). At the low dose, infection prevalence in the RB-C was 75% and in the DR-K was 29% and 9%, for the replicate groups. At the high parasite dose, prevalence in the RB-C increased (96%, 100%) but remained the low for the DR-K (13%, 33%). Clinical signs were not detected in either species at 5 months postexposure. Mean lesion scores for the RB-C (2.6, 2.4) were higher than the DR-K (0.8, 0.5) at the low exposure dose. With an increase in exposure dose, the mean lesion scores of the RB-C (4.2, 3.0) were again higher than the mean lesion scores for the DR-K (0.1, 0.8). The mean spore concentrations for the DR-K were lower than those for the RB-C, for both exposure doses. For both strains, the mean spore concentration was greater at the higher dose than at the lower dose. The unexposed control groups for this species did not show any evidence of infection by M. cerebralis.

For cranial spore concentrations, the ANOVA showed a significant effect due to the difference in fish species ($P = 0.0175$). There were nonsignificant effects on the concentration of spores, due to the parasite dose ($P = 0.0893$) and interaction between fish species and dose ($P = 0.3116$).

Chinook Salmon

No signs of infection, as measured by presence of spores, clinical disease signs, or histopathology, were detected in Deschutes River chinook salmon (DR-C) at either the high or low dose challenge, despite the high infection prevalence and disease severity observed in the simultaneously exposed RB-C (Table 2).

DISCUSSION

Our experimental challenges showed a range of susceptibility to M. cerebralis among the salmonid species indigenous to the Deschutes River, Oregon. As demonstrated in other studies (O'Grodnick 1979; Hedrick et al. 1999a, 1999b, 2001a, 2001b), rainbow trout were the most susceptible species. The susceptibility of the wild Deschutes River rainbow trout tested in this study was not significantly different from that of the control rainbow trout strain, although decreased clinical signs and mean spore count (at the high dose) and lesion severity (at the low dose) suggested a decreased sensitivity.

Results of challenges with steelhead in this study contrasted with other reports where this species was found highly susceptible (Hedrick et al. 2001a, 2001b). Although differences in mean spore concentration between the steelhead and rainbow trout were only supported statistically for the 1999 brood, we believe that these differences were also present in the 2000 brood, but that significance was masked by a tank effect in the rainbow trout. One principle difference between studies, that might account for this result, is the size at exposure. In this study, the fish were considerably larger (1.0 g in 1999; 0.3 g in 2000) than those used in the studies by Hedrick et al. (2001a, 2001b), where fish ranged from 0.06 g to 0.14 g. The effects of size on infection severity are well-documented (Hoffman and Byrne 1974; Halliday 1976; Markiw 1992), and it would be expected that the larger steelhead in this study would show fewer signs of clinical disease than smaller steelhead. However, when compared with simultaneously exposed rainbow trout of approximately the same size, the Deschutes River steelhead appear to be more resistant to M. cerebralis.

The only published report on susceptibility of sockeye salmon, or its land-locked form the kokanee, is a study by O'Grodnick (1979), comparing the susceptibility of various salmonids to whirling disease. The age and size of fish at exposure was not specified, and exposure consisted of a natural challenge to unknown numbers of infectious units. In that exposure, the number of fish infected and the presence of clinical signs was similar between the sockeye salmon and control rainbow trout; however, mean spore concentrations were lower for the sockeye. In this study, the size at exposure for both the rainbow trout control and the kokanee salmon was at least 1.5 g, and this may explain the absence of clinical disease signs. However, for all measures of infection severity (excluding clinical signs), kokanee salmon were more resistant than the rainbow trout control.

Results of exposures of chinook salmon also differ from those of Hedrick et al. (2001b), which demonstrated infection levels comparable to those seen in rainbow trout. Again, one difference between these studies is the fish size at challenge. In the study by Hedrick et al. (2001b), fish were exposed at 0.25 g; fish in the present study were 1.0 g. The challenge dose in that study was also slightly higher (2,500 triactinomyxons/fish). In the only other published study of susceptibility of chinook salmon to M. cerebralis, O'Grodnick (1979) demonstrated an intermediate resistance, although the size at exposure was not stated.

One explanation for the disparity, in the reported susceptibilities of chinook salmon and steelhead, is that different species develop resistance at different rates, perhaps depending on cartilage ossification. Another possibility is that resistance developed against the myxozoan Ceratomyxa shasta might have resulted in a partial resistance against M. cerebralis infection, a hypothesis that was investigated by Hedrick et al. (2001a). Ceratomyxa shasta is enzootic throughout the Crooked and Deschutes Rivers and is found in portions of the Metolius River (Ratliff 1981). The parasite acts as a strong selection factor, and most of the salmonids in the Deschutes system have developed a high level of resistance. The rainbow trout tested in the study by Hedrick et al. (2001a) were of a hatchery strain originating from wild fish collected in the lower portions of the Deschutes River. Although not significantly significant, results in that study were similar to those seen here, with the lower Deschutes River rainbow trout strain showing fewer clinical signs, lower lesion scores, and a decreased mean spore concentration, compared with the control strain. However, in our study, differences in susceptibility were less apparent in the native rainbow trout than among the other species. That this was not demonstrated for all species from this watershed makes this C. shasta resistance hypothesis unlikely; however, it seems plausible that resistance develops by different mechanisms between species. It also seems unusual that chinook salmon from this river system would have developed so complete a resistance, and we feel that this challenge should be repeated before drawing any conclusions.

These data will play a critical role in assessing the potential risks involved in the proposed fish passage plans (Ratliff et al. 1998) and contribute to our current understanding of susceptibility to M. cerebralis, as it relates to fish age and species. Results of these challenges demonstrate that Deschutes river salmonids show variable susceptibility to M. cerebralis and suggest that if the parasite became established in this system there would be some risk of infection and disease for these populations. The most susceptible were the native rainbow trout, which are resident both above and below the PRB dam. We have established the presence of two key components necessary for the parasite to establish in this river: the alternate host T. tubifex and the

susceptible salmonid host. However, other ecological factors are likely to contribute to the ability of the parasite to establish and persist in this system, and further investigation of these other components is essential to determine the risk.

ACKNOWLEDGMENTS

This work was supported with funds from Portland General Electric and the National Partnership on the Management of Wild and Native Coldwater Fisheries. The Oregon Department of Fish and Wildlife generously provided fish for the study. This is Oregon Agricultural Station technical paper 11839.

REFERENCES

Andree, K. B., E. MacConnell, and R. P. Hedrick. 1998. A nested polymerase chain reaction for the detection of genomic DNA of Myxobolus cerebralis in rainbow trout Oncorhynchus mykiss. Diseases of Aquatic Organisms 34:145–154.

El-Matbouli, M., R. W. Hoffman, H. Schoel, T. S. McDowell, and R. P. Hedrick. 1999. Whirling disease: host specificity and interaction between the actinosporean stage of Myxobolus cerebralis and rainbow trout Oncorhynchus mykiss. Diseases of Aquatic Organisms 35:1–12.

Groot, C., and L. Margolis, editors. 1991. Pacific salmon life histories. University of British Columbia Press, Vancouver, British Columbia.

Halliday, M. M. 1976. The biology of Myxosoma cerebralis: the causative organism of whirling disease of salmonids. Journal of Fish Biology 9:339–357.

Hedrick, R. P., M. El-Matbouli, M. A. Adkinson, E. MacConnell. 1998. Whirling disease: re-emergence among wild trout. Immunological Reviews 166:365–376.

Hedrick, R. P., T. S. McDowell, M. Gay, G. D. Marty, M. P. Georgiadis, and E. MacConnell. 1999a. Comparative susceptibility of rainbow trout (Oncorhynchus mykiss) and brown trout (Salmo trutta) to Myxobolus cerebralis the cause of salmonid whirling disease. Diseases of Aquatic Organisms 37:173–183.

Hedrick, R. P., T. S. McDowell, K. Mukkatira, M. P. Georgiadis, and E. MacConnell. 1999.b. Susceptibility of selected inland salmonids to experimentally induced infections with Myxobolus cerebralis, the causative agent of whirling disease. Journal of Aquatic Animal Health 11:330–339.

Hedrick, R. P., T. S. McDowell, K. Mukkatira, M. P. Georgiadis, and E. MacConnell. 2001a. Salmonids resistant to Ceratomyxa shasta are susceptible to experimentally induced infections with Myxobolus cerebralis. Journal of Aquatic Animal Health 13:35–42.

Hedrick, R. P., T. S. McDowell, K. Mukkatira, M. P. Georgiadis, and E. MacConnell. 2001b. Susceptibility of three species of anadromous salmonids to experimentally induced infections with *Myxobolus cerebralis*, the causative agent of whirling disease. Journal of Aquatic Animal Health 13:43–50.

Höfer, B. 1903. Ueber die Drehkrankheit der Regenbogenforelle. Allgemeinen Fischerei-Zeitung 28:7–8.

Hoffman, G. L., and C. J. Byrne. 1974. Fish age as related to susceptibility to *Myxosoma cerebralis*. Progressive Fish-Culturist 36:151.

Hoffman, G. L., and R. E. Putz. 1969. Host susceptibility and the effect of aging, freezing, heat, and chemicals on the spores of *Myxosoma cerebralis*. Progressive Fish-Culturist 31:35–37.

Lorz, H. V., A. Amandi, C. R. Banner, and J. S. Rohovec. 1989. Detection of *Myxobolus (Myxosoma) cerebralis* in salmonid fishes in Oregon. Journal of Aquatic Animal Health 1:217–221.

Markiw, M. E. 1992. Experimentally induced whirling disease I. Dose response of fry and adults of rainbow trout exposed to the triactinomyxon stage of *Myxobolus cerebralis*. Journal of Aquatic Animal Health 4:40–43.

Nehring, R. B., and P. G. Walker. 1996. Whirling disease in the wild: the new reality in the intermountain west. Fisheries 21:28–32.

O'Grodnick, J. J. 1979. Susceptibility of various salmonids to whirling disease (*Myxosoma cerebralis*).

Transactions of the American Fisheries Society 108:187–190.

Ratliff, D. E. 1981. *Ceratomyxa shasta*: Epizootiology in chinook salmon of central Oregon. Transactions of the American Fisheries Society 110:507–513.

Ratliff, D., C. Fagan, M. Riehle, M. Powell, S. Carlon, P. Lickwar, A. Stuart, and J. Eisner. 1998. Alternatives for renewing fish passage at the Pelton Round Butte hydroelectric project Madras, Oregon. Interagency Fisheries Technical Subcommittee. Portland General Electric Co., Portland, Oregon.

Statsci. 2000. S-PLUS Statistical software, MathSoft, Inc., Seattle, Washington.

Thoesen, J. C., editor. 1994. Suggested procedures for the detection and identification of certain finfish and shellfish pathogens. 4th edition, Version 1. Fish Health Section, American Fisheries Society.

Thompson, K. G., R. B. Nehring, D. C. Bowden, and T. Wygant. 1999. Field exposure of seven species or subspecies of salmonids to *Myxobolus cerebralis* in the Colorado River, Middle Park, Colorado. Journal of Aquatic Animal Health 11:312–329.

Vincent, E. R. 1996. Whirling disease and wild trout: the Montana experience. Fisheries 21:32–33.

Yasutake, W. T., and J. H. Wales. 1983. Microscopic anatomy of salmonids: An atlas. Washington, D.C.

American Fisheries Society Symposium 29:125–134, 2002
© 2002 by the American Fisheries Society

Laboratory Studies Indicating that Living Brown Trout *Salmo trutta* Expel Viable *Myxobolus cerebralis* Myxospores

R. Barry Nehring*

Colorado Division of Wildlife, 2300 S. Townsend Avenue, Montrose, Colorado 81401, USA

Kevin G. Thompson[1]

Colorado Cooperative Fish and Wildlife Research Unit, Room 201 Wagar Building
Colorado State University, Fort Collins, Colorado 80523, USA

Karen A. Taurman and David L. Shuler

Colorado Division of Wildlife, 317 W. Prospect Street, Fort Collins, Colorado 80526, USA

ABSTRACT. In Colorado, Windy Gap Reservoir is a focus of *Myxobolus cerebralis* infectivity of greater intensity than may be explained by the potential contribution of *M. cerebralis* myxospores by dead fish. One mechanism that would help explain this situation is the expulsion of viable *M. cerebralis* myxospores by living infected fish. We conducted laboratory experiments to see if *Tubifex tubifex*, purged of infection by incubation at 26°C for a minimum of 30 d, could become reinfected by exposure to feces and wastes from aquaria containing *M. cerebralis*-infected brown trout *Salmo trutta*. In two separate experiments, replicate experimental units of *T. tubifex* were thoroughly infected in this manner. By comparison, evidence of infection in negative control replicates was much weaker in both experiments. It is possible that the purging process used to remove initial infection was not 100% effective, yet the differences between experimental and negative control replicates were dramatic. Positive control replicates, intentionally exposed to harvested myxospores of *M. cerebralis*, became heavily infected in both experiments. These results strongly support the hypothesis that brown trout are capable of expelling viable *M. cerebralis* myxospores.

Windy Gap Reservoir is a shallow, 42-ha impoundment on the Colorado River in Middle Park, Colorado. Completed in 1985, the reservoir supports a sparse salmonid population that has never been enhanced by stocking nor exploited by anglers. *Myxobolus cerebralis* actinospores have been observed in the reservoir effluent at long-term densities about 30 times greater than at two sample locations that are short distances above the reservoir (Thompson and Nehring 2000). We estimated the total actinospore production from Windy Gap Reservoir from 1 April 1997 to 31 March 1998 based on density estimates from 47 filtration occasions during that period. Allowing for a long-term average filtration efficiency of 50%, and using United States Geological Survey water records to determine mean discharge for the time interval surrounding each filtering occasion, we estimated that nearly 960 billion actinospores came out of the reservoir during the 12-month period. A similar exercise for 1998–1999 resulted in an estimate of 1.8 trillion actinospores.

Infected fish from state sources have never been stocked in the drainage upstream of the reservoir. We are unaware of infected fish stocked in the drainage by private vendors since the original introduction in the Colorado drainage in 1987. However, the reservoir sustains a dense population of aquatic oligochaetes. Zendt and Bergersen (2000) estimated there were more than 47,500 oligochaetes/m² in Windy Gap Reservoir, of which 73% were *Tubifex tubifex*. About 1.2% of the *T. tubifex* were infected with *M. cerebralis*. These circumstances cause considerable wonder regarding the origin of the myxospores required to sustain such an infection rate in the reservoir and the level of actinospores observed in the reservoir discharge.

If all myxospores must originate from carcasses of dead fish, a large number of carcasses would be required to produce the actinospores observed in the reservoir effluent. Using a population proportion of 73% for *T. tubifex* (Zendt and Bergersen 2000), an estimate of 27,000 myxospores per brown trout *Salmo trutta* upstream of Windy Gap Reservoir (author's unpublished data), and 457 as the maximum estimated number of actinospores produced

*Corresponding author: barry.nehring@state.co.us
[1]Present address: Colorado Division of Wildlife, 2300 S. Townsend Avenue, Montrose, Colorado 81401, USA

for each myxospore (this study), 107,000 to 200,000 dead trout would be required each year in this small reservoir to maintain the level of actinospore production estimated in 1997 and 1998. Even halving the numbers of fish to account for the fact that not all myxospores are contained in head tissues, the estimate of dead fish required remains substantial. One possible alternative explanation is that during the spring runoff myxospores are transported from upstream locations into Windy Gap Reservoir, where they settle out and provide for infection of the abundant *T. tubifex* population.

Currently, most investigators hold that myxospores of *M. cerebralis* are released from host fish only when they die and decompose or are eaten by predators (e.g., see Baldwin et al. 1998; El-Matbouli et al. 1999a). However, others suggest trout are capable of expelling viable myxospores. Uspenskaya (1957), reporting on early investigations of *M. cerebralis* in the former Soviet Union, claimed to have observed mature myxospores in "almost all organs and tissues of diseased trout" at certain times in the course of infection. Rydlo (1971) also investigated this phenomenon, and found "sporadic" occurrence of myxospores in intestine and liver, and skin of "one-summer" rainbow trout *Oncorhynchus mykiss*. Taylor and Haber (1974) observed *M. cerebralis* myxospores in opercular cysts on Lahontan cutthroat trout *O. clarki henshawii*, suggesting that rupture of these cysts would provide a method of transmission of the parasite from live infected fish. These early reports, along with the dynamic of actinospore production observed in Windy Gap Reservoir, led us to reconsider the hypothesis that viable *M. cerebralis* myxospores may be expelled or shed from living brown trout.

Our study objective was to examine the hypothesis that brown trout expel or shed viable *M. cerebralis* myxospores. We pursued this objective by conducting long-term laboratory studies.

METHODS

Experiment 1

More than 400 brown trout fingerlings, exposed to actinospores of *M. cerebralis* in the Colorado River from the time of hatching in 1998, were collected from the wild in September 1998. The fish were held in floating flow-through cages in the Colorado River and trained to an artificial diet. On 22 March 1999, 394 surviving brown trout were moved to the

Colorado Division of Wildlife Aquatic Toxicology Research laboratory in Fort Collins, Colorado, and equally distributed (n = 65 or 66) among six clean glass 90-L aquaria containing no substrate. Aquaria were supplied with dechlorinated tap water at 0.33 L/min, temperature ranged from 9.3°C to 19°C depending on season of the year. The fish were almost 1 year of age when transferred, mean length was 79 mm (n = 21, SE = 2.12). A sample of fish, obtained when they were moved to the laboratory, was examined by the plankton centrifuge method (O'Grodnick 1975) on individual heads to establish that *M. cerebralis* myxospores were present.

The brown trout were fed a commercial trout diet at a rate of 2.3% body weight per day. The feed rate decreased as the fish grew (Piper et al. 1982). Any moribund fish were removed daily to prevent dead brown trout from contributing myxospores to the aquaria. Once each week, all the aquaria were cleaned by siphoning the accumulated fecal material and other wastes off the bottom. This material was used to feed experimental lots of oligochaete worms containing *T. tubifex* so that any expelled myxospores in the material would be available for ingestion.

The commercial fish food was sampled 25 times during November and December 1999 and tested by polymerase chain reaction (PCR) using a single round modification (see Schisler et al. 2001) of the test developed by Andree et al. (1998) to see whether it might contain *M. cerebralis* DNA. At the testing laboratory, the entire sample (fish food in 70% EtOH) was spun at 1000 × g for 5 min, after which all supernatant was aspirated off and discarded. About 100 mg of the pellet was transferred to a microfuge tube, spun at 14,000 × g for 1 min, then resuspended in 200 μL lysis buffer. Total DNA was extracted using a spin-column DNA purification procedure (DNeasy Tissue kit, Qiagen, Valencia, California).

Sixteen samples were taken from centrifuged aquarium waste products from 18 June to 5 November 1999 and subjected to the PCR test. At the testing laboratory, waste samples were gently mixed; then, about 0.5 mL was transferred to a microfuge tube. The microfuge tubes were spun at 14,000 × g for 1 min; then, the supernatant was removed. The resulting pellet was air dried for 15 min at 55°C, then heated in a microwave oven (700 watt) on high for 1 min, and finally resuspended in 200 μL lysis buffer. Total DNA was extracted using a spin-column DNA purification procedure.

The strength of PCR gel signals were independently graded by two reviewers on a five-point scale ranging from negative (0) to very strong positive (4) by comparison to previously established strong positive and weak positive control preparations (Schisler et al. 2001). Scores were consistently in close agreement between the two judges; rare discrepancies were resolved by mutual viewing and discussion, to reach agreement. The lowest positive signal was termed "weak positive" (1) and required the use of an ultra-sensitive fluorescent DNA stain (SYBR Green) to be observed. For this reason, in this study we did not consider a sample unquestionably positive unless it scored one of the higher positive ratings (2, 3, or 4).

A mixed culture of aquatic oligochaetes originating from Mount Whitney Hatchery and containing *T. tubifex* was acquired from the University of California-Davis. The worms were purged of any previous *M. cerebralis* infection by incubation at 26°C for 30 d (El-Matbouli et al. 1999b) prior to shipment to Colorado. The worms were divided into nine lots of about 5.5 g wet weight each. Three lots each were randomly assigned as replicates to each of three groups: negative control, positive control, and experimental.

The experiment was initiated on 1 April 1999. All worms were held in 500 mL of aerated, dechlorinated tap water in an incubator at 15°C. All were fed Spirulina discs (dehydrated algae) and Algamac 2000, a commercially prepared diet for tubificid worms. Each positive control replicate was seeded with 1.4 million *M. cerebralis* myxospores harvested from wild rainbow trout *Oncorhynchus mykiss* held in the Colorado River at the same location as the brown trout. For eight weeks, the experimental worms were fed fecal and waste material obtained directly from the aquaria containing the infected brown trout. However, the worms became stressed due to the biological oxygen demand created by the amount of waste material in the small volume of water. From week 9 until the end of the study, all worm replicates were held in larger containers with 3.8 L of water. The aquarium wastes used to inoculate the experimental worms were centrifuged to minimize the amount of organic material while retaining any myxospores. The resulting concentrated material was then rehydrated and divided equally among the three experimental replicates. All of the larger containers were placed in a water bath at 15°C.

The water from each worm container was filtered weekly and examined for *M. cerebralis* actinospores by dissecting microscope at 40–80×. To prevent cross-contamination, the positive control, negative control and experimental groups were filtered on different days. A clean Pecap® (AREA, Inc., Homestead, Florida) 20-µm mesh screen was used for each filtration. Screens were soaked overnight in a bleach solution before a different group was filtered. Siphon tubes used for withdrawing the water were dedicated to individual replicates. A subsample of each filtrate was obtained for PCR testing for the presence of genomic DNA of the parasite, and results were graded on the same five-point scale described earlier. Once actinospores were observed, producing replicates were filtered, enumerated and tested by PCR twice weekly.

On 10 occasions from March 1999 through January 2000, 15 to 50 brown trout (Table 1) were sacrificed to determine prevalence of infection and average concentrations of myxospores in cranial tissues by the plankton centrifuge method (O'Grodnick 1975). From September 1999 through January 2000, additional fish were collected on each occasion and analyzed by the pepsin-trypsin digest (PTD) method (Markiw and Wolf 1974). For occasions when samples were analyzed by both methods, comparisons were made of the spore concentrations obtained by 2-sample t-test on natural log-transformed values, allowing for unequal variances. Test statistics yielding $P \leq 0.05$ were considered statistically significant.

Experiment 2

Methods for the second experiment were the same with the following exceptions. A new worm culture was obtained from Windy Gap Reservoir and the upper Colorado River, and was purged of infection by 45 d incubation at 26°C. The culture was enhanced by the addition of worms from two of the negative control groups used in the first experiment. The combined worms were divided into three replicates each for positive control and experimental lots, each containing 4.3 g wet weight of worms, and one negative control containing 3.3 g wet weight. The positive control lots were each inoculated with about 112,200 myxospores harvested from cranial tissues of *M. cerebralis*-infected brown trout, about 8% of the dosage per positive control replicate used in the first experiment. We also retained and included one of the negative control replicates from the previous experiment (California worms), and treated it in the same fashion as the other negative control.

Table 1. Prevalence of infection and mean myxospore concentration in brown trout heads sampled on 10 occasions from March 1999 through January 2000, during the first laboratory experiment. Mean myxospore concentration is for all heads in the sample.

Date	N	Prevalence (%)		Mean myxospore concentration	
		Plankton Centrifuge	PTD	Plankton Centrifuge	PTD
03/26/99	15	67		2,100	
04/10/99[a]	18	94		5,600	
06/26/99	15	100		10,300	
07/23/99	15	93		12,900	
08/28/99	15	40		5,300	
09/26/99	15[b]	93	47	42,000	7,000
10/10/99	15[b]	13	87	500	15,700
11/12/99	20[b]	75	45	8,800	5,900
12/20/99	15[b]	80	53	12,400	15,600
01/03/00	50[b]	76	92	19,000	29,100

[a]The April sample was composed of mortalities retained from 1 April through 25 April 1999 (mean date of death, 10 April).
[b]N for each method.

The infected brown trout were distributed equally among aquaria and inventoried weekly prior to cleaning. If any fish were unaccounted for since the previous inventory, or if any fish had died in an aquarium before being removed, the wastes from that aquarium were discarded for a period of 30 d before being used again. The brown trout used in this experiment were collected from the Williams Fork River (enzootic for M. cerebralis) in late June 1999. They were placed in sentinel cages and held below Windy Gap Dam for 30 d to ensure that they had a significant exposure to M. cerebralis actinospores. After the 30-d exposure, they were moved to Fort Collins and held in the same fashion as the fish in the first experiment. They were further exposed to the M. cerebralis actinospores produced by the positive control replicates of the first experiment from 7 September to 1 October 1999. At the beginning of the experiment, we had 143 brown trout (mean length 125 mm, SE 4.88, n = 15) distributed as equally as possible in seven aquaria. These fish provided fecal material and aquaria waste for the experimental replicates. The negative and positive control lots of T. tubifex received feces and aquarium wastes collected from aquaria of brown trout not exposed to M. cerebralis that were held in the lab as part of an unrelated study. Samples of aquarium wastes and feces from both the infected (n = 70) and uninfected fish (n = 30) were periodically tested by the PCR technique for M. cerebralis DNA. The second experiment began on 23 February 2000.

RESULTS

Experiment 1

Plankton centrifuge analysis showed that the sample of brown trout taken just prior to the beginning of the experiment was 67% positive for M. cerebralis, with an overall mean concentration of 2,100 myxospores per fish (Table 1).

Actinospores were first detected by microscope in water from positive control replicates between 15 July and 20 July 1999, 106 to 111 d post exposure. Positive PCR signals occurred in one of the positive control replicates on 8 April and 7 May. All other water samples analyzed up to one week prior to the onset of actinospore production tested negative by PCR for parasite DNA. In all three positive control replicates, positive PCR signals were consistently noted, beginning one week before (two replicates) to one week after (one replicate) the first actinospores were observed by microscope. Actinospore production rapidly increased to a high level, with the three replicates commonly producing an average of more than 10^6 actinospores every three to four days, through the months of August and September (day 123 to day 183; Figure 1). The level of production gradually tapered off thereafter. Total actinospore production among the three positive control groups through the end of December 2000 was estimated at 68 million, 56 million, and 59 million. Trace levels of actinospores were still being produced in the

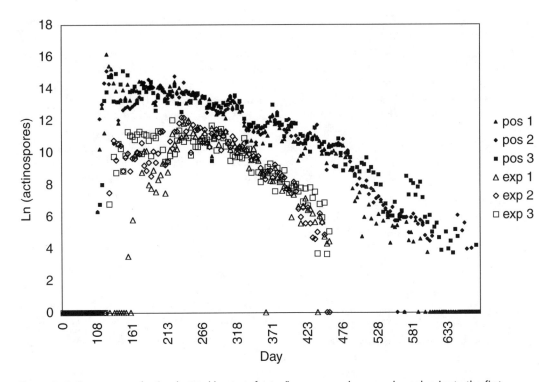

Figure 1. Actinospore production (natural log-transformed) as measured once each week prior to the first microscope observation of actinospores and twice each week thereafter from the three positive control replicates (solid symbols) and the three experimental replicates (open symbols) of worms held in the first experiment.

positive control replicates on 5 January 2001, 540 d after initial release began.

Among the three experimental groups receiving fecal material, actinospores first appeared in water from two replicates on 3 August (125 d post-exposure to aquarium wastes) and in the third replicate on 30 August 1999 (152 d post-exposure to aquarium wastes). Positive PCR tests in all three replicates coincided with the first observation of multiple actinospores by dissecting microscope. While positive control replicates reached peak production almost immediately after initial observation of actinospores, the experimental replicates did not reach peak production until 80–100 d after initial observation of actinospores (Figure 1). Total production among the three experimental replicates through the end of June 2000 was estimated at 2.3 million, 3.2 million, and 3.0 million actinospores. At this time (332 d after the first actinospores were observed), spore production had dropped below 50 per d. We ceased monitoring the experimental replicates for actinospore release in July 2000.

Actinospores were never observed by weekly microscope examination of concentrated filtrates (n = 147) from the negative control replicates through 16 February 2000, when two ceased to be monitored. The third replicate was retained and used as a negative control for the second experiment. On five occasions, samples from negative control filtrates tested positive for M. cerebralis by PCR. Four occurred in September 1999, the fifth in December 1999. All three replicates tested positive by PCR at least once.

Periodic testing of fish samples to confirm myxospore presence showed that the fish remained infected through January 2000 (Table 1). Prevalence and mean concentration in cranial tissues sampled was highly variable over the 10-month period. Means differed significantly (P < 0.05) between the plankton centrifuge and PTD methods in September, October and January. Neither method consistently showed a higher prevalence or mean concentration than the other.

No genomic DNA of M. cerebralis was detected in the 25 fish food samples tested by PCR.

Neither did any of the 16 aquarium waste samples test definitively positive by PCR.

Eighty-four fish died during the experiment; an additional 290 were sampled at various times for testing cranial myxospore concentrations. Fifteen were sampled for histological examination. This totals 389 fish, and because the 100-fish sample in January consisted of all remaining fish, there was a discrepancy of five fish. One fish was confirmed as lost to predation in September 1999. The fate of the other four fish is unknown.

Experiment 2

A 15-fish sample from the brown trout collected on 16 February 2000 tested positive by PTD for the presence of M. cerebralis myxospores, 7 d before the experiment started (Table 2). Prevalence of infection was 93%, mean myxospore concentration was 60,000, and estimates of myxospores in individual heads ranged from 1000 to 444,000 among those that contained spores.

The first actinospores were detected by microscope in the positive control replicates on 11 May, 78 d post exposure. Positive PCR signals (2 or greater) were first observed on 2 May in one replicate and on 11 May in the other replicates. Water samples from all three positive control replicates thereafter consistently tested positive by PCR through 2 November 2000. The number of actinospores observed rapidly increased to a high level. These replicates commonly produced an average of more than 10^6 actinospores every three to four days through mid-August (day 81 to day 180; Figure 2). The level of production remained high through early November 2000. Total production among the three positive control groups was estimated at 50 million, 52 million, and 46 million.

Among the three experimental replicates receiving waste material from the aquaria containing M. cerebralis-infected brown trout, the first actinospores were seen between 29 June and 13 July (127–141 d post-exposure to aquarium wastes, Figure 2). The first and third replicates produced actinospores consistently after the first observation, but none were seen in the second replicate from 20 July until 31 August, after which it too produced actinospores consistently. Positive PCR tests for water samples drawn from all three replicates coincided with the first observation of actinospores by dissecting microscope. A delay of 80 d or more between onset of actinospore release and apparent peak production was observed (Figure 2). Total estimated actinospore production among the three experimental replicates through 8 January 2001 was estimated at 539 thousand, 197 thousand, and 372 thousand actinospores.

No actinospores were observed by weekly microscope examination of the negative control replicate that was carried over from the first experiment. It tested positive by PCR on 13 September 2000 and negative on 95 other occasions. Triactinomyxon-form actinospores, morphologically similar to those of M. cerebralis by visual inspection (but not measured), were observed from the new negative control replicate on eight occasions. All of these observations occurred between 3 May and 12 July 2000, and microscope counts ranged from 2 to 36 for total individual sample volumes of 0.8 ml. All of these samples were tested by PCR, as well as 45 others in which no actinospores were observed by microscope, and none yielded a positive signal.

Seventy waste samples from aquaria containing infected brown trout and 30 from aquaria containing uninfected brown trout were tested for the presence of M. cerebralis genomic DNA by PCR. Only two samples from aquaria containing infected fish gave a positive result. None from the uninfected fish aquaria tested positive.

The brown trout remained infected with M. cerebralis through October 2000 (Table 2). There was considerable variability in prevalence of

Table 2. Prevalence of infection and mean myxospore concentration in brown trout heads sampled on four occasions from February 2000 through October 2000, during the second laboratory experiment. Mean myxospore concentration is for all heads in the sample.

Date	N	Method	Prevalence (%)	Mean myxospore concentration
02/16/00	15	PTD	93	60,000
03/16/00	15	PTD	53	10,400
07/14/00	15	Plankton centrifuge	67	14,900
10/31/00	16	Plankton centrifuge	88	19,500

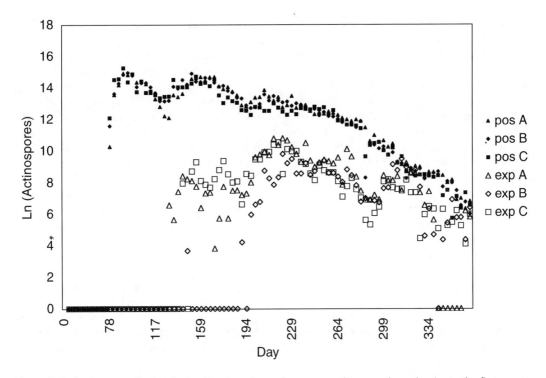

Figure 2. Actinospore production (natural log-transformed) as measured once each week prior to the first microscope observation of actinospores and twice each week thereafter from the three positive control replicates (solid symbols) and the three experimental replicates (open symbols) of worms held in the second experiment.

infection and mean cranial myxospore concentration over time.

All weekly fish inventories concurred with the numbers of fish expected to be present.

DISCUSSION

It is apparent that the experimental replicates of worms in each of the experiments encountered viable M. cerebralis myxospores after being exposed to the wastes collected from the aquaria containing M. cerebralis-infected brown trout. All six experimental replicates produced actinospores that were demonstrated by PCR to be those of M. cerebralis. Markiw (1986) showed that actinospores should appear about 104 to 113 d post exposure at 12.5°C. It should happen somewhat faster at 15°C, if some worms became infected at the very beginning of the experiment. Actinospores in experimental replicates were first seen, in the first experiment, at 125 d and, in the second experiment, at 127 d. In contrast, actinospores were first observed in the positive control replicates 106 d and 78 d post-exposure to myxospores in the first and second experiments. This indicates that there was a delay

of perhaps several weeks before there was a measurable number of worms infected in the experimental replicates. The pattern of actinospore production in experimental replicates of both experiments also indicates an increasing incidence of infection over several weeks; otherwise, actinospore production should have peaked suddenly as it did in the positive control replicates.

It is possible that the myxospore exposure in experimental replicates occurred in the first experiment because of predation. There were four fish unaccounted for at the end of that experiment and one known predation victim. However, the only documented predation occurred in September 1999, approximately 60 d after the onset of actinospore production in the experimental replicates. No predation occurred in the second experiment, yet the T. tubifex in the three experimental replicates again became infected with M. cerebralis.

Actinospore production estimates were higher for the experimental replicates in the first experiment compared with the second. Two possible explanations for this result are the numbers of brown trout used as the source of infectivity, and

differences in the procedures used to expose experimental worms to the waste and fecal material. We
started with 394 brown trout in the first experiment but only 143 for the second experiment.
However, our initial estimates of the level of infection in the brown trout were much higher in the
second experiment compared with the first, so this
seems to be an inadequate explanation unless myxospores are not shed in direct proportion to the
level of infection in individual fish. Alternatively,
in the first experiment, the waste was fed directly
to the worms for the first 54 d of the experiment.
Afterward, we centrifuged the waste to reduce the
amount of organic matter added to the experimental worm containers. In the second test, all waste
material was centrifuged from the beginning of the
experiment to prevent elevated biological oxygen
demand in the experimental containers. The cen-
trifuging process may have resulted in the loss of
myxospores much greater than we expected, thereby reducing the number of myxospores introduced
into the experimental worm replicates.

Total estimated actinospore production in the
positive control groups did not differ greatly
between the two experiments. It was estimated to
be 183 million in the first experiment and 148 million in the second. This is somewhat surprising
since positive control replicates in the first experiment received 12.5 times the number of myxospores, than did the positive controls in the second experiment. However, these results are not
unlike those of Markiw (1986). In two experiments, using three myxospore dosages that differed
by 10 fold, Markiw documented only a 1.9-fold difference in total estimated actinospore production
between the lowest and highest dose. Also, the
positive control replicates in the second experiment were fed aquarium wastes, and the additional
organic input may have had an influence on the
numbers of actinospores produced.

No actinospores were ever observed in negative controls in the first experiment, but each the
three replicates tested positive by PCR for M. cerebralis at least once during the experiment. We sur-
mise that this could happen because the purging
procedure may not be 100% effective in eliminating M. cerebralis infection in the worms. Nevertheless, comparison of experimental replicates with
negative control replicates in both experiments
clearly demonstrates there was more infection
among the experimental worms than could be
explained by incomplete purging of any previous

infection. Although Gilbert and Granath (2001)
reported persistent infection in T. tubifex, and
recurring actinospore release more than 600 d after
initial exposure, we observed one negative control
replicate for more than 700 d without observing
actinospores. This suggests that the purging process
was highly effective in this study.

The observation of triactinomyxon-form actinospores in the new negative control replicate of
the second experiment was puzzling. The identity of
these actinospores remains unknown, because none
of the filtrates that contained them tested positive
by PCR for M. cerebralis. Our experience with hundreds of PCR samples matched with microscope
actinospore counts leads us to conclude that the
probability of having eight positive samples test
PCR-negative is very small. We estimated this
probability based on the binomial model (Ott 1993,
p. 147–149) for these samples in two groups (range
of microscope count 2–4, n = 5; and 14–36, n = 3).
The probability of five failures in five trials is 0.099
(low-count group), and the probability of three failures in three trials is 0.0003 (high-count group).
Again, the most plausible explanation is that the
purging process (El-Matbouli et al. 1999b) was not
quite 100% effective. However, in this case, the
recurring infection apparently was not M. cerebralis.

Although the infection of the worms in the
experimental replicates by exposure to aquarium
wastes was successful in both experiments, the
exact process by which myxospores were shed or
expelled by the fish remains unknown. We had
very little success in establishing the presence of
M. cerebralis myxospores by PCR in the aquarium
wastes used to expose the worms. Only two times in
either experiment did aquarium waste samples test
positive for M. cerebralis by PCR. However,
because of the large volume of aquarium waste
involved, the density of myxospores in any given
week's worth of waste may not have been large. It
is also possible that PCR tests of aquarium wastes
were inhibited by the amount of organic material
present, although evidence of such inhibition was
not noted by the laboratory conducting the testing.

All positive control replicates contained the
same mass weight of worms and, within each
experiment, all were dosed with about the same
number of myxospores. Assuming each replicate
contained approximately the same percentage of
T. tubifex worms that were susceptible to M. cerebralis, it is possible to estimate the yield of actinospores per myxospore dose. Among the three

positive control replicates in the first experiment, the mean actinospore yield was 42 per myxospore (range 38.6 to 46.7). In the second experiment, the mean yield was 457 actinospores per myxospore (range 412 to 486). In three similar experimental exposures, Markiw (1986) reported yield estimates of 3.1, 9.0, and 78.4 actinospores per myxospore dose. Our experimental replicates also contained the same mass weight of worms as the positive controls within each experiment. Therefore, the estimates of mean yield of actinospores per myxospore dose from the positive control replicates can be used to back calculate a crude estimate of the number of myxospores in each experimental replicate. In the first experiment, this results in estimates of 51,500, 75,800, and 70,700 myxospores among the experimental replicates. In the second experiment, it results in estimates of 1,200, 400, and 800 myxospores.

Possible avenues of myxospore escape from living host fish may include the rupture of spore-containing cysts formed on the operculum (Taylor and Haber 1974). These investigators identified such cysts in cutthroat trout and identified myxospores within them as those of M. cerebralis. We found a high degree of correlation in the magnitude of cranial myxospore concentration and the presence of myxospores in the pectoral fins of brown trout used in the second experiment. Myxospores of M. cerebralis were detected in the pectoral fins of 4 of 15 fish sacrificed on 14 July 2000. Three of the four fish had cranial myxospore concentrations ranging from 47,000 to 70,000. In contrast, myxospores were detected in pectoral fins of only 1 of 11 fish with cranial myxospore concentrations ranging from 0 to 17,000. These data, together with the findings of Taylor and Haber (1974) demonstrate that myxospores of M. cerebralis occur in peripheral tissues where they may well be shed through abrasion of epidermal surfaces. Myxospores residing in pectoral fin tissues might also be released by the practice of fin nipping among fish.

Whether other avenues exist will require further investigation accomplished with more sophisticated tools than experimental exposures. However, advanced tools are available now that were not available when Uspenskaya (1957) and Rydlo (1971) reported finding M. cerebralis myxospores in organs and soft tissues. For example, the in situ hybridization (ISH) protocol (Antonio et al. 1998) can be used to locate spores in peripheral tissues and organs where they may be passively shed.

ACKNOWLEDGMENTS

Portions of this work were funded by the National Partnership for the Management of Wild and Native Coldwater Fisheries, the Whirling Disease Foundation, Colorado Trout Unlimited and several of its chapters, and National Trout Unlimited. Some of the oligochaetes used in this study were supplied by T. McDowell and R. Hedrick. We thank C. Wilson and three anonymous reviewers for comments that improved the manuscript.

REFERENCES

Andree, K. B., E. MacConnell, and R. P. Hedrick. 1998. A nested polymerase chain reaction for the detection of genomic DNA of Myxobolus cerebralis in rainbow trout Oncorhynchus mykiss. Diseases of Aquatic Organisms 34:145–154.

Antonio, D. B., K. B. Andree, T. S. McDowell, and R. P. Hedrick. 1998. Detection of Myxobolus cerebralis in rainbow trout and oligochaete tissues by using a nonradioactive in situ hybridization (ISH) protocol. Journal of Aquatic Animal Health 10:338–347.

Baldwin, T. J., and six coauthors. 1998. Distribution of Myxobolus cerebralis in salmonid fishes in Montana. Journal of Aquatic Animal Health 10:361–371.

El-Matbouli, M., R. W. Hoffman, H. Schoel, T. S. McDowell, and R. P. Hedrick. 1999a. Whirling disease: host specificity and interaction between the actinosporean stage of Myxobolus cerebralis and rainbow trout Oncorhynchus mykiss. Diseases of Aquatic Organisms 35:1–12.

El-Matbouli, M., T. S. McDowell, D. B. Antonio, K. B. Andree, and R. P. Hedrick. 1999b. Effect of water temperature on the development, release and survival of the triactinomyxon stage of Myxobolus cerebralis in its oligochaete host. International Journal for Parasitology 29:627–641.

Gilbert, M. A., and W. O. Granath, Jr. 2001. Persistent infection of Myxobolus cerebralis, the causative agent of salmonid whirling disease, in Tubifex tubifex. Journal of Parasitology 87:101–107.

Markiw, M. E. 1986. Salmonid whirling disease: dynamics of experimental production of the infective stage—the triactinomyxon spore. Canadian Journal of Fisheries and Aquatic Sciences 43:521–526.

Markiw, M. E., and K. Wolf. 1974. Myxosoma cerebralis: Isolation and concentration from fish skeletal elements—sequential enzymatic digestions and purification by differential centrifugation. Journal of the Fisheries Research Board of Canada 31:15–20.

O'Grodnick, J. 1975. Whirling disease Myxosoma cerebralis spore concentration using the continuous plankton centrifuge. Journal of Wildlife Diseases 11:54–57.

Ott, L. R. 1993. An introduction to statistical methods and data analysis, 4th edition. Duxbury Press, Belmont, California.

Piper, R. G., I. B. McElwain, L. E. Orme, J. P. McCraren, L. G. Fowler, and J. R. Leonard. 1982. Fish hatchery management. United States Fish and Wildlife Service, Washington, D.C.

Rydlo, M. 1971. Evidence of *Myxosoma cerebralis* spores in various organs of rainbow trout afflicted with whirling disease. Österreichs Fischerei 24 (7):97–99.

Schisler, G. S., E. P. Bergersen, P. G. Walker, J. Wood, and J. K. Epp. 2001. Comparison of single-round polymerase chain reaction (PCR) and pepsin-trypsin digest (PTD) methods for detection of *Myxobolus cerebralis*. Diseases of Aquatic Organisms 45:109–114.

Taylor, R. E. L., and M. H. Haber. 1974. Opercular cyst formation in trout infected with *Myxosoma cerebralis*. Journal of Wildlife Diseases 10:347–351.

Thompson, K. G., and R. B. Nehring. 2000. A simple technique used to filter and quantify the actinospore of *Myxobolus cerebralis* and determine its seasonal abundance in the Colorado River. Journal of Aquatic Animal Health 12:316–323.

Uspenskaya, A. V. 1957. The ecology and spreading of the pathogen of trout whirling disease—*Myxosoma cerebralis* (Hofer, 1903, Plehn, 1905) in the fish ponds of the Soviet Union. Pages 47–55 *in* G. K. Petrushevskii, editor. Parasites and diseases of fish. Bulletin All-Union Scientific Research Institute of Fresh-water Fisheries. Leningrad (In Russian) English translation: Office of Technical Services, OTS 60-51169, Washington D.C.

Zendt, J. S., and E. P. Bergersen. 2000. Distribution and abundance of the aquatic oligochaete host *Tubifex tubifex* for the salmonid whirling disease parasite *Myxobolus cerebralis* in the upper Colorado River basin. North American Journal of Fisheries Management 20:502–512.

American Fisheries Society Symposium 29:135–141, 2002
© 2002 by the American Fisheries Society

Effects of *Myxobolus cerebralis* Infection on Juvenile Spring Chinook Salmon in the Lostine River, Oregon

TODD A. SANDELL

Hatfield Marine Science Center, Oregon State University, Newport, Oregon 97365, USA

HARRIET V. LORZ, SARAH A. SOLLID AND JERRI L. BARTHOLOMEW*

Center for Salmon Disease Research and the Department of Microbiology
Oregon State University, Corvallis, Oregon 97331, USA

ABSTRACT. The potential for *Myxobolus cerebralis*, the cause of salmonid whirling disease, to affect resident populations of spring chinook salmon *Oncorhynchus tshawytscha* in the Lostine River, Oregon, was investigated in this study. Spring chinook salmon and rainbow trout *O. mykiss* fry were held in the Lostine River for 14 d in late March 1999, when resident chinook salmon alevins naturally emerge. After exposure, fry were held in pathogen-free water in the laboratory. The prevalence of infection at 5 months postexposure, as determined by PCR, was equivalent in both species (37.5% and 41%, respectively). Only rainbow trout developed cranial lesions (average lesion severity 0.4 on a 5-point scale; 4 of 10 fish examined were positive), and no spores were detected in homogenates of cartilage from fish of either species. Comparison of data on chinook salmon spawning sites (1996–2000) with known distribution of *M. cerebralis* in the Lostine River demonstrated that the majority of chinook salmon spawn in the middle section of the river, where levels of *M. cerebralis* exposure were reduced. Results of this study indicate that juvenile chinook salmon may become infected with *M. cerebralis*, when naturally exposed to the parasite, but suggest that the timing and location of their emergence may mitigate the negative impacts of *M. cerebralis* infection in this river.

Myxobolus cerebralis was first documented in northeastern Oregon, U.S.A. in 1986 (Lorz et al. 1989), but the effects of the parasite on anadromous salmonid populations remain poorly understood. A number of anadromous species are susceptible to infection by *M. cerebralis*, including species that are currently or historically indigenous to this region: steelhead *Oncorhynchus mykiss* and chinook *O. tshawytscha*, coho *O. kisutch*, and sockeye *O. nerka* salmon (Hoffman and Putz 1969; Horsch 1987; O'Grodnick 1979; El-Matbouli et al. 1992; Baldwin et al. 1998; Hedrick et al. 2001). Differences in susceptibility among salmonid species have been demonstrated in natural exposures (O'Grodnick 1979; El-Matbouli et al. 1992; Markiw 1992; Hedrick et al. 1999a, 1999b), and controlled laboratory exposure studies showed that juvenile chinook salmon and steelhead are highly susceptible to *M. cerebralis* (Hedrick et al. 2001). Evidence of variations in susceptibility among strains of the same species were demonstrated in a recent study by

Thompson et al. (1999); these researchers reported differences in the susceptibility of four subspecies of cutthroat trout *O. clarki* (*bouvieri, stomias, virginalis,* and *pleuriticas*). However, variations in susceptibility among strains of other species, including chinook salmon, have not been clearly demonstrated.

Rainbow trout that are heavily infected with *M. cerebralis* may develop clinical signs of whirling disease, including spinal, cranial, and opercular deformities, black tails, and whirling behavior. High mortality may result from severe infections in fry; 100% mortality was reported by 12 d postexposure in rainbow trout sac fry (2 d old), exposed to 1,000 triactinomyxons (TAMs) per fish (Markiw 1991) in the laboratory. Although mortality is not always an outcome of exposure, it is presumed that sublethal *M. cerebralis* infections are detrimental to the development and survival of these infected fish. Decreased swimming performance was reported among experimentally infected rainbow trout, and researchers speculate that infected free ranging fish would be less able to escape predation and compete for food and other resources (Eileen K. N. Ryce, Montana Cooperative Fishery Research Unit, Montana State University,

---

*Corresponding author. bartholj@orst.edu

135

personal communication). In other laboratory studies, clinical disease and reduced weight gains were reported in experimentally infected chinook salmon (exposed to 2,500 triactinomyxons/fish) and steelhead (exposed to 200 triactinomyxons/fish) in comparison with uninfected, age-matched control fish (Hedrick et al. 2001).

A recent study by the authors (Sandell et al. 2001) described the distribution of M. *cerebralis* in the Lostine River, a tributary of the Wallowa River (Snake River basin; Figure 1). This river is typical of streams originating in the mountains of this region, with high gradients in the upper reaches and lower gradients where it flows through the valley floor. Sentinel rainbow trout became infected (as measured by the presence of parasite DNA) throughout the main stem of the river (river km 0–29), even when exposure cages were placed in areas where the oligochaete host *Tubifex tubifex* was not readily detected. The findings of that study generated interest in the effects of M. *cerebralis* on indigenous anadromous species. The potential effect of the parasite on chinook salmon in the

Lostine River was evaluated in this study, using infection data from sentinel exposure of chinook salmon fry, locations of chinook salmon spawning redds, and the known distribution and seasonal occurrence of M. *cerebralis* in the river.

METHODS

Fish

Rainbow trout were obtained as eyed eggs (Mt. Lassen Trout Farm, Red Bluff, California) and were hatched and reared in specific pathogen-free (SPF) water at 13°C. After absorption of the egg yolk, fry were fed a commercial trout food ad libitum. Chinook salmon were obtained from Rapid River Hatchery (Riggins, Idaho) at 6 weeks of age. The two species were closely age-matched at the time of exposure, although the rainbow trout (average weight 0.74 g; 1,860 total temperature units) were larger than the chinook salmon (average weight 0.59 g; 1,781 total temperature units) because of differences in feeding behavior while in captivity.

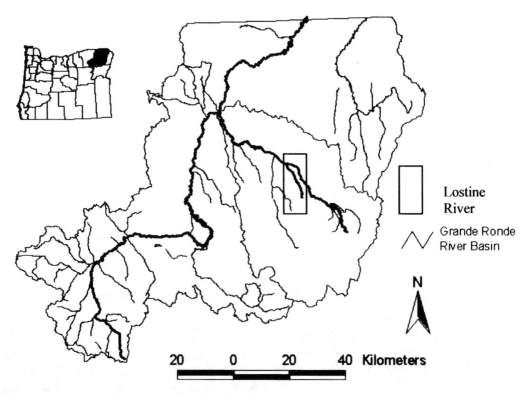

Figure 1. Map showing location of the Lostine River, Oregon. Locations of study sites, as measured from the river's mouth: site 1 river kilometer (km) - 0.8; site 2 - km 11.3; site 3 - km 15.3; site 4 - km 18.5; site 5 - km 29.

Natural Exposure

Both species were transported to the exposure site on the Lostine River in aerated, water-filled coolers. On 26 March 1999, 100 chinook salmon and 100 rainbow trout fry were placed in adjacent sentinel cages at km 0.8 on the Lostine River, as described by Sandell et al. (2001). Because the number of available chinook salmon fry available was severely limited, they were exposed only at the site where exposed fish had the highest prevalence of infection in 1998–1999 (Sandell et al. 2001). Water temperature was measured at the site using a Corning M-90 Checkmate portable meter (Corning Inc., Corning, New York). After the 14 d exposure, 34 rainbow trout fry were sampled to determine that the parasite was present at the time of exposure. The fish were killed with an overdose (500 mg/L) of tricaine methanesulfonate (Finquel, Argent Laboratories, Redmond, Washington) and whole alevins were placed in tissue lysis buffer (Qiagen, Valencia, California) for PCR analysis, as described in Sandell et al. (2001). No chinook salmon were sampled at this time, due to their limited number. The remaining fry of both species were returned to the laboratory and maintained for 5 months in SPF well water at 13°C on a commercial trout diet. During this time the fish were observed daily, and clinical signs of whirling disease were recorded. Control groups of 25 unexposed fish from the same cohort of each species were maintained in SPF well water at 13°C.

Assessment of Infection and Disease Severity

After 5 months, the surviving fry were killed with a lethal dose of anesthesia, as described above. The heads were divided sagitally, with half analyzed for the presence of M. *cerebralis* DNA by PCR (Andree et al. 1998) and spore enumeration. The remaining half heads were placed in tissue cassettes (Fisher Scientific, Pittsburgh, Pennsylvania) and stored in 10% buffered formalin. After PCR assay of all fish, fixed tissues from 10 M. *cerebralis*-positive fry from each group were sent to the Washington Animal Disease Diagnostic Laboratory (Pullman, Washington) for histological analysis. Infection severity was expressed using a scale from 0 (none) to 5 (severe). Description of the histopathology represented by this scale is given by Hedrick et al. (1999b). Spores were enumerated as described in Sandell et al. (2001).

Chinook Salmon Redd Counts

Data on chinook salmon redd numbers and locations were obtained from biologists with the Oregon Department of Fish and Wildlife (Pat Keniry and Tim Hoffnagle, personal communication). Spawning salmon surveys were conducted three times per year on the Lostine River between August and October during 1996–2000. The number of redds per reach were averaged for this analysis. River reaches defined in the spawning survey corresponded with the five sentinel exposure sites used to determine the distribution of M. *cerebralis* (Sandell et al. 2001). Estimated elevations were obtained from a USGS topographical map, centered on the town of Lostine, Oregon.

Distribution of *Myxobolus cerebralis* in the Lostine River

Studies characterizing the distribution and seasonal prevalence of M. *cerebralis* infection in the Lostine River for 1998–1999 were described by Sandell et al. (2001). Briefly, Mt. Lassen rainbow trout were obtained as eyed eggs, placed in sentinel cages at five exposure sites between river km 0 and 29 (Figure 1), and exposed for 14 d after hatching. The prevalence of infection was determined by detection of M. *cerebralis* DNA, using a nested PCR assay (Andree et al. 1998). The data reported here are the average prevalence of infection among sentinel rainbow trout exposed at each site in 1998–1999. These data were used to select the location of the exposure site in 1999.

Statistical Analysis

Data on the presence or absence of M. *cerebralis* DNA by PCR were analyzed using a comparison of binomial proportions; significance was defined as $P < 0.05$ (Ramsey and Schafer 1997).

RESULTS

Natural Exposure

The water temperature at river km 0.8, at the time of exposure in March 1999, was 9.3°C. After the 14 d exposure, 43% of the rainbow trout fry analyzed by PCR were positive for M. *cerebralis* DNA. We detected no significant differences in the prevalence of infection between the fish species at 5 months postexposure; 31.6% of the rainbow trout (6 of 19) and 37.5% of the chinook salmon (9 of 24) were positive for M. *cerebralis* DNA. No spores were detected from homogenized samples of head cartilage in either group, and none of the fish developed signs of whirling disease. Cranial lesions were detected in 4 of 10 rainbow trout fry examined; the mean lesion score among those fish examined histologically was 0.4. No lesions were detected in the chinook salmon. There was no evidence of M. *cerebralis* in the unexposed control fish of either species by PCR, detection of spores, or histological lesions.

Chinook Salmon Redd Counts

Although chinook salmon have historically spawned throughout the Lostine River, data on redds from the last 5 years (1996–2000) indicate that most salmon spawned in the middle section of the river, between river km 12 to 16 (Figure 2). This area corresponds to the exposure site at river km 15.3, where the average prevalence of infection in the sentinel fry was 10.4%. The elevation in this river section transitions from the low gradient of the lower river to the steep gradient of the upper river (Figure 3).

DISCUSSION

The declines of rainbow trout populations in the Rocky Mountain states, attributed in part to whirling disease (Nehring and Walker 1996; Vincent 1996; Baldwin et al. 1998), have led to reexamination of the potential impact of M. *cerebralis* on other fish populations outside these states. In the Columbia River basin, anadromous salmonid populations have declined for numerous reasons, including harvest, barriers to fish passage, and habitat destruction (National Resource Council [NRC] 1996; National Oceanic and Atmospheric Administration [NOAA] 1999). The detection of M. *cerebralis* in all of the contiguous states in this region (Bergersen and Anderson 1997) raises questions about the potential role of the parasite in salmonid population declines.

A previous study demonstrated that M. *cerebralis* was present throughout the main stem of the Lostine River, with infection levels highest in the lower river and during the late summer (August to September; Sandell et al. 2001). To determine if chinook salmon are exposed to a significant level of parasite exposure, chinook salmon and rainbow trout fry were held in the lower Lostine River during March, when they naturally emerge from the gravel. At 5 months after the exposure, the percentages of chinook salmon and rainbow trout that were positive for M. *cerebralis* DNA were similar (37.5 and 41%, respectively), suggesting that the

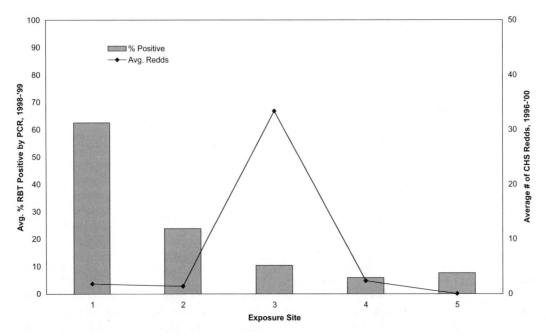

Figure 2. Average prevalence of *Myxobolus cerebralis* infection among rainbow trout (1998–1999) and average number of chinook salmon redds (1996–2000) at Lostine River exposure sites.

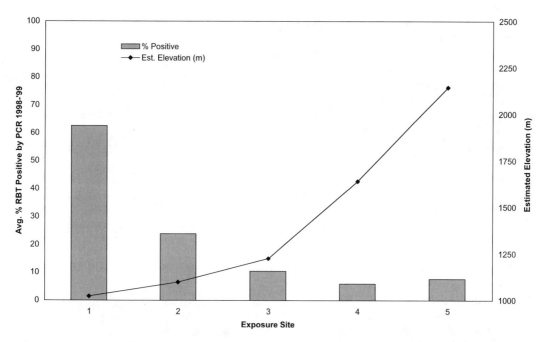

Figure 3. Average prevalence of *Myxobolus cerebralis* infection among rainbow trout (1998–1999) and estimated river elevation (meters) of Lostine River exposure sites.

two species are equally likely to become infected. However, clinical disease signs did not develop in either species, and presence of parasite DNA did not correlate with histological lesions or spore numbers. Only rainbow trout exhibited histologic evidence of infection, and lesion severity was low, which suggests that these fish were not exposed to a high infectious dose.

Other studies on the susceptibility of chinook salmon to whirling disease have demonstrated that this species is susceptible, but the conditions of these challenges varied greatly and are difficult to compare with the present study. In a controlled laboratory challenge, Hedrick et al. (2001) demonstrated that chinook salmon (0.25 g) were highly susceptible when challenged with a single high dose of triactinomyxons (2,500 triactinomyxons per fish). O'Grodnick (1979) reported that chinook salmon (obtained from the Michigan Department of Natural Resources; specific strain and size not identified) were less severely affected than rainbow trout in a natural exposure; however, parasite dose was not quantified. The chinook salmon used in this study were larger (0.59 g) than those used in the study by Hedrick et al. (2001), and this may have contributed to the lack of clinical signs. However, the infectious dose was an important variable that could not be quantified in this study. There is little data on what parasite dose constitutes a "natural exposure" in rivers where *M. cerebralis* is enzootic. Attempts were made to quantify triactinomyxon numbers in the Lostine River during periods of peak infection, using the filtration technique of Thompson and Nehring (2000). No triactinomyxons were detected from any sample, suggesting that, even at the time of maximum infection prevalence, resident fish in the Lostine River receive a lower level of exposure than occurs in other enzootic areas where population impacts have been reported.

In addition to emerging in March and April, when parasite exposure is reduced, the majority of chinook salmon alevins in the Lostine River are also likely to hatch upstream of those exposure sites where the greatest prevalence of *M. cerebralis* infection occurred (river km 0.8–11.3; Figures 1–3). The combination of emergence timing and location may prevent chinook salmon alevins from becoming infected during their most susceptible life stage (Markiw 1991). For these reasons, it appears that populations of chinook salmon in the Lostine River are at a lower risk of being heavily impacted by *M. cerebralis* than rainbow trout and steelhead, which emerge later in the year and in the lower river.

There are, however, several aspects of chinook salmon biology that may influence the effects of *M. cerebralis* on these populations. Spring chinook salmon are typically present in the river as juveniles for 1 year before smolting and migrating out of the system. The effects of this prolonged exposure are unknown. The effect of *M. cerebralis* infection on chinook salmon and steelhead smolts during the par–smolt transformation and adaptation to salt water is also undocumented. The impact of infection on juveniles at this stage may be pronounced, as the transformation causes physiological stress that has been shown to be immune-suppressive (Angelidas et al. 1987; Maule et al. 1987, 1989; Wang and Belosevic 1995). For these reasons, further studies examining the effects of a long-term natural exposure on chinook salmon juveniles and smolts, as well as possible susceptibility differences between salmonid strains native to the Columbia and Snake River basins, are needed to more accurately predict the impact of *M. cerebralis* on salmonid populations in this region.

ACKNOWLEDGMENTS

This work was funded by the National Partnership on the Management of Wild and Native Coldwater Fisheries and by Portland General Electric Co. We thank the Willett, Wynan, and Link families for allowing us access to their property, while conducting this research. We are also indebted to the Idaho Department of Fish and Game for the chinook salmon used in this study, to Ron Hedrick and Terri McDowell for cultures of *T. tubifex*, to Pat Keniry and Tim Hoffnagle (Oregon Department of Fish and Wildlife) for data on chinook salmon redds in the Lostine River, and to Don Stevens for his assistance in collecting and maintaining specimens, This is Oregon Agricultural Station technical paper 11838.

REFERENCES

Andree, K. B., E. MacConnell, and R. P. Hedrick. 1998. A nested polymerase chain reaction for the detection of genomic DNA of *Myxobolus cerebralis* in rainbow trout *Oncorhynchus mykiss*. Diseases of Aquatic Organisms 34:145–154.

Angelidas, P., F. Baudin-Laurencin, and P. Youinou. 1987. Stress in rainbow trout, *Salmo gairdneri*: effects upon phagocyte chemiluminescence, circulating leucocytes and susceptibility to *Aeromonas salmonicida*. Journal of Fish Biology 21 (Supplement A):113–122.

Baldwin, T. J., J. E. Peterson, G. C. McGhee, K. D. Staigmiller, E. S. Motteram, C. C. Downs, and D. R.

Stanek. 1998. Distribution of *Myxobolus cerebralis* in salmonid fishes in Montana. Journal of Aquatic Animal Health 10:361–371.

Bergersen, E. P., and D. E. Anderson. 1997. The distribution and spread of *Myxobolus cerebralis* in the United States. Fisheries 22:6–7.

El-Matbouli, M., T. Fischer-Scherl, and R. W. Hoffman. 1992. Present knowledge on the life cycle, taxonomy, pathology, and therapy of some *Myxosporea* spp. important for freshwater fish. Annual Review of Fish Diseases 367–402.

Hedrick, R. P., T. S. McDowell, M. Gay, G. D. Marty, M. P. Georgiadis, and E. MacConnell. 1999a. Comparative susceptibility of rainbow trout *Oncorhynchus mykiss* and brown trout *Salmo trutta* to *Myxobolus cerebralis*, the cause of salmonid whirling disease. Diseases of Aquatic Organisms 37:173–183.

Hedrick, R. P., T. S. McDowell, K. Mukkatira, and M. P. Georgiadis. 1999b. Susceptibility of selected inland salmonids to experimentally induced infections with *Myxobolus cerebralis*, the causative agent of whirling disease. Journal of Aquatic Animal Health 11:330–339.

Hedrick, R. P., T. S. McDowell, K. Mukkatira, and M. P. Georgiadis. 2001. Susceptibility of three species of anadromous salmonids to experimentally-induced infections with *Myxobolus cerebralis* the causative agent of whirling disease. Journal of Aquatic Animal Health 13:43–50.

Hoffman, G. L., and R. E. Putz. 1969. Host susceptibility and the effect of aging, freezing, heat, and chemicals on spores of *Myxosoma cerebralis*. The Progressive Fish-Culturist 35–37.

Horsch, C. M. 1987. A case history of whirling disease in a drainage system: Battle Creek drainage of the upper Sacramento River basin, California, USA. Journal of Fish Diseases 10:453–460.

Lorz, H. V., A. Amandi, C. R. Banner, and J. S. Rohovec. 1989. Detection of *Myxobolus* (*Myxosoma*) *cerebralis* in salmonid fishes in Oregon. Journal of Aquatic Animal Health 1:217–221.

Markiw, M. E. 1991. Whirling Disease: earliest susceptible age of rainbow trout to the Triactinomyxid of *Myxosoma cerebralis*. Aquaculture 92:1–6.

Markiw, M. E. 1992. Salmonid whirling disease. U.S. Fish and Wildlife Service, Fish and Wildlife Leaflet 17.

Maule, A. G., C. B. Schreck, and S. L. Kaattari. 1987. Changes in the immune system of coho salmon (*Oncorhynchus kisutch*) during the parr-to-smolt transformation and after implantation of cortisol. Canadian Journal of Fisheries and Aquatic Sciences 44:161–166.

Maule, A. G., R. A. Tripp, S. L. Kaattari, and C. B. Schreck. 1989. Stress alters immune function and disease resistance in chinook salmon (*Oncorhynchus tshawytscha*). Journal of Endocrinology 120:135–142.

Nehring, R. B., and P. G. Walker. 1996. Whirling disease in the wild: the new reality in the intermountain West. Fisheries 21:28–30.

NOAA (National Oceanic, and Atmospheric Administration). 1999. Endangered and threatened species:

threatened status for three chinook salmon evolutionarily significant units (ESUs) in Washington. Portland, Oregon.

NRC (National Research Council), Committee on Protection and Management of Pacific Northwest Anadromous Salmonids. 1996. Upstream: salmon and society in the Pacific Northwest. National Academy Press, Washington, D.C.

O'Grodnick, J. J. 1979. Susceptibility of various salmonids to whirling disease (*Myxosoma cerebralis*). Transactions of the American Fisheries Society 108:187–190.

Ramsey, F. L., and D. W. Schafer. 1997. The statistical sleuth: a course in methods of data analysis. Duxbury Press, Belmont, California.

Sandell, T. A., Lorz, H. V., Stevens, D. G., and J. L. Bartholomew. 2001. *Myxobolus cerebralis* dynamics in the Lostine River, Oregon: implications for resi-

dent and anadromous salmonids. Journal of Aquatic Animal Health 13:142–150.

Thompson, K. G., R. B. Nehring, D. C. Bowden, and T. Wygant. 1999. Field exposure of seven species or subspecies of salmonids to *Myxobolus cerebralis* in the Colorado River, Middle Park, Colorado. Journal of Aquatic Animal Health 11:312–329.

Thompson, K. G., and R. B. Nehring. 2000. A simple technique used to filter, and quantify the actinospore of *Myxobolus cerebralis* and determine its seasonal abundance in the Colorado River. Journal of Aquatic Animal Health 12:316–323.

Vincent, E. R. 1996. Whirling disease and wild trout: the Montana experience. Fisheries 21:32–33.

Wang, R., and M. Belosevic. 1995. The *in vitro* effects of estradiol and cortisol on the function of a long-term goldfish macrophage cell line. Developmental and Comparative Immunology 19:327–336.

Section 5

Ecology

American Fisheries Society Symposium 29:145–166, 2002
© 2002 by the American Fisheries Society

The Ecology of *Myxobolus cerebralis*

BILLIE L. KERANS*

Department of Ecology, Montana State University –Bozeman, Bozeman, Montana 59717, USA
Phone: (406) 994-3725; Fax: (406) 994-3190; Email: bkerans@montana.edu

ALEXANDER V. ZALE

Montana Cooperative Fishery Research Unit, USGS
Department of Ecology, Montana State University –Bozeman, Bozeman, Montana 59717, USA
Phone: (406) 994-2380; Fax: (406) 994-7479; Email: zale@montana.edu

ABSTRACT. The myxosporean parasite *Myxobolus cerebralis* is the causative agent of salmonid whirling disease. Containing its spread and limiting its effects in the Intermountain West will require judicious management programs, but such actions await a comprehensive understanding of the biology and ecology of this parasite and its hosts and how these elements interact; we do not yet know the weaknesses of this organism. To better guide efforts aimed at such an understanding, we assembled available information on the ecology of the parasite, organizing it into a conceptual model of its life cycle, to help foster understanding, focus future research, and lead eventually to a mathematical model for evaluating control measures. *Myxobolus cerebralis* has a complex life cycle with two obligate hosts, a salmonid fish and the oligochaete *Tubifex tubifex*, parasitized by the myxosporean and the actinosporean, respectively, and two infective "spore" stages, the myxospore and the triactinomyxon. This complexity is enhanced by the variable suitability of multiple salmonid species to serve as hosts, varying host suitability of genetic variants of *T. tubifex*, relatively recent introduction of *M. cerebralis* to North America, and unique traits of the parasite that preclude easy classification into conventional modeling categories. Much is known about the anatomy and function of myxospores and triactinomyxons from laboratory studies, but information on their distribution, abundance, and dispersal in natural systems is limited and based on indirect observations. Similarly, we understand development of the parasite within its hosts and resulting pathologies well but know little about host immune reactions and other mechanisms controlling proliferation within hosts or how environmental factors affect these defenses. Population-level effects on fish in natural systems have been quantified only rarely, where good prewhirling disease data exist, and effects on *T. tubifex* populations are unknown. Most rates and frequencies needed to infer relationships and model system dynamics have not been directly quantified in natural systems, but rapid progress is being made. Larger issues, including effects of *M. cerebralis* on community dynamics and ecosystem structure and function, have yet to be explored.

Whirling disease is the common name of the disorder caused by the parasite *Myxobolus cerebralis* (Myxozoa: Myxosporea) that has been implicated in severe declines of some wild rainbow trout *Oncorhynchus mykiss* populations in the Intermountain West of the United States in the 1990s (Nehring and Walker 1996; Vincent 1996) and threatens others. Managers are interested in containing the spread of the parasite and ameliorating its effects on salmonid hosts. However, design and execution of rational management strategies will first necessitate a comprehensive understanding of the biology and ecology of this parasite and its hosts

and how these elements interact. Only when we understand how and why *M. cerebralis* has evolved to do what it does to complete its life cycle can we look for ways to thwart and disrupt that cycle.

Unfortunately, the ecology of *M. cerebralis* in the Intermountain West is complex and will be difficult to understand comprehensively for several reasons. First, *M. cerebralis* has a complicated life cycle with two obligate hosts—fish and oligochaetes—and two intermediary spore stages—the triactinomyxon and the myxospore (Hedrick et al. 1998). Model simulations show that dynamics of apparently simple systems that include linked species can be complex even in the absence of any outside environmental effects, with host and parasite population

---

*Corresponding author

stability, oscillations, and extinctions all as viable outcomes (e.g., Anderson and May 1986). Second, a number of salmonid species and genetic variants of *Tubifex tubifex* (the only oligochaete known to host *M. cerebralis*; Wolf et al. 1986) of varying degrees of susceptibility to *M. cerebralis* (Hedrick et al. 1998; Stevens et al. 2001) coexist in many rivers of the Intermountain West, making the dynamics of the host-parasite relationship less predictable. Third, population dynamics of invading species are usually difficult to predict, and *M. cerebralis* has only recently invaded North America (Hoffman 1970, 1990). Nonindigenous parasites have been known to invade and persist in some regions but not others, and often, there is a considerable delay between invasion and expansion to levels at which populations of the hosts are affected (Anderson and May 1986). Therefore, sufficient time has likely not yet passed to definitively ascertain where *M. cerebralis* has or has not become established in some locations in the Intermountain West. Where nonindigenous parasites do persist, their population dynamics are often erratic, exhibiting wide swings in abundance and prevalence especially during the early stages of establishment (Anderson and May 1986). Moreover, some of the fish hosts of *M. cerebralis* are also nonindigenous to most or all of the Intermountain West, and their range expansion is similarly unpredictable. The combination of interactions among a minimum of three species, varying susceptibilities among hosts, and the nonindigenous nature of the parasite and some hosts makes comprehensive understanding of the ecology of this host–parasite relationship difficult.

Nevertheless, the complexities associated with *M. cerebralis* in the Intermountain West do not mean that we will never be able to understand the host–parasite relationship well enough to manage the system. However, the complexities do suggest that we must place this host–parasite relationship into its underlying evolutionary and ecological context to begin to understand it. We can thereby compartmentalize the relationship into discrete and comprehensible components. Conceptualization of the life cycle of *M. cerebralis*, including interactions with other organisms, followed by development of a mathematical model quantifying components and their rates in the entire system, can provide a means to understanding the evolutionary and ecological relationships between this parasite and its hosts. This strategy has helped clarify other host–parasite relationships to identify

important components for which information is unavailable and thereby better focus research to evaluate the potential efficacy of management strategies and to enhance understanding of the underlying processes of invasion, persistence, and spread of the parasites (Anderson 1994; Cherry et al. 1998; McKenzie 2000).

In this paper, we 1) describe the life cycle and taxonomy of *M. cerebralis* from an evolutionary ecology perspective, 2) examine the general life history strategies of parasites and how the characteristics of *M. cerebralis* fit these strategies, 3) conceptualize the life cycle of *M. cerebralis* and place existing information into this framework—and identify needed information that does not yet exist—to facilitate future development of an explanatory mathematical model, and 4) discuss how parasitism of salmonids and worms by *M. cerebralis* has the potential to indirectly affect the other organisms in the community and how other organisms could influence parasitism by *M. cerebralis* of salmonids and worms. We thereby hope to enhance our understanding of invasion and spread of *M. cerebralis* and its effects on salmonid populations in the Intermountain West and thereby facilitate management.

LIFE CYCLE, TAXONOMY, AND EVOLUTION

Myxobolus cerebralis is a member of the protistan Phylum Myxozoa, but recent evidence links the Myxozoa to both the radial (Siddall et al. 1995) and the bilateral (Smothers et al. 1994; Anderson et al. 1998) metazoans. *Myxobolus cerebralis* has a complex life cycle with two hosts, a fish and an oligochaete parasitized by the myxosporean and the actinosporean stages, respectively, and two infective "spores," the myxospore and the triactinomyxon (Figure 1). This life cycle was first documented by Maria Markiw and Ken Wolf (Markiw and Wolf 1983; Wolf and Markiw 1984), and the conspecificity of the spores has since been verified using genetic tests (Andree et al. 1997).

The linkage of the myxosporean and actinosporean stages of *M. cerebralis* was the first evidence for the redundancy of the two classes of the Myxozoa, the Myxosporea, which typically parasitize fish (Lom 1987), and the Actinosporea, which typically parasitize oligochaetes (Janiszewska 1955, 1957; Marques 1984). The two-host life cycle has since been documented for 24 other myxosporeans

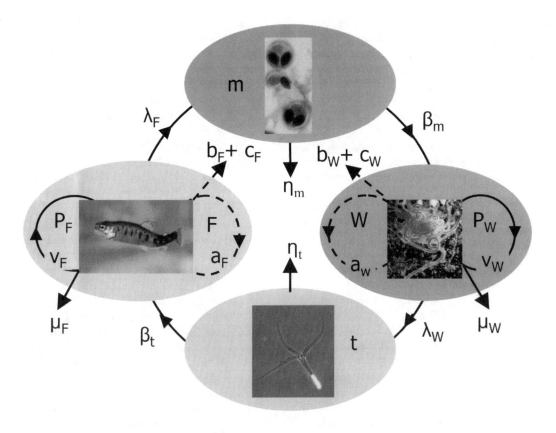

Figure 1. The conceptualized life cycle of *Myxobolus cerebralis* (solid lines) and its hosts (dashed lines). *Myxobolus cerebralis* resides in four compartments in the worm host (P_w), the fish host (P_f) and in the environment as a triactinomyxon (t) or myxospore (m). The parameters are described in detail in the text, but in general they describe the transitions between fish and worm hosts and the environment, and birth and death of parasites, fish, and worms.

that infect freshwater fish and worms and one that infects fish and bryozoans (Kent et al. 2001). Thus, the taxonomic affiliation of the Myxosporea and Actinosporea is currently under revision, and it has been proposed that they be united into one taxon (Kent et al. 1994; Siddall et al. 1995). About 1,300 species of myxosporeans have been described (Lester et al. 1999), and only 59 actinosporeans have been described or partially described as species (Lom et al. 1997). It is unknown if all myxosporeans have the two-host life cycle because life cycles are known for only about 1% of the described myxosporeans (Lester et al. 1998), and 44 described species of actinosporeans have no known associated myxosporeans (Lom et al. 1997). New actinosporeans are regularly reported (e.g., Xiao and Desser 1998a, 1998b, 1998c). Many myxosporeans likely remain undiscovered because parasites of many tropical and subtropical teleosts are poorly known

(Lester et al. 1998).

Conclusive understanding of why the life cycle of M. *cerebralis* evolved as it did will depend on where and in what order the two-host life cycle evolved. We do not know if the Myxozoa first parasitized worms or fish or if they originated in marine or freshwater systems. Compelling evidence for both worms and fish as the primitive host exists (Kent et al. 1994; Siddall et al. 1995; Diamant 1997; Lester et al. 1998, 1999; Kent and Lom 1999) and is at present equivocal and inconclusive, in our opinion. Although all myxosporeans with known associated actinosporeans inhabit freshwater, some actinosporeans infect marine worms (Hallett et al. 1995, 1998; Roubal et al. 1997; Koie 2000), and some myxosporeans parasitize marine fish (Alvarez-Pettitero and Sitja-Bobadilla 1993; Diamant 1997). It is unknown whether any of the marine actinosporeans and myxosporeans exhibit the two-host life cycle. Among marine myxospore-ans, only the life cycle of Myxidium leei has been documented, and it exhibits direct fish-to-fish transmission without an associated actinosporean (Diamant 1997, 1998).

Two-host life cycles are thought to have evolved from single-host life cycles to bridge eco-logical gaps in the single-host life cycle (Schmidt and Roberts 1977). If primitive myxozoans first par-asitized worms in the ocean, which seems plausible given that gametic fusion occurs in the worm host (El-Matbouli and Hoffmann 1998; El-Matbouli et al. 1998) and that the Myxozoa are possibly related to the predominantly-marine Cnidaria (Siddall et al. 1995), then perhaps the two-host life cycle evolved as a mechanism for the parasite to keep its offspring near potential worm hosts upon invading fluvial freshwater systems. Progeny of nonmotile freshwater invertebrates released into the water col-umn tend to be carried downstream away from suit-able hosts and habitats, but parasitism of a mobile fish host, especially one that tends to move upstream to spawn, would negate such displace-ment. An example of this is glochidial parasitism of freshwater fish by unionid mussels (Graf and O'Foighil 2000). Conversely, if marine fish were the primitive host, which seems plausible given the many myxosporeans known to parasitize marine fish, then adaptation to use a sedentary worm as an alternative host might help M. cerebralis bridge the temporal gap in freshwater between when fish hosts are most likely to die and release the parasite (after spawning) and when susceptible and abundant hosts (offspring of these same fish) are available months later following emergence.

Myxobolus cerebralis almost certainly evolved in Eurasia (Hoffman 1970, 1990) and, thus, is an invasive organism in the United States. Hoffman (1970) proposed that M. cerebralis co-evolved with brown trout Salmo trutta, based on historical records (whirling disease was known only from Germany, France, and Denmark prior to 1952; Hoffman 1970) and the resistance brown trout dis-play towards this parasite; brown trout can be infected (Halliday 1976; O'Grodnick 1979) but become diseased only when exposed to high tri-actinomyxon densities (Hedrick et al. 1999a). If this hypothesis of a Eurasian origin for M. cerebralis is true, it probably co-evolved with only brown trout and Atlantic salmon S. salar in North and Baltic sea drainages. This contention is supported by the fact that the resistance of Atlantic salmon to

M. cerebralis is similar to that of brown trout (V. Blazer, National Fish Health Research Laboratory, personal communication). In addition, rainbow trout have become established in only a few loca-tions in the North and Baltic sea drainages of northern Europe, despite many attempts to intro-duce them and continued opportunities for inva-sion from hatcheries (Hindar et al. 1996; Saegrov et al. 1996; Landergren 1999), presumably because of the presence of M. cerebralis (A. Zvirgzds, Lat-vian Academy of Sciences, personal communica-tion). Myxobolus cerebralis was likely not native to the Danube drainage, which flows southeast into the Black Sea. The huchen Hucho hucho, a salmonid endemic to the Danube drainage, lacks resistance to M. cerebralis (El-Matbouli et al. 1992). In addition, reproducing populations of rainbow trout flourish in headwater tributaries of the Danube (Weiss and Schmutz 1999). However, reproducing populations of rainbow trout also exist in the north-flowing Alpine Rhine and its tributar-ies, but only in the headwaters above Lake Con-stance (P. Armin, Swiss Federal Institute for Envi-ronmental Science and Technology, personal com-munication). An alternate explanation is that M. cerebralis is present in these drainages, and rainbow trout have developed resistance to it through expo-sure over many generations.

MATHEMATICAL MODELS AND LIFE HISTORY CHARACTERISTICS OF PARASITES

Parasites encompass a diverse array of organisms with widely differing characteristics and ecologies. Nevertheless, categorization of parasites into two very broad groups, the microparasites and the macroparasites, has been useful in achieving under-standing of their effects on hosts and the develop-ment and simplification of mathematical models that differ substantially for microparasites and macroparasites (May and Anderson 1979). The microparasites are typically viruses, bacteria, and protozoans, whereas the macroparasites include parasitic helminths and arthropods (Anderson and May 1979, 1991; May and Anderson 1979; Dobson and Hudson 1986).

"Microparasites … are characterized by small size, short generation times, extremely high rates of direct reproduction within the host, and a tenden-cy to induce immunity to reinfection in those hosts that survive the initial onslaught. The duration of infection is typically short in relation to the life span of the host and is therefore transient in nature" (Anderson and May 1979; Table 1). Microparasitic infections can have various degrees of pathogenicity, but because they multiply rapidly within the host, the degree of pathogenicity does not depend on the initial abundance of parasites entering the host (Dobson and Hudson 1986; Anderson and May 1991). On the other hand, "macroparasites … tend to have much longer generation times than microparasites, and direct multiplication within the host is either absent or occurs at a low rate" (Anderson and May 1979). Both the immune response elicited by macroparasites, and their pathogenicity, generally depend on the number of parasites present in a given host (May and Anderson 1979; Dobson and Hudson 1986). Macroparasitic infections, therefore, tend to be of a persistent nature with the host being continuously reinfected.

How does M. cerebralis fit into these categories? In some ways, it possesses the characteristics of a microparasite. It is much smaller than either host. The stages of the parasite vary between about 3 and 80 µm within the hosts (El-Matbouli et al. 1995; El-Matbouli and Hoffmann 1998). The myxospores are about 10 µm in diameter, whereas the processes of the triactinomyxons are about 200 µm long (Lom and Hoffman 1971; El-Matbouli and Hoffmann 1998). The parasite goes through several replication cycles in both hosts (asexual in fish, sexual and asexual in worms), such that its capacity for reproduction is much higher than that of either host (El-Matbouli et al. 1995; El-Matbouli and Hoffmann 1998). Maturation to mature transmission (spore) stages occurs in about 3–4 months in both fish and worms (El-Matbouli et al. 1995; El-Matbouli and Hoffmann 1998), whereas T. tubifex can live up to 2 years in the laboratory (Gilbert and Granath 2001), and most salmonids have life cycles that extend over 3–6 years but can be much longer (Scott and Crossman 1973).

Myxobolus cerebralis also possesses attributes characteristic of macroparasites (Table 1). Some immunity does develop in rainbow trout, but it is weak (E. K. N. Ryce, A. V. Zale, and E. MacConnell, unpublished data), and worms that have been

Table 1. Life history characteristics of *Myxobolus cerebralis*, microparasites and macroparasites. This table was adapted from one presented in Dobson and Hudson (1986) and it contains material from Anderson and May (1979, 1991) and May and Anderson (1979).

Life history characteristics	Microparasite	Macroparasite	*Myxobolus cerebralis* in worms	*Myxobolus cerebralis* in fish
Body size of parasite relative to host	Much smaller than host	Smaller than host	Much smaller than host	Much smaller than host
Intrinsic growth rate of parasite population relative to host population	Much faster than host	Faster than host	Much faster than host	Much faster than host
Life span of parasite relative to host	Much shorter	Shorter	Much shorter	Much shorter
Immunity in hosts	Typically induces long-term immunity in hosts that survive	Typically immunity depends on parasite abundance and is short-lived	No immunity?	Some immunity
Duration of infection	Tends to be short relative to host life span	Tends to be persistent, long relative to host life span	Long and persistent	Long and persistent
Pathogenicity	Mild to deleterious	Depends on parasite abundance	Depends on parasite abundance	Depends on parasite abundance

purged of infection can be reinfected (El-Matbouli et al. 1999). Infection in both hosts is persistent (Hedrick et al. 1998; El-Matbouli et al. 1998; Gilbert and Granath 2001), and pathogenicity is dependent on parasite abundance in both fish and worms (Hedrick et al. 1999a; B. L. Kerans, C. Rasmussen, R. Stevens and J. Winton, unpublished data). The myxospore can lie dormant for years if not decades (Halliday 1976), thereby achieving a lifespan longer than either host. Modeling the infection dynamics of M. cerebralis will, therefore, require a hybrid model that includes characteristics of both microparasite and macroparasite models.

The dynamics of microparasitic infections in hosts are typically modeled by dividing the host population into susceptible, infected, and immune classes, with the assumption that the host population size remains constant (Anderson and May 1979). Such models will probably not be helpful when examining M. cerebralis in the Intermountain West because rainbow trout populations have clearly declined in some cases, and thus, the assumption that host population size is constant has been violated (May and Anderson 1979). Models for macroparasites include host population size as a dynamic variable (May and Anderson 1979) and, thus, are more appropriate for modeling M. cerebralis infections in natural waters. A model with host population sizes as dynamic variables can take into account the fact that pathogenicity of M. cerebralis to both hosts depends on parasite abundance, that hosts lack strong immunity, and that infection is persistent. A full model for M. cerebralis might include equations describing population dynamics of the parasite within both hosts and for both transmission stages (four linked equations). In addition, models of the population dynamics of both hosts could be included. Whether all equations will be needed to adequately describe parasite population dynamics will depend on the time scale of the dynamics of the various stages (see Table 2 in May and Anderson 1979). For example, if the dynamics of the transmission stages in the environment occur on much shorter time scales than the dynamics of the parasite in both hosts, then they can be assumed to be at some equilibrium value set by prevailing conditions, and separate equations for them may be collapsed out of the modeling exercise.

A four-equation model of M. cerebralis that included the population dynamics of the parasite in both hosts and for both transmission stages using some realistic assumptions about the system (i.e.,

ignoring fish predation on worms and assuming mortality of myxospores was negligible) was developed by B. R. Cherry and G. Smith (University of Pennsylvania, personal communication). Their preliminary analysis suggested that reducing abundance of T. tubifex might not be an effective strategy for reducing parasite numbers, but that increasing worm mortality might cause the parasite to decline. This result can be explained because with negligible mortality all myxospores are eventually found by T. tubifex, so the density of worms is unimportant; however, increasing mortality of infected T. tubifex decreases the number of parasites that actually survive the developmental period. Cherry and Smith also suggested that if the assumption of negligible myxospore death rate is violated, for example, if a substantial proportion of myxospores are de-activated by other resistant worms, then T. tubifex density may become important. Although these indications were preliminary, they are an example of the information that becomes available when models are used to examine the relative importance of various components. The following conceptual model of the M. cerebralis life cycle highlights what is known and unknown to aid mathematical modelers and others interested in understanding M. cerebralis.

CONCEPTUAL MODEL OF THE LIFE CYCLE

Figure 1 depicts the conceptualized life cycle of M. cerebralis and its hosts. The parasite's life cycle is depicted by the solid lines, and the hosts' life cycles are depicted by the dashed lines. The parasite resides in four compartments (shown as ovals), including the fish, the worm, or in the environment as the triactinomyxon or myxospore transmission stages. Abundances of the worm (W), the fish (F), the parasite within the worm (P_W) and within the fish (P_F), the triactinomyxon (t), and the myxospore (m) will depend to various degrees on a number of parameters (Anderson and May 1979; May and Anderson 1979). We present these parameters to focus attention on information that will be needed for modeling; the mathematical form of each will depend on the specific model. If hosts and parasites are modeled in continuous time (i.e., differential equations), then the parameters are expressed as rates (per individual per time). If the models have discrete time steps, then the parameters are expressed as numbers per individual.

The abundances of triactinomyxons (t) and myxospores (m) in the environment are positive functions of λ_W or λ_F (the numbers of triactino-myxons or myxospores produced by an individual parasite that has successfully invaded the worm and fish hosts, respectively) and the number of parasites within hosts (P_W, P_F). Triactinomyxon and myxospore abundances decrease when they become deactivated (η_t or η_m, the proportions of triactino-myons or myxospores that are deactivated within the environment) and when they are transmitted to the hosts. Parasite abundances in fish (P_F) and worm (P_W) populations increase as a function of the susceptibility of the hosts (β_t, β_m), the abundances of infective stages in the environment (t, m), and the abundances of worms (W) and fish (F). The proportion of contacts between hosts and infective stages that result in a parasite being transmitted are described by β_t and β_m (i.e., susceptibility of hosts to infection). Parasite numbers in fish and worm populations are reduced as the result of physical factors or perhaps as a result of the immune response of hosts. The proportion of parasites that die within the worm and fish are described by μ_W and μ_F, respectively. Within both hosts, the parasite also replicates (ν_W, ν_F). The numbers of transmission stages produced by each parasite (λ) are presumably functions of ν and μ. The populations of fish (F) and worms (W) increase as a function of the number of offspring produced by a fish (a_F) and a worm (a_W) and decrease as a function of the proportion of fish and worms removed by mortality (b_F, b_W). In addition, populations of fish and worms are reduced by c_F and c_W, which describe the negative effects of the parasite populations on the host populations.

Myxospores in the Environment

Myxospores are released into sediments upon the death of infected salmonids (Hedrick et al. 1998). Abundance of myxospores in the environment (m) will be a function of the production of myxospores by an individual parasite in fish (λ_F, see "Production of Myxospores"), the deactivation rate of myxospores in the environment (η_m or, alternatively, survival rate), and the number of parasites in fish (P_F).

Only indirect information is available about the abundance of myxospores in the environment. In laboratory experiments, we were able to infect *T. tubifex* in only one of 16 trials using natural sediments collected from three side channels of the Madison River, Montana (B. L. Kerans, unpublished data), where whirling disease has caused

severe declines in rainbow trout abundances (Vincent 1996). However, naturally infected worms were collected from these channels (R. C. Krueger and B. L. Kerans, unpublished data). These data suggest that myxospores are widespread; however, their availability is low or patchy even in areas where whirling disease has caused severe declines in rainbow trout abundances. Techniques for enumeration of myxospores in natural sediments have been developed (Lemmon and Kerans 2001) but they are probably not sensitive enough to detect the low numbers of myxospores that the above experiment suggested are available in streams.

Survival of myxospores in the environment appears to be fairly high because of their hard shell. Long viability of myxospores is a bet-hedging strategy that insures that they will survive at least as long as their hosts. After release from fish, they can survive for at least 5 months at 13°C (El-Matbouli and Hoffmann 1991), with anecdotal evidence that they can survive up to 30 years (see references cited in Halliday 1976). Myxospores can survive passage through the guts of birds (Taylor and Lott 1978) and fish and freezing at −20°C for 3 months (El-Matbouli and Hoffmann 1991). However, myxospores may possibly be deactivated when consumed by nonsusceptible tubificids (M. El-Matbouli, University of Munich, M. Gay, T. S. McDowell, M. P. Georgiadis and R. P. Hedrick, University of California-Davis, personal communication). In natural waters, *T. tubifex* is often found in association with other tubificid species (Brinkhurst 1970). In the Intermountain West, that association is often with *Limnodrilus hoffmeisteri*, *Rhyacodrilus* spp., and *Ilyodrilus templetoni* (Zendt and Bergersen 2000; B. L. Kerans, unpublished data). Whether other nonsusceptible tubificids or other organisms can deactivate myxospores in natural waters is unknown. However, *T. tubifex* can be highly infected (~30%) when it is only a small component of the worm assemblage and *Rhyacodrilus* spp. is the dominant tubificid (R. C. Krueger and B. L. Kerans, unpublished data).

No direct evidence exists concerning dispersal of myxospores after they are released from the fish host. Because of their small size, myxospores are probably dispersed passively by water currents, as are other fine sediments within streams. Consequently, they probably are deposited in backwaters and other areas where water velocities are low. In addition, their hard shell may allow them to survive dispersal out of the water.

Transmission of M. cerebralis to T. tubifex

The transmission parameter, β_m, quantifies the ability of myxospores to infect the worm host (i.e., host susceptibility). Tubificids come in contact with and ingest myxospores as they burrow in sediments, consuming particles and associated bacteria (Coler et al. 1967; Brinkhurst and Chua 1969; Wavre and Brinkhurst 1971). Because myxospores must come in contact with susceptible worms for the parasite to be transmitted, their spatial overlap is critical.

Currently, *Tubifex tubifex* is the only tubificid that is known to host *M. cerebralis* (Wolf et al. 1986; Hedrick et al. 1998). In the laboratory, experimentally exposed isolates of *L. hoffmeisteri* and *I. templetoni* did not produce triactinomyxons (B. L. Kerans, C. Rasmussen, R. Stevens, and J. R. Winton, unpublished data). However, geographic and genetic isolates of *T. tubifex* may have differential susceptibility to *M. cerebralis*. For example, a genetically distinct isolate of *T. tubifex* has been detected (Beauchamp et al. 2001; C. Rasmussen, Western Fisheries Research Center, personal communication) that appears resistant to *M. cerebralis* (R. Hedrick, personal communication). Other groups of *T. tubifex* can be distinguished genetically, and it is not clear at this time whether they are different species or subspecies (Beauchamp et al. 2001; C. Rasmussen, personal communication). Such discussions have long surrounded the ubiquitous *T. tubifex*, as other authors have found genetic isolates of the species that often have disparate environmental requirements (e.g., Anlauf 1994, 1997; Anlauf and Neumann 1997; Sturmbauer et al. 1999). Moreover, some of these genetically distinct groups have differential ability to produce triactinomyxons (Stevens et al. 2001); however, it is unclear whether the differential triactinmyxon production occurs because of differential susceptibilities or immune responses (i.e., resistance) of the variants (see "Myxobolus cerebralis in Worm Populations"). It is possible that variation in susceptibility of genetic isolates of the *T. tubifex* complex within natural streams may play a role in the severity of *M. cerebralis* infections within natural streams.

Although the specific form of β_m will depend on the model, a rough estimate could be found in experiments where a fixed number of *T. tubifex* were exposed to a fixed number of myxospores. The proportion of worms infected would indicate their susceptibility. In experiments where 100 worms were exposed to 350 myxospores per worm, the percentage of infected worms (as indicated by the number producing triactinomyxons) varied between 6% and 33% (T. Waldrop, V. Blazer, D. Smith, B. Schill, and C. Densmore, National Fish Health Research Laboratory, personal communication). The variation in percentage infected was related to temperature and sediment type (see "Myxobolus cerebralis in Worm Populations") and may have been a function of myxospore encounter rate. Similar studies, using constant doses of myxospores and comparing the proportion of infected worms across the genetic variants of *T. tubifex*, could aid in our understanding of their differential susceptibility.

In the natural environment, on large spatial scales, the overlap between myxospores and *T. tubifex* is probably patchy. *Tubifex tubifex* and some other tubificids such as *L. hoffmeisteri* are cosmopolitan organisms, occurring in the fine benthic sediments of freshwaters worldwide (Brinkhurst and Jamieson 1971; Brinkhurst 1996). However, *T. tubifex* is not ubiquitous—not all freshwater environments contain populations, and where the species occurs, densities can vary from less than 100 worms/m^2 to hundreds of thousands of worms/m^2 (e.g., >600,000/m^2; Sarkka 1987). Moreover, tubificids are normal components of natural freshwater ecosystems and contribute significantly to the mixing of sediments (Robbins et al. 1989), breakdown of coarse particulate organic matter (e.g., leaves, Chauvet et al. 1993), and the cycling of nutrients and organic matter (Chatarpaul et al. 1979; Robbins et al. 1989) in lentic and lotic systems. *Tubifex tubifex* appears to be common in the Intermountain West and has been found in streams where rainbow trout populations have experienced severe declines as a result of whirling disease (Zendt and Bergerson 2000; R. C. Krueger and B. L. Kerans, unpublished data), as well as those where they have not (B. L. Kerans, unpublished data).

Within infected streams, the spatial overlap between *T. tubifex* and myxospores is probably extensive. *Tubitex tubifex* is a habitat generalist (e.g., Juget and Lafont 1994) and extremely tolerant of a wide range of environmental conditions (e.g., Chapman et al. 1982). In streams, it can be found in silty, backwater areas, in pools, and associated with leaf litter and is less likely to be abundant in rocky areas with high water velocities in the main channel (Lazim and Learner 1987; Chauvet et al. 1993; Juget and Lafont 1994; Dumnicka 1994). *Tubitex tubifex* and other tubificids tend to be most abundant in the top 10 cm of substrate, although they are sometimes found deeper (Newrkla and Wijegoonawardana 1987; Lazim et al. 1989; Robbins et al. 1989). Often, high population densities occur under eutrophic conditions (e.g., Sarkka 1987). Zendt and Bergerson (2000) found the highest relative abundance of *T. tubifex* in areas where riparian zones were heavily disturbed. Myxospores are probably not very abundant in natural waters, but they are probably passively dispersed to areas of low water velocity and fine sediments—the best habitat for worms.

Myxobolus cerebralis in Worm Populations

The rate of increase of parasites within the worm host population will partially depend on the replication (v_W) and mortality (μ_W) rates of the parasite within individual worms, as well as the transmission parameter, β_m, and the abundances of worms (W) and myxospores (m) (e.g., $\beta_m mW$). Thus, the dynamics of the parasite within worms, the abundance of susceptible worms and myxospores, and the ability of myxospores to infect worms will all play a role in the persistence and increase of the parasite population in worms.

El-Matbouli and Hoffmann (1998) have documented the process of myxospore attachment, sporoplasm penetration, and subsequent parasitic development in *T. tubifex*. After ingestion by *T. tubifex*, the myxospore attaches to the gut epithelium by extruded polar filaments. The sporoplasm of the myxospore penetrates between the gut epithelial cells. The parasite undergoes schizogony (multiple mitotic divisions), gametogony (formation of haploid gametocytes), and sporogony (divisions that produce the triactinomyxon) all between the cells of the gut epithelium. Thus, *T. tubifex* is the definitive host for *M. cerebralis* because meiosis and gametic fusion occur within the worm host. Moreover, studies of nuclear DNA concentrations support this contention (El-Matbouli et al. 1998). The final developmental stage is a pansporocyst, which contains eight folded triactinomyxons. Interestingly, development is asynchronous, so that mature pansporocysts and immature stages are present simultaneously (El-Matbouli and Hoffmann 1998).

The parasite clearly proliferates within the worm; however, proliferation rate varies among worm variants. When 100 *T. tubifex* were challenged with 50 myxospores per worm in laboratory experiments, worms from the Mount Whitney Fish Hatchery in California produced 45 times more tri-

actinomyxons than myxospores received (Stevens et al. 2001). However, *T. tubifex* from the Madison River, Montana, only produced six times more triactinomyxons than myxospores at the same myxospore dose. Variation in proliferation among *T. tubifex* isolates may be produced by differential parasite replication (v_W) and mortality (μ_W) rates. In addition, the worm immune system may cause differential mortality in the various stages (i.e., schizogony, gametogony, sporogony). Invertebrates do have immune systems that frequently cause mortality of parasites (Bayne 1991), but little is known about how worm immune systems respond to *M. cerebralis*.

Water temperature influences the development time of the parasite to the triactinomyxon stage after penetration of the sporoplasm into the gut lining. *Tubitex tubifex* infected and reared at 8°C in the laboratory first released triactinomyxons at 170 d postexposure (1,360°C days; R. Stevens and B. L. Kerans, unpublished data), whereas those reared at 15°C first produced triactinomyxons after 89–90 d (1,320–1,335°C days; El-Matbouli and Hoffmann 1998; R. Stevens and B L. Kerans, unpublished data). Moreover, both water temperature and sediment type seem to influence prevalence of infection in *T. tubifex*. Infection prevalence was about 22.2% at 17°C, 14.6% at 13°C, and 11.4% at 9°C, in the laboratory (T. Waldrop, V. Blazer, D. Smith, B. Schill, and C. Densmore, personal communication). No infected worms were detected at 20°C. Although the mechanisms producing the differences in prevalence of infection are unknown, a combination of temperature-mediated functions—both development rate of the parasite within the host and worm activity rate, which affects the contact rate between tubificids and myxospores—between 9°C and 17°C, appears plausible.

Prevalence of *M. cerebralis* infection in worms in natural bodies of water is typically quite low, even in streams where whirling disease has caused significant mortality in rainbow trout. Two methods have been used to estimate these percentages. First, the polymerase chain reaction (PCR) developed by Andree et al. (1998) has been used to detect *M. cerebralis* DNA in worms. Using this method or modifications, the percentage of infected worms with morphology similar to *T. tubifex* is usually much less than 10% (Rognlie and Knapp 1998; Zendt and Bergersen 2000). The second method examines the proportion of field-collected worms of morphology similar to *T. tubifex* that release triactinomyxons.

Using this method, prevalence of infection in worms was estimated at 3.8% in the upper Colorado River in Colorado (K. A. Beauchamp, M. J. Gay, G. O. Kelley, R. P. Hedrick, University of California-Davis, M. El-Matbouli, University of Munich, R. D. Kathman, Aquatic Resources Center, and R. B. Nehring, Colorado Division of Wildlife, personal communication) and less than 10% in the upper Madison River in Montana (R. C. Krueger and B. L. Kerans, unpublished data).

Examining prevalence of M. cerebralis infection in T. tubifex collected from natural waters is problematic and by nature imprecise. Typically, only a small percentage of worms can be identified to species because species-level identification requires worms to be sexually mature, and mature worms are typically rare (Brinkhurst 1996; Kathman and Brinkhurst 1998). Because T. tubifex is often found in association with other worm species of similar external morphology (chaetal form), species can easily be confused when immature (Brinkhurst 1996). Moreover, as stated above, the percentage of worms releasing triactinomyxons is usually very low. This combination often results in a very low probability that an identified T. tubifex will be found to be infected. For example, Rognlie and Knapp (1998) examined the M. cerebralis infection rate in 704 worms from Willow Creek, Montana, where rainbow trout were highly infected (Baldwin et al. 2000). Eighteen worms were found to be infected using PCR; however, only two of the infected worms were positively identified as T. tubifex. One way to correct for this problem is to determine the relative abundances of T. tubifex and other morphologically similar species using the mature specimens present. Unidentified immatures are then assigned to species based on the relative abundances of the mature specimens. This method assumes similar age-specific mortality of the different worms. Such corrections can result in profound shifts in the estimate of the prevalence of infection in T. tubifex. For example, when the percentage of worms releasing triactinomyxons was determined for identified T. tubifex and all morphologically similar immatures, prevalence of infection was typically much less than 10% in Madison River side channels (R. C. Krueger and B. L. Kerans, unpublished data). When the percentage of T. tubifex releasing triactinomyxons was estimated by correcting for the relative abundance of other species within the assemblage, the prevalence of infection in T. tubifex in single collections rose to about 30% in some locations.

Worms may be infected with several myxozoans and, therefore, produce actinospores of more than one species that could be confused with M. cerebralis triactinomyxons. For example, worms collected from the upper Colorado River released four actinospore types (G. O. Kelley, M. Gay, T. S. McDowell, K. A. Beauchamp, K. B. Andree and R. P. Hedrick, University of California-Davis, M. El-Matbouli, and R. B. Nehring, personal communication). Separation of actinospores is based on subtle morphological differences (Lom et al. 1997), although PCR tests are available for corroborative identification of M. cerebralis triactinomyxons.

Correlations among the prevalence of infection in T. tubifex and abiotic and biotic characteristics in natural waters are just beginning to be examined. In a study of six side channels of the Madison River in Montana, the percentage of T. tubifex releasing triactinomyxons varied between less than 1% and 18% across sites. These percentages were positively correlated to the relative abundance of fine sediments and their content of organic material and negatively correlated to water velocity and temperature (R. C. Krueger and B. L. Kerans, unpublished data). Percentages of infected T. tubifex also varied seasonally, but unpredictably.

Effects of M. cerebralis Infections on Tubificid Populations

Harboring parasites can often cause negative effects to individuals (i.e., disease), which may cause changes in host population dynamics (Dobson and Hudson 1986; Minchella and Scott 1991). These negative effects, such as depressed individual growth, decreased reproductive rate, and mortality, are summarized by the parameter c_W, which quantifies how individual parasites depress the population growth of worms. Infected worms can live for almost 2 years in the laboratory (Gilbert and Granath 2001); thus, direct mortality associated with M. cerebralis infection appears to be low. However, the intestines of infected worms contain masses of M. cerebralis cells that cause compression and hypotrophy of epithelial cells (El-Matbouli and Hoffmann 1998), suggesting that the growth and reproduction of infected worms might be impaired. Infection by M. cerebralis caused reductions in population growth rates, individual growth, reproduction, and some mortality in T. tubifex in laboratory experiments (Stevens et al. 2001). Furthermore, worm population growth rates were negatively correlated to the total numbers of triactinomyxons produced, strongly suggesting that the level of parasite propagation directly influenced the success of the *T. tubifex* populations (B. L. Kerans, C. Rasmussen, R. Stevens, and J. R. Winton, unpublished data).

Production of Triactinomyxons

The triactinomyxon production parameter (λ_w) describes the number of triactinomyxons produced by a myxospore after it has successfully invaded the worm host. This parameter is a function of υ_w and μ_w. Individual triactinomyxons leave worms through the anus and are released directly into the water column (El-Matbouli and Hoffmann 1998). Worms also produce fecal packets in the laboratory, each containing many triactinomyxons (Gilbert and Granath 2001), but the normalcy and significance of such packets are unknown.

Only indirect information is available about the number of triactinomyxons that an individual myxospore can produce. When 100 *T. tubifex* from the Mount Whitney Fish Hatchery in California were exposed to 5,000; 50,000; and 100,000 myxospores, they produced on average 228,333; 383,417; and 233,250 triactinomyxons, respectively (Stevens et al. 2001). Thus, if we assume that all myxospores were encountered by worms and successfully invaded worms, each myxospore produced 45, 8, and 2 triactinomyxons, respectively. Number of triactinomyxons produced was not related to the abundances of myxospores (at least between 50 and 1,000 per worm). Such a relationship is to be expected if individual parasites are able to proliferate until the capacity of individual worms is attained, suggesting that the proliferation rate of *M. cerebralis* in worms may be related to the capacity of the individual worms to support parasites and not to any innate capacity of the parasite to replicate. Moreover, threshold myxospore doses required to infect this variant of *T. tubifex* are clearly lower than 50 myxospores per worm. In a recently completed experiment, the percentage of infected Mount Whitney *T. tubifex* was low at doses of one and 5 myxospores per worm, intermediate at 10 myxospores per worm, and 100% when myxospore doses were 50 per worm or greater (J. Zickovich, L. Steinbach, B. L. Kerans, and E. MacConnell, unpublished data).

Environmental conditions affect production of triactinomyxons in the laboratory. Worms exposed to myxospores and held at different temperatures produced more triactinomyxons at 13°C than at 9°C and 17°C in mud or sand (T. Waldrop, V. Blazer, D. Smith, B. Schill, and C. Densmore, personal communication). Worms in leaf litter produced fewer triactinomyxons than those in mud or sand, at all temperatures. No triactinomyxons were produced when worms were held at 20°C. Moreover, duration of triactinomyxon release can be abbreviated by increasing temperature during only the release period (El-Matbouli et al. 1999). The variation in production of triactinomyxons across temperatures suggests that an optimum temperature might exist for triactintomyxon production; however, whether the optimum temperature varies among *T. tubifex* variants is unknown.

Worms do not die when triactinomyxons are released (El-Matbouli and Hoffmann 1998; Gilbert and Granath 2001). Worms exposed to myxospores only once in the laboratory released triactinomyxons twice over a 20-month period, with individual worms releasing similar numbers of triactinomyxons during the first and second release periods (Gilbert and Granath 2001). In addition, all variants of *T. tubifex* are not equal triactinomyxon producers. Geographic and genetic isolates of morphologically-identified *T. tubifex* exhibited differences in triactinomyxon production that varied over several orders of magnitude when challenged with similar numbers of myxospores (Stevens et al. 2001; B. L. Kerans, C. Rasmussen, R. Stevens, and J. R. Winton, unpublished data).

Triactinomyxons in the Environment

Abundance of triactinomyxons in the environment (t) is a function of the production of triactinomyxons by myxospores in worms (λ_w), the number of parasites in worms (P_w), and the deactivation rate of triactinomyxons in the environment (η_t, or alternatively, survival rate). After leaving its worm host, the triactinomyxon is presumed to drift passively in the water column until it comes in contact with a fish.

Detection and enumeration of triactinomyxons in streams is hampered by their apparent low densities and the lack of good quantification techniques. Triactinomyxons were detected by concentrating them using 20-μm mesh screen in the upper Colorado River drainage, where rainbow trout populations have declined (Thompson and Nehring 2000). Densities were greatest in summer and early autumn and lowest during spring. However, densities were highly variable and low (maximum density about 21 triactinomyxons/L). Polymerase chain reaction (PCR) verification of field samples yielded 31% false negative and 9% false positive error rates for presence/absence (Thompson and Nehring 2000). B. Allen and E. P. Bergerson (Colorado Cooperative Fish and Wildlife Research Unit, personal communication) also found low abundances of triactinomyxons in the Cache La Poudre River using a similar filtration technique, which they concluded was not sensitive for detecting low densities of triactinomyxons. Screen filtration efficacy tests, using known numbers of triactinomyxons, determined the technique is imprecise and inaccurate (F. T. Barrows, Bozeman Fish Technology Center, personal communication); some triactinomyxons became entangled in the mesh and could not be rinsed off, while others passed through the mesh. Ongoing trials at the Bozeman Fish Technology Center show packed-bed filtration provides accurate and precise estimates of triactinomyxon densities in the laboratory and field (F. T. Barrows, personal communication).

Comparative *in situ* exposure studies of sentinel trout in infected streams provide an indirect measure of the number of triactinomyxons in the environment. They show high levels of temporal and spatial variability in infection rates and severities in streams in Oregon, Idaho, Montana, and Colorado (Thompson et al. 1999; Vincent 2000; Sandell et al. 2001; Hiner and Moffitt 2001; D. R. Burton, C. Hogge, and K. Johnson, Idaho Department of Fish and Game, personal communication; R. B. Nehring, K. G. Thompson, T. Wygant, Colorado Divison of Wildlife, and D. C. Bowden, Colorado State University, personal communication). However, patterns are emerging where streams have been monitored for several years at several locations. For example, maximum infections of sentinel trout occurred from May through July at the same locations in the Madison River in Montana over several years (Vincent 2000).

The temporal and spatial variability in infection prevalence and disease severity in sentinel fish has led to investigations of abiotic and biotic factors that drive the patterns. Richard Vincent (Montana Fish, Wildlife and Parks, personal communication) has hypothesized that water temperature drives the seasonal patterns seen on the Madison River. In Willow Creek, Montana, the percentage of sentinel rainbow trout infected increased linearly with increasing water temperatures between 6°C and 12°C (Baldwin et al. 2000).

Correlates to spatial variation in prevalence and disease severity in sentinel salmonids are also being examined. Conductivity was the only water quality parameter that correlated to disease severity in the Lostine River in Oregon (Sandell et al. 2001). Disease severity in sentinel salmonids was positively correlated to the abundance of oligochaetes and chironomids in the South Fork of the Boise River in Idaho (Hiner and Moffitt 2001). The disease severities in sentinel fish in side channels of the Madison River were positively correlated to the percentages of fine sediments and the abundances of *T. tubifex* producing triactinomyxons and negatively correlated to water temperatures and velocities (R. C. Krueger and B. L. Kerans, unpublished data).

After entering the water column, triactinomyxons survive only for a short duration, relative to the length of time myxospores can survive (i.e., hours or days versus months or years; Markiw 1992b). Several environmental variables influence their survival. Triactinomyxon survival at 15 d declined from 80% at 6°C to 50% at 20°C (El-Matbouli et al. 1999). Triactinomyxons were killed by desiccation (room temperature for 1 h) and freezing (−20°C; E. J. Wagner, M. Smith, C. Wilson, Utah Division of Wildlife Resources, and D. Roberts, J. Austin, Utah State University, personal communication). Salinities greater than 20‰ killed 80% of triactinomyxons within 1 h. Fewer triactinomyxons died within 24 h at salinities between 5 and 15‰ (M. A. Smith, A. Howa, E. J. Wagner, Utah Division of Wildlife Resources, personal communication). Predation by ostracods, amphipods, and cladocerans reduced triactinomyxon numbers in laboratory cultures (R. Arndt and E. J. Wagner, Utah Divison of Wildlife Resources, personal communication; D. Gustafson, Montana State University, personal communication).

How far triactinomyxons disperse is not well known. They are neutrally buoyant and can remain suspended in the water column for more than 24 h (E. J. Wagner, R. Arndt, and M. Smith, personal communication). Given their small size and buoyancy, one would expect triactinomyxons to be transported at the rate of water velocity or just slightly slower; thus, in 24 h, they could be many kilometers downstream. Triactinomyxon densities declined rapidly over short distances below a fish-rearing unit on the Cache la Poudre River, Colorado, suggesting that triactinomyxons may not disperse far from their source (B. Allen and E. P. Bergersen, personal communication). Indirect evidence also suggests that triactinomyxons do not disperse far. Infection severity of sentinel rainbow trout can change appreciably within a few kilometers, suggesting that triactinomyxons are removed from the water column or deactivated quickly.

Transmission of *M. cerebralis* to Salmonids

The parameter, β_t, quantifies the ability of triactinomyxons to infect the fish host (i.e., host susceptibility). Salmonids come into contact with triactinomyxons floating in the water column. Consequently, spatial and temporal overlap between susceptible fish and triactinomyxons is critical.

Many species of salmonids can be infected by *M. cerebralis* in the Intermountain West, including brown trout, rainbow trout, cutthroat trout *O. clarki*, brook trout *Salvelinus fontinalis*, bull trout *S. confluentus*, and mountain whitefish *Prosopium williamson* (Hedrick et al. 1998, 1999a, 1999b). Moreover, in other regions, anadromous coho salmon *O. kisutch*, chinook salmon *O. tshawytscha*, steelhead *O. mykiss* (Hedrick et al. 2001) and Atlantic salmon (V. Blazer, personal communication) are also susceptible. Only Arctic grayling *Thymallus arcticus* (Hedrick et al. 1999b) and lake trout *S. namaycush* (V. Blazer, personal communication) are invulnerable. Field exposures of salmonids typically give similar results, although the percentage of infected brown trout is often lower than that of rainbow trout (Thompson et al. 1999; Baldwin et al. 2000). These results suggest that almost all salmonids (and all ages and sizes of rainbow trout; Markiw 1992a; E. K. N. Ryce, E. MacConnell and A. V. Zale, unpublished data) provide appropriate habitat for *M. cerebralis*, although parasite proliferation after infection and, thus, host resistance to disease, may differ (see "*Myxobolus cerebralis* in Fish Populations"). Distantly related salmonids such as Artic grayling are apparently so dissimilar to *Salmo* that they do not provide good habitat.

Because triactinomyxons are short lived, both spatial and temporal overlap between susceptible salmonids and the triactinomyxon transmission stage are critical to the persistence of the parasite in natural systems. Close proximity of both hosts, when triactinomyxons are released, increases probability of transmission from the worm to the fish and reduces duration that the fragile triactinomyxon is drifting and vulnerable to damage, predation, and senescence. Young salmonids typically inhabit shallow, low-velocity, lateral habitats, in spring and summer. These are also the habitats where *T. tubifex* densities are highest (B. L. Kerans, unpublished data) and where myxospore-laden carcasses

of adult fish that died after spawning are likely to be deposited. Targeting juvenile salmonids is adaptive because they are much more abundant than older cohorts, they have a high cartilage:bone ratio, and they possess little innate resistance and no acquired immunity (E. K. N. Ryce, A. V. Zale, and E. MacConnell, unpublished data). Spawning and rearing areas of rainbow trout in the Madison River, Montana, overlap with some of the locations where sentinel fish exhibit the most severe infections (Downing 2000) and where densities of *T. tubifex* releasing triactinomyxons are high (R. C. Krueger and B. L. Kerans, unpublished data).

Temporal overlap between susceptible salmonids and triactinomyxons also appears to be significant. Infection rates of salmonids in natural waters of the Intermountain West appear to peak in spring and summer (Baldwin et al. 2000; E. R. Vincent, personal communication). This time likely corresponds to the time in northern Europe when age-0 *Salmo* are abundant in nursery habitats, possess maximum amounts of cartilage to support *M. cerebralis* trophozoites, and are large enough and sufficiently developed to be able to avoid becoming impaired by disease. Because brown trout can become diseased if exposed to large numbers of triactinomyxons shortly after emergence (Hedrick et al. 1999a) and would, therefore, likely die in natural waters before maturation of myxospores was achieved, later release of triactinomyxons would tend to increase reproductive success of *M. cerebralis*. Unfortunately, this timing appears to be especially detrimental to spring spawning rainbow and especially cutthroat trout, which are much younger and less developed at this time and, therefore, suffer high mortality.

Such variability may offer possible strategies to ameliorate the effects of the parasite on rainbow trout populations, by reducing the probability of contact between susceptible salmonids and triactinomyxons, in either time or space. The consistency of the temporal patterns of disease severity in sentinel rainbow trout from year to year on the Madison River has led researchers to suggest that an earlier spawning and rearing strain of rainbow trout might be able to avoid the triactinomyxon stage, at least in part, by emerging earlier (D. C. Downing, T. E. McMahon, B. L. Kerans, and E. R. Vincent, unpublished data). Physical conditions that vary from year to year may also ameliorate the effects of the parasite. For example, the abundances of age-1 Madison River rainbow trout in 1999 and 2000 were similar to abundances before whirling disease was detected (Clancey and Downing 2001). This recruitment surge may be related to unusually high flows during the time when young rainbow trout were rearing during the previous year (E. R. Vincent, personal communication), or it could be unrelated to any physical features—resulting from only dynamics of the host and parasite interaction. High flows might destroy triactinomyxons or dilute their numbers, thereby reducing contact between triactinomyxons and susceptible salmonids. Moreover, consistency in the spatial patterns of disease severity in sentinel rainbow trout also suggests that some areas of the Madison River would be better rearing areas for small, susceptible fish. In fact, even during the years of highest rainbow trout population declines, some trout survived, perhaps because they reared in areas where triactinomyxon densities were low.

Myxobolus cerebralis in Fish Populations

The rate of increase of parasites within the fish host population will partially depend on the replication (v_F) and mortality (μ_F) rates of the parasite within individual fish, the susceptibility of individual fish (β_t), and the abundances of fish (F) and triactinomyxons (t) (e.g., $\beta_t tF$). Thus, the dynamics of the parasite within fish, the abundances of susceptible fish and triactinomyxons, and the ability of triactinomyxons to infect fish will all play a role in the persistence and increase of the parasite population in salmonids.

When a triactinomyxon encounters a fish, it releases its sporoplasm into the epidermis of the skin, fins, gills, or buccal cavity (El-Matbouli et al. 1995). The parasite migrates from the epidermis through nervous tissue to the cartilage. During this migration, M. cerebralis is in presporogonic stages wherein proliferation takes place. After a period of continued presporogonic development in the cartilage, sporogony begins with multiple mitotic divisions and differentiation. The final developmental stage is the elliptical myxospore, which is covered by two hard valves.

The parasite proliferates (v_F) within the fish host, but susceptible salmonid species have differential immune responses to M. cerebralis, as judged by myxospore production. For example, in laboratory studies where salmonids were exposed to similar numbers of triactinomyxons, rainbow trout produced many more myxospores than brown trout (Hedrick et al. 1999a), bull trout (Hedrick et al. 1999b), and coho salmon (Hedrick et al. 2001).

Myxospore numbers in westslope and Yellowstone cutthroat trout and chinook salmon were similar to rainbow trout (Hedrick et al. 1999b, 2001). Although they are not well understood, host cellular or humoral responses, or both, are activated when the sporoplasms migrate from the epidermis to the nervous tissue, and presumably, those cells that do not enter the nervous tissue are destroyed (μ_F; Hedrick et al. 1998). Moreover, the host's cellular response is stimulated during the active phase of cartilage destruction, and lesions form in affected areas (Hedrick et al. 1998). Although not all aspects of the differential immune responses of salmonids are understood, these results suggest that the immune response of brown trout, which co-evolved with M. cerebralis, is stronger than that of many other species. Perhaps the immune responses of bull trout and coho salmon were developed in response to other myxosporeans.

Development of the parasite to the myxospore stage after exposure to the triactinomyxon stage takes about 90 d at 16°C (El-Matbouli et al. 1995; Hedrick et al. 1998) and is temperature-dependent. Development of the first recognizable myxospores, though perhaps immature, occurred at 120, 101, and 52 d at 7°C, 12°C, and 17°C, respectively (Halliday 1973).

The prevalence of M. cerebralis in salmonids seems to be patchy across streams in the Intermountain West. Salmonids infected with M. cerebralis were found in 230 water bodies in 21 of the 22 major drainages in Montana (Baldwin et al. 1998). Prevalence of infection varied from 0% to more than 50%, among drainages and species of trout. Variation in prevalence among drainages may result from the time the parasite has been in the various drainages or perhaps from the reduced suitability of some streams for the persistence of the parasite. Determination of the prevalence of infection using wild caught fish can be problematic. For example, across drainages in Montana, prevalence of infection was higher in brook trout and brown trout than in rainbow trout (Baldwin et al. 1998). This pattern may result from sampling bias toward survivors, as rainbow trout are more likely to die from the disease than the other species (Baldwin et al. 1998).

Few studies have examined the physical conditions that relate to infection prevalence in wild salmonids. Temperature did not predict prevalence of infection in the Salt River drainage in Wyoming in 1997, but percentage of fine sediments and distance to mainstem did (D. J. Isaak and W. A.

Hubert, Wyoming Cooperative Fish and Wildlife Research Unit, personal communication). They inferred that M. cerebralis may have been in the early stages of invasion in this drainage.

Effects of M. cerebralis Infections on Salmonid Populations

The negative effects of whirling disease on individual salmonids, including "tail-chasing" behavior (from which the disease gets its common name), skeletal deformities, blackened tails, and death, have been widely documented (e.g., Hedrick et al. 1998). These negative effects could result in depressed individual growth, reproduction, and mortality and are summarized by the parameter c_F.

The pathology of whirling disease in rainbow trout is well known. Briefly, trophozoites (i.e., feeding stages) of M. cerebralis feed upon cartilage and destroy the structural framework of the skeleton, often leaving fish permanently disfigured (Hedrick et al. 1998). Recently, Rose et al. (2000) showed that M. cerebralis causes spinal cord constriction and brain stem compression, which, they argue, cause the whirling behavior and periods of immobility.

Negative effects to individual fish in the laboratory depend on the species of salmonid (Hedrick et al. 1998), the dose of triactinomyxons the fish received (Hedrick et al. 1998; E. K. N Ryce, E. Mac-Connell, and A. V. Zale, unpublished data), and the age and size of the fish (E. K. N. Ryce, E. Mac-Connell, and A. V. Zale, unpublished data). Brown trout show fewer clinical signs and much less severe lesions than rainbow trout in comparative laboratory trials (Hedrick et al. 1999a) and in field exposures of sentinel fishes (Thompson et al. 1999; Baldwin et al. 2000). Thus, whirling disease is not as severe in brown trout, which is thought to have co-evolved with M. cerebralis, as it is in rainbow trout. Lesions decreased in severity for other salmonids in the following order: chinook salmon (similar to rainbow trout), westslope and Yellowstone cutthroat trout, coho salmon, and bull trout (Hedrick et al. 1999b, 2001). Both infected chinook salmon and rainbow trout weighed less than their uninfected counterparts (Hedrick et al. 2001). For both rainbow and brown trout, disease severity (as measured by number and severity of lesions) was positively correlated to triactinomyxon dose (Hedrick et al. 1999a). Moreover, severity of whirling disease pathology is dependent upon the age of rainbow trout when first exposed to the triactinomyxon stage; effects of the disease are substantially reduced when first exposure

occurs after 9 weeks post hatch (E. K. N. Ryce, E. MacConnell, A. V. Zale, unpublished data).

In the wild, these individual effects have translated into population-level effects, in some cases. Rainbow trout abundances have declined in the Madison River, Montana, and the upper Colorado River, Colorado (Nehring and Walker 1996; Vincent 1996). In addition, brown trout populations have been little affected on the Madison River (Vincent 1996); however, they have declined in the upper Colorado River (Nehring and Walker 1996). Modin (1998) inferred little effect on rainbow trout in California streams. Thus, detection of the effects of M. cerebralis has been variable, which perhaps has more to do with lack of good pre-whirling disease population data than with lack of effects of the parasite.

Production of Myxospores

The myxospore production parameter (λ_F) describes the rate at which a triactinomyxon produces myxospores once it is able to successfully invade the fish host. This parameter will be a function of v_F and μ_F. Myxospores are released into the environment with the death of infected salmonids (Hedrick et al. 1998).

The decomposition of one rainbow trout carcass can release more than a million myxospores (Hedrick et al. 1999a) because M. cerebralis has a high replication rate in rainbow trout. When rainbow trout were exposed to a dose of 10 triactinomyxons/fish, the number of myxospores per head averaged between 28,000 and 78,000. Thus, if we assume that all triactinomyxons produced myxospores, each triactinomyxon produced between 2,800 and 7,800 myxospores in only the heads. Production of myxospores in rainbow trout appears to be dose-related (Hedrick et al. 1999a). Perhaps an upper limit to myxospore production occurs; however, it appears that the dose of triactinomyxons producing such a limit has not been studied.

Hedrick et al. (1999a) showed that brown trout produced fewer myxospores at a given triactinomyxon dose than rainbow trout, but nevertheless, they produced several thousand myxospores. Other species of salmonids also produced thousands of myxospores (Hedrick et al. 1999b). Moreover, infected salmonids may release myxospores before death (R. B. Nehring, K. A. Taurman, and D. L. Shuler, Colorado Division of Wildlife, and K. G. Thompson, Colorado Cooperative Fish and Wildlife Research Unit, personal communication). Such "live" release of spores is common among myxozoans that parasitize gill filaments (Eszterbauer et al. 2000). Thus, "live" release of M. cerebralis myxospores deserves further study.

Because myxospores are presumably carried by fish until their death, any factor which affects the survival of fish will influence myxospore deposition. Infection reduces the stamina of rainbow trout (E. K. N. Ryce, E. MacConnell, and A. V. Zale, unpublished data) and, because salmonids are under stress when spawning, it is possible that infection by M. cerebralis enhances the already increased mortality rate of salmonids following spawning. Myxospores would thereby be released near salmonid spawning and rearing areas. Dead fish carcasses attract decomposing organisms (Wipfli et al. 1998), which probably promote the release of myxospores into the environment (i.e., accelerate decomposition of bone). The worms eat myxospores, become parasitized, and release triactinomyxons the following spring when young juvenile fish are abundant. Death of infected juvenile fish could also release myxospores; however, if death occurs before the parasite has completed development, or in an environment without the worm hosts, these parasites will be lost from the system.

Transportation of fish between drainages and movement of fish within infected drainages probably play important roles in dispersal of the parasite. Myxospores are especially prone to interdrainage dispersal by humans. The original invasion to North America was caused by importation of frozen rainbow trout from Europe (Hoffman 1970, 1990). Stocking infected fish in reservoirs may increase myxospore burden and subsequent triactinomyxon release (R. B. Nehring and K. G. Thompson, personal communication). Little is known about how myxospores are spread within drainages, but infected fish probably distribute the parasite. Many salmonids are highly mobile, especially anadromous, fluvial, and adfluvial forms.

COMMUNITY ECOLOGY

Myxobolus cerebralis and its hosts are embedded in aquatic communities consisting of many other organisms that may interact with the parasite and its fish and worm hosts. The life cycle of M. cerebralis in only one salmonid host and one genetic variant of T. tubifex is depicted in Figure 1. To reflect the fact that M. cerebralis is embedded in the natural environment with many susceptible salmonid species and

genetic variants of T. tubifex, Figure 1 could be modified by adding vertical stacks of different salmonids and genetic variants of T. tubifex, each with differential parasite proliferation rates. Because M. cerebralis has little genetic diversity (Andree et al. 1999), these multiple stacks of salmonid and T. tubifex hosts contribute to only one pool of myxospores and triactinomyxons within the environment. Many salmonids that are "acceptable" habitat for M. cerebralis can be sites for parasite proliferation and sources of myxospores in infected drainages. For example, mountain whitefish are plentiful in the Madison River, yet we know little about their population dynamics and whether they make a significant contribution to the myxospore burden and, consequently, to the disease in wild rainbow trout. If other susceptible salmonids make significant contributions to the myxospore burden in infected drainages, then information on their ecology would be needed to provide insights and information useful for ameliorating whirling disease in rainbow trout.

The negative effects of M. cerebralis on salmonid and worm hosts have the potential to indirectly affect aquatic community structure by altering the outcome of interactions among the hosts and other organisms (Minchella and Scott 1991). For example, reduction in rainbow trout abundances in areas where M. cerebralis infections occur could open the community to population increases of other fish species that are more resistant to infection and compete with rainbow trout for food and space. Complex interactions could also occur within the worm assemblage because T. tubifex is negatively affected by M. cerebralis infections. Associations between T. tubifex and other tubificid species appear to be mutualistic (Brinkhurst and Kennedy 1965; Brinkhurst et al. 1972; Chua and Brinkhurst 1973; Milbrink 1993). We found that the negative effects of the presence of myxospores on T. tubifex were negated if T. tubifex was grown in culture with L. hoffmeisteri (B. L. Kerans and J. C. Lemmon, unpublished data), presumably because L. hoffmeisteri consumed myxospores, which were then unavailable to T. tubifex. Although it is unknown whether such interactions occur in the natural environment, these results suggest that the interactions among M. cerebralis, T. tubifex, and other tubificids are complex. Examining interactions among salmonids, tubificids, and M. cerebralis will provide information about how M. cerebralis affects communities and ecosystem structure and function.

WHAT DO WE KNOW AND WHAT INFORMATION IS NEEDED?

Our conceptual modeling exercise clarifies those aspects of the life cycle of M. *cerebralis* that we understand and can quantify, as well as those that we do not. For example, we know that myxospores are highly resistant to the range of physical features that commonly occur in streams in the Intermountain West and can remain viable for long periods when outside of the fish host. Triactinomyxons are much more delicate than myxospores, can be deactivated rather easily by chemical, physical, and biotic processes, and survive only a short time under natural field conditions. In contrast, we have only indirect information on myxospore (worm exposure to natural sediments) and triactinomyxon (sentinel fish studies) distributions and abundances in nature. This information suggests that distributions are spatially variable and abundances are low, even within systems where whirling disease has caused population declines of rainbow trout. More detailed information on distribution, abundance, dispersal, and the abiotic and biotic processes that affect myxospores and triactinomyxons in natural systems is needed. Development of new quantification techniques that determine myxospore and triactinomyxon abundances, with the accuracy and precision needed for measurements in natural waters, will aid these efforts.

Tubifex tubifex is likely the only oligochaete host species of M. *cerebralis*, although geographic and genetic variants of T. *tubifex* are either differentially susceptible, resistant, or both. In contrast, many salmonid species in the Intermountain West can be infected by M. *cerebralis*. Further studies of comparative susceptibilities of genetic variants of T. *tubifex* and salmonid species may aid in determining the evolutionary pathways of M. *cerebralis* and other myxozoans. Fish of any age and size can be infected, but whether developmental stages of worms are differentially susceptible to M. *cerebralis* is unknown. In natural waters, the distributions of myxospores, T. *tubifex*, triactinomyxons, and susceptible salmonids and their overlap in space and time are just beginning to be examined, and some patterns are emerging. Further examinations of the seasonal and spatial patterns of triactinomyxon release and its overlap with susceptible salmonids and genetic variants of T. *tubifex* in infected systems are needed to determine the generality of these patterns. Moreover, we do not know how the presence of different genetic variants of T. *tubifex* within M. *cerebralis*-infected streams relates to triactinomyxon production and disease in wild salmonids.

Attachment to, and development of, the parasite within the worm and fish hosts is well documented. Temperature affects these processes. Proliferation rates of M. *cerebralis* vary among geographic and genetic isolates of T. *tubifex* and among different salmonid species. The mechanisms producing such variation in proliferation, including how the immune systems of worms and fish respond to infection, are not well understood. Examination of the mechanisms producing differential proliferation among worms and salmonid species may provide insights into the evolution of the parasite and potential control mechanisms. Prevalence of infection in worms seems to be low, even in areas where whirling disease has devastated rainbow trout populations; however, prevalence of infection in T. *tubifex* can be high when it is a small component of the worm assemblage. Typically, prevalence of infection in salmonids is higher than prevalence of infection in worms. Determining the prevalences of infection in worm and fish hosts in natural waters poses difficult technical problems, and these proportions should be treated with caution. More research is needed on the prevalence of infection in salmonids and T. *tubifex* and explanatory environmental correlates.

Myxobolus cerebralis infections negatively affect individual salmonids, worms, and their populations. The pathology of whirling disease in salmonids is well known, but effects of M. *cerebralis* on worms are just beginning to be examined. The negative effects are positively correlated to the number of infective stages (triactinomyxons and myxospores) produced in both worms and fish. Experimental evidence suggests that brown trout are more resistant to disease than rainbow trout. The resistance of other salmonids varies. Moreover, larger, older rainbow trout are also more resistant to the disease than smaller, younger fish. The effects of M. *cerebralis* on individual rainbow trout have been detected at the population level in at least two natural systems where pre-whirling disease data were good. In other systems, where data are less complete, the effects of the disease on fish are more equivocal. Information is needed on the effects of whirling disease on other salmonid populations. How the negative effects of M. *cerebralis* on worms that have been shown in the laboratory translate to assemblages in natural waters is unknown. The low prevalence of infection that often occurs in worms might suggest that these effects are slight in natural waters; however, because *T. tubifex* can sometimes be highly infected, such effects may indeed be important under some circumstances.

Myxobolus cerebralis proliferates in both the worm and salmonid hosts; however, the number of myxospores produced by an individual triactinomyxon that has successfully invaded a fish host and the number of triactinomyxons produced by an individual myxospore that has successfully invaded a worm host have only been deduced indirectly. Myxospore production is positively correlated to the doses of triactinomyxons with which fish are challenged, but preliminary evidence suggests that *M. cerebralis* proliferation within the worm host is limited by host capacity. To date, the "threshold" myxospore dose, required to initiate infection, has been determined for only one of the geographic variants of *T. tubifex*. Experiments that directly examine the ability of individual myxospores to produce infection in individual worms of different genetic or geographic variants have not been done and may clarify the factors that are related to proliferation. Individual *T. tubifex* can produce several thousand triactinomyxons and can produce at least twice during their lifetime as the result of one infection, and production is temperature dependent. Individual salmonids can produce millions of myxospores, which are thought to be released with death of the fish, although the possibility of release of myxospores before death warrants further examination. Clearly, we have learned much but have much more to learn.

REMEDYING THE INFORMATION DEFICIENCIES

The information deficiencies we have identified in this review can be remedied through rigorous and integrated laboratory and field research. Great progress has been made in recent years in the laboratory. On the other hand, field investigations prerequisite to design and execution of rational management and control programs have been hampered by the complex life cycle of *M. cerebralis*, its obligate and optional interactions with numerous species, and complexities associated with its recent introduction to North America. Logistical constraints such as the lack of appropriate instrumentation, inadequate or absent methods testing, and limited funding opportunities have also impeded progress. Despite these constraints, much excellent and informative field research has been conducted. Conversely, some field investigations have suffered from poor sampling

designs, lack of clearly defined hypotheses, use of untested methodologies and analytical techniques, and infidelity to rigorous protocols, thereby yielding ambiguous findings. At this juncture, it is clear that whirling disease in the Intermountain West is a much more complicated and difficult problem than we, at first, thought. The quick and simplistic studies that we designed before we knew better, typically modeled after investigations of bacterial and viral diseases in domestic animals or routine fish-population monitoring programs, were destined to provide equivocal and unreliable conclusions. We now need to recognize fully the complexity of the problem confronting us and respond to it through the conduct of exact, rigorous, and defensible field science coupled with integrated laboratory experiments.

ACKNOWLEDGMENTS

We thank T. McMahon, C. Rasmussen, L. Steinbach, J. Bartholomew, and two anonymous reviewers for their comments on the manuscript, and R. Stevens and E. K. N. Ryce for photographs. We thank all of our colleagues who confirmed the many personal communications. The Montana Cooperative Fishery Research Unit is jointly sponsored by the U.S. Geological Survey, Montana State University, and the Montana Department of Fish, Wildlife and Parks. The heavy pall of the terrorist attacks in New York, Virginia, and Pennsylvania shrouded the world as we completed revisions on this article in September 2001. May the memory of the victims of this profound tragedy persist eternally.

REFERENCES

Alvarez-Pettitero, P., and A. Sitja-Bobadilla. 1993. Pathology of Myxosporea in marine fish culture. Diseases of Aquatic Organisms 17:229–238.

Anderson, C. M., E. U. Canning, and B. Okamura. 1998. A triploblast origin for the Myxozoa? Nature (London) 392:346–347.

Anderson, R. M. 1994. Mathematical studies of parasitic infection and immunity. Science 264:1884–1886.

Anderson, R. M., and R. M. May. 1979. Population biology of infectious diseases: Part I. Nature (London) 280:361–367.

Anderson, R. M., and R. M. May. 1986. The invasion, persistence and spread of infectious diseases within animal and plant communities. Philosophical Transactions of the Royal Society of London B 314:533–568.

Anderson, R. M., and R. M. May. 1991. Infectious diseases of humans. Oxford University Press, Oxford.

Andree, K. B., S. J. Gresoviac, and R. P. Hedrick. 1997. Small subunit ribosomal RNA sequences unite alternate actinosporean and myxosporean stages of *Myxobolus cerebralis* the causative agent of whirling disease in salmonid fish. Journal of Eukaryotic Microbiology 44:208–215.

Andree, K. B., E. M. MacConnell, and R. P. Hedrick. 1998. A nested polymerase chain reaction for the detection of *Myxobolus cerebralis* in rainbow trout. Diseases of Aquatic Organisms 34:145–154.

Andree, K. B., M. El-Matbouli, R. W. Hoffman, and R. P. Hedrick. 1999. Comparison of 18S and ITS-1 rDNA sequences of selected geographic isolates of *Myxobolus cerebralis*. International Journal for Parasitology 29:771–775.

Anlauf, A. 1994. Some characteristics of genetic variants of *Tubifex tubifex* (Müller, 1774) (Oligochaeta: Tubificidae) in laboratory cultures. Hydrobiologia 278:1–6.

Anlauf, A. 1997. Enzyme variability of *Tubifex tubifex* (Müller) (Oligochaeta: Tubificidae) and seven other tubificid species. Archiv für Hydrobiologie 139:83–100.

Anlauf, A., and D. Neumann. 1997. The genetic variability of *Tubifex tubifex* (Müller) in 20 populations and its relation to habitat type. Archiv für Hydrobiologie 139:145–162.

Baldwin, T. J., J. E. Peterson, G. C. McGhee, K. D. Staigmiller, E. S. Motteram, C. C. Downs, and D. R. Stanek. 1998. Distribution of *Myxobolus cerebralis* in salmonid fishes in Montana. Journal of Aquatic Animal Health 10:361–371.

Baldwin, T. J., E. R. Vincent, R. M. Silflow, and D. Stanek. 2000. *Myxobolus cerebralis* infection in rainbow trout (*Oncorhynchus mykiss*) and brown trout (*Salmo trutta*) exposed under natural stream conditions. Journal of Veterinary Diagnostic Investigation 12:312–321.

Bayne, C. J. 1991. Invertebrate host immune mechanisms and parasite escapes. Pages 299–315 *in* C. A. Toft, A. Aeschlimann, and L. Bolis, editors. Parasite-host associations. Coexistence or conflict? Oxford University Press, Oxford.

Beauchamp, K. A., R. D. Kathman, T. S. McDowell, and R. P. Hedrick. 2001. Molecular phylogeny of tubificid oligochaetes with special emphasis on *Tubifex tubifex* (Tubificidae). Molecular Phylogenetics and Evolution 19:216–224.

Brinkhurst, R. O. 1970. Distribution and abundance of tubificid (Oligochaeta) species in Toronto Harbour, Lake Ontario. Journal of the Fisheries Research Board of Canada 27:1961–1969.

Brinkhurst, R. O. 1996. On the role of tubificid oligochaetes in relation to fish disease with special reference to the Myxozoa. Annual Review of Fish Diseases 6:29–40.

Brinkhurst, R. O., and K. E. Chua. 1969. Evidence of interspecific interactions in the respiration of tubificid oligochaetes. Journal of the Fisheries Research Board of Canada 26:2659–2668.

Brinkhurst, R. O., K. E. Chua, and N. K. Kaushick. 1972. Interspecific interactions and selective feeding of tubificid oligochates. Limnology and Oceanography 17:122–133.

Brinkhurst, R. O., and C. R. Kennedy. 1965. Studies on the biology of Tubificidae (Annelida: Oligochaeta) in a polluted stream. Journal of Animal Ecology 34:429–443.

Brinkhurst, R. O., and B. G. M. Jamieson. 1971. Aquatic oligochaeta of the world. University of Toronto Press, Toronto.

Chapman, P. M., M. A. Farrell, and R. O. Brinkhurst. 1982. Relative tolerance of selected aquatic oligochaetes to individual pollutants and environmental factors. Aquatic Toxicology 2:47–67.

Chatarpaul, L., J. B. Robinson, and N. K. Kaushik. 1979. Role of tubificid worms in nitrogen transformations in stream sediment. Journal of the Fisheries Research Board of Canada 36:673–678.

Chauvet, E., N. Giani, and M. O. Gessner. 1993. Breakdown and invertebrate colonization of leaf litter in two contrasting streams–significance of oligochaetes in a large river. Canadian Journal of Fisheries and Aquatic Sciences 50:488–495.

Cherry, B. R., M. J. Reeves, and G. Smith. 1998. Evaluation of bovine viral diarrhea virus control using a mathematical model of infection dynamics. Preventive Veterinary Medicine 33:91–108.

Chua, K. E., and R. O. Brinkhurst. 1973. Evidence of interspecific interactions in the respiration of tubificid oligochaetes. Journal of the Fisheries Research Broad of Canada 30:617–622.

Clancey, P., and D. Downing. 2001. Madison River/Ennis Reservoir fisheries and Madison River drainage westslope cutthroat trout conservation and restoration program. Montana Fish, Wildlife and Parks, Ennis, Montana.

Coler, R. A., H. B. Gunner, and B. M. Zuckerman. 1967. Selective feeding of tubificids on bacteria. Nature (London) 216:1143–1144.

Diamant, A. 1997. Fish-to-fish transmission of a marine myxosporean. Diseases of Aquatic Organisms 30:99–105.

Diamant, A. 1998. Red drum *Sciaenops ocellatus* (Sciaenidae), a recent introduction to Mediterranean mariculture, is susceptible to *Myxidium leei* (Myxosporea). Aquaculture 162:33–39.

Dobson, A. P., and P. J. Hudson. 1986. Parasites, disease and the structure of ecological communities. Trends in Ecology and Evolution 1:11–15.

Downing, D. C. 2000. Spawning and rearing ecology of Madison River rainbow trout in relation to whirling disease infection risk. Master's thesis. Montana State University, Bozeman.

Dumnicka, E. 1994. Habitat preferences of invertebrates (especially Oligochaeta) in a stream. Acta Hydrobiologia 36:91–101.

El-Matbouli, M., and R. W. Hoffmann. 1991. Effects of freezing, aging, and passage through the alimentary canal of predatory animals on the viability of *Myxobolus cerebralis* spores. Journal of Aquatic Animal Health 3:260–262.

El-Matbouli, M., T. Fischer-Scherl, and R. W. Hoffmann. 1992. Present knowledge of the life cycle, taxonomy, pathology, and therapy of some Myxosporea spp. important for freshwater fish. Annual Review of Fish Diseases 3:367–402.

El-Matbouli, M., and R. W. Hoffmann. 1998. Light and electron microscopic studies on the chronological development of Myxobolus cerebralis to the actinosporean stage in Tubifex tubifex. International Journal for Parasitology 28:195–217.

El-Matbouli, M., R. W. Hoffmann, and C. Mandok. 1995. Light and electron microscopic observations on the route of the triactinomyxon-sporoplasm of Myxobolus cerebralis from epidermis into rainbow trout (Onchorhynchus mykiss). Journal of Fish Biology 46:919–935.

El-Matbouli, M., T. W. Holstein, and R. W. Hoffmann. 1998. Determination of nuclear DNA concentration in cells of Myxobolus cerebralis and triactinomyxon spores, the causative agent of whirling disease. Parasitological Research 84:694–699.

El-Matbouli, M., T. S. McDowell, D. B. Antonio, K. B. Andree, and R. P. Hedrick. 1999. Effect of water temperature on the development, release and survival of the triactinomyxon stage of Myxobolus cerebralis in its oligochaete host. International Journal for Parasitology 29:627–641.

Eszterbauer, E., C. Szekely, K. Molnar, and F. Baska. 2000. Development of Myxobolus bramae (Myxosporea: Myxobolidae) in an oligochaete alternate host, Tubifex tubifex. Journal of Fish Diseases 23:19–25.

Gilbert, M. A., and W. O. Granath, Jr. 2001. Persistent infection of Myxobolus cerebralis, the causative agent of salmonid whirling disease, in Tubifex tubifex. Journal of Parasitology 87:101–107.

Graf, D. L., and D. O'Foighil. 2000. The evolution of brooding characters among the freshwater pearly mussels (Mollusca: Bivalvia: Unionoidea) of North America. Journal of Molluscan Studies 66:157–170.

Hallett, S. L., C. Erseus, and R. J. G. Lester. 1995. An actinosporean from an Australian marine oligochaete. Bulletin of the European Association of Fish Pathologists 15:168–171.

Hallett, S. L., P. J. O'Donoghue, and R. J. G. Lester. 1998. Structure and development of a marine actinosporean, Sphaeractinomyxon ersei n. sp. (Myxozoa). Journal of Eukaryotic Microbiology 45:142–150.

Halliday, M. M. 1973. Studies on Myxosoma cerebralis, a parasite of salmonids. II. The development and pathology of Myxosoma cerebralis, in experimentally infected rainbow trout (Salmo gairdneri) fry reared at different water temperatures. Nordisk Veterinaermedicin 25:349–358.

Halliday, M. M. 1976. The biology of Myxosoma cerebralis: the causative organism of whirling disease. Journal of Fish Biology 9:339–357.

Hedrick, R. P., M. El-Matbouli, M. A. Adkinson, and E. MacConnell. 1998. Whirling disease: re-emergence among wild trout. Immunological Reviews 166:365–376.

Hedrick, R. P., T. S. McDowell, M. Gay, G. D. Marty, M. P. Georgiadis, and E. MacConnell. 1999a. Compar-

ative susceptibility of rainbow trout Oncorhynchus mykiss and brown trout Salmo trutta to Myxobolus cerebralis, the cause of salmonid whirling disease. Diseases of Aquatic Organisms 37:173–183.

Hedrick, R. P., T. S. McDowell, K. Mukkatira, M. P. Georgiadis, and E. MacConnell. 1999b. Susceptibility of selected inland salmonids to experimentally induced infections with Myxobolus cerebralis, the causative agent of whirling disease. Journal of Aquatic Animal Health 11:330–339.

Hedrick, R. P., T. S. McDowell, K. Mukkatira, M. Georgiadis, and E. MacConnell. 2001. Susceptibility of three species of anadromous salmonids to experimentally induced infections with Myxobolus cerebralis, the causative agent of whirling disease. Journal of Aquatic Animal Health 13:43–50.

Hindar, K., I. A. Fleming, N. Jonsson, J. Breistein, H. Saegrov, E. Karlsbakk, M. Gammelsaeter, and B. O. Dønnum. 1996. Regnbueørret i Norge: forekomst, reproduksjon og etablering. Norsk Institutt for Naturforskning Oppdragsmelding 454.

Hiner, M., and C. M. Moffitt. 2001. Variation in Myxobolus cerebralis infections in field-exposed cutthroat and rainbow trout in Idaho. Journal of Aquatic Animal Health 13:124–132.

Hoffman, G. L. 1970. Intercontinental and transcontinental dissemination and transfaunation of fish parasites with emphasis on whirling disease (Myxosoma cerebralis). Pages 69–81 in S. F. Snieszko, editor. Symposium on diseases of fishes and shellfishes. American Fisheries Society, Special Publication 5, Bethesda, Maryland.

Hoffman, G. L. 1990. Myxobolus cerebralis, a worldwide cause of salmonid whirling disease. Journal of Aquatic Animal Health 2:30–37.

Janiszewska, J. 1955. Actinomyxidia. Morphology, ecology, history of investigations, systematics, development. Acta Parasitologica Pollonia 2:405–437.

Janiszewska, J. 1957. Actinomyxidia II. New systematics, sexual cycle, description of new genera and species. Zoologica Pollonica 8:3–33.

Juget, L., and M. Lafont. 1994. Theoretical habitat templets, species traits, and species richness: aquatic oligochaetes in the upper Rhone River and its floodplain. Freshwater Biology 31:327–340.

Kathman, R. D., and R. O. Brinkhurst. 1998. Guide to the freshwater oligochaetes of North America. Aquatic Resources Center, College Grove, Tennessee.

Kent, M., and J. Lom. 1999. Can a new species of Myxozoa be described based solely on their actinosporean stage? Parasitology Today 15:472–473.

Kent, M. L., L. Margolis, and J. O. Corliss. 1994. The demise of a class of protists: taxonomic and nomenclatural revisions proposed for the protist phylum Myxozoa Grass, 1970. Canadian Journal of Zoology 72:932–937.

Kent, M. L., K. B. Andree, J. L. Bartholomew, M. El-Matbouli, S. S. Desser, R. H. Devlin, S. W. Feist, R. P. Hedrick, R. W. Hoffmann, J. Khattra, S. L. Hallett,

R. J. G. Lester, M. Longshaw, O. Palenzeula, M. E. Siddall, and C. Xiao. 2001. Recent advances in our knowledge of the Myxozoa. Journal of Eukaryotic Microbiology 48:395–413.

Koie, M. 2000. First record of an Actinosporean (Myxozoa) in a marine polychaete annelid. Journal of Parasitology 86:871–872.

Landergren, P. 1999. Spawning of anadromous rainbow trout, *Oncorhynchus mykiss* (Walbaum): a threat to sea trout, *Salmo trutta* L., populations? Fisheries Research 40:55–63.

Lazim, M. N., and M. A. Learner. 1987. The influence of sediment composition and leaf litter on the distribution of tubificid worms (Oligochaeta). Oecologia 72:131–136.

Lazim, M. N., M. A. Learner, and S. Cooper. 1989. The importance of worm identity and life history on determining the vertical distribution of tubificids (Oligochaeta) in riverine mud. Hydrobiologia 178:81–91.

Lemmon, J. C., and B. L. Kerans. 2001. Extraction of whirling disease myxospores from sediments using the plankton centrifuge and sodium hexametaphosphate. Intermountain Journal of Sciences 7:57–62.

Lester, R. J. G., S. L. Hallett, M. El-Matbouli, and E. U. Canning. 1998. The case for naming actinosporeans using the zoological code. Parasitology Today 14:476–477.

Lester, R. J. G., S. L. Hallett, M. El-Matbouli, and E. U. Canning. 1999. Can a new species of Myxozoa be described based solely on their actinosporean stage? Reply. Parasitology Today 15:508.

Lom, J. 1987. Myxosporea: a new look at long-known parasites of fish. Parasitology Today 3:327–332.

Lom, J., and G. L. Hoffman. 1971. Morphology of the spores of *Myxosoma cerebralis* (Hofer, 1903) and *M. cartilaginis* (Hoffman, Putz, and Dunbar, 1965). Journal of Parasitology 57:1302–1308.

Lom, J., J. McGeorge, S. W. Feist, D. Morris, and A. Adams. 1997. Guidelines for the uniform characterization of the actinosporean stages of parasites of the phylum Myxozoa. Diseases of Aquatic Organisms 30:1–9.

Markiw, M. E. 1992a. Experimentally induced whirling disease I. Dose response of fry and adults of rainbow trout exposed to the triactinomyxon stage of *Myxobolus cerebralis*. Journal of Aquatic Animal Health 4:40–43.

Markiw, M. E. 1992b. Experimentally induced whirling disease II. Determination of longevity of the infective triactinomyxon stage of *Myxobolus cerebralis* by vital staining. Journal of Aquatic Animal Health 4:44–47.

Markiw, M. E., and K. Wolf. 1983. *Myxosoma cerebralis* (Myxozoa: Myxosporea) etiologic agent of salmonid whirling disease requires tubificid worm (Annelida: Oligochaeta) in its life cycle. Journal of Protozoology 30:561–564.

Marques, A. 1984. Contribution a la connaissance des actinomyxidies: ultrastructure, cycle biologique, system-atique. Doctoral dissertation. Universite des Sciences et Techniques du Languedoc, Montpellier, France.

May, R. M., and R. M. Anderson. 1979. Population biology of infectious diseases: Part 2. Nature(London) 280:455–461.

McKenzie, F. E. 2000. Why model malaria? Parasitology Today 16:511–516.

Milbrink, G. 1993. Evidence for mutualistic interactions in freshwater oligochaete communities. Oikos 68:317–322.

Minchella, D. J., and M. E. Scott. 1991. Parasitism: a cryptic determinant of animal community structure. Trends in Ecology and Evolution 6:250–254.

Modin, J. 1998. Whirling disease in California: a review of its history, distribution, and impacts, 1965–1997. Journal of Aquatic Animal Health 10:132–142.

Nehring, R. B., and P. G. Walker. 1996. Whirling disease in the wild: the new reality in the Intermountain West. Fisheries 21(6):28–30.

Newrkla, P., and N. Wijegoonawardana. 1987. Vertical distribution and abundance of benthic invertebrates in profundal sediments of Mondsee, with special reference to oligochaetes. Hydrobiologia 155:227–234.

O'Grodnick, J. J. 1979. Susceptibility of various salmonids to whirling disease (*Myxosoma cerebralis*). Transactions of the American Fisheries Society 108:187–190.

Robbins, J. A., T. Keilty, D. S. White, and D. N. Edgington. 1989. Relationships among tubificid abundances, sediment composition, and accumulation rates in Lake Erie. Canadian Journal of Fisheries and Aquatic Sciences 46:223–231.

Rognlie, M. G., and S. E. Knapp. 1998. *Myxobolus cerebralis* in *Tubifex tubifex* from a whirling disease epizootic in Montana. Journal of Parasitology 84:711–713.

Rose, J. D., G. S. Marrs, C. Lewis, and G. Schisler. 2000. Whirling disease behavior and its relation to pathology of brain stem and spinal cord in rainbow trout. Journal of Aquatic Animal Health 12:107–118.

Roubal, F. R., S. L. Hallet, and R. J. G. Lester. 1997. First record of triactinomyxon actinosporean in marine oligochaete. Bulletin of the European Association of Fish Pathologists 17:83–85.

Sarkka, J. 1987. The occurrence of oligochaetes in lake chains receiving pulp mill waste and their relation to eutrophication on the trophic scale. Hydrobiologia 155:259–266.

Saegrov, H., K. Hindar, and K. Urdal. 1996. Natural reproduction of anadromous rainbow trout in Norway. Journal of Fish Biology 48:292–294.

Sandell, T. A., H. V. Lorz, D. G. Stevens, and J. L. Bartholmew. 2001. Dynamics of *Myxobolus cerebralis* in the Lostine River, Oregon: implications for resident and anadromous salmonids. Journal of Aquatic Animal Health 13:142–150.

Schmidt, G. D., and L. S. Roberts. 1977. Foundations of parasitology. C. V. Mosby, St. Louis, Missouri.

Scott, W. B., and E. J. Crossman. 1973. Freshwater fishes of Canada. Fisheries Research Board of Canada, Bulletin 184, Ottawa.

Siddall, M. E., D. S. Martin, D. Bridge, S. S. Desser, and D. K. Cone. 1995. The demise of a phylum of protists: phylogeny of Myxozoa and other parasitic Cnidaria. Journal of Parasitology 81:961–967.

Smothers, J. F., C. D. von Dohlen, L. H. Smith, Jr., and R. D. Spall. 1994. Molecular evidence that the myxozoan protists are metazoans. Science 265:306–308.

Stevens, R., B. L. Kerans, J. C. Lemmon, and C. Rasmussen. 2001. The effects of *Myxobolus cerebralis* myxospore dose on triactinomyxon production and biology of *Tubifex tubifex* from two geographic regions. Journal of Parasitology 87:315–321.

Sturmbauer, C., G. B. Opadiya, H. Niederstätter, A. Riedmann, and R. Dallinger. 1999. Mitochondrial DNA reveals cryptic oligochaete species differing in cadmium resistance. Molecular Biology and Evolution 16:967–974.

Taylor, R. L., and M. Lott. 1978. Transmission of salmonid whirling disease by birds fed trout infected with *Myxosoma cerebralis*. Journal of Protozoology 25:105–106.

Thompson, K. G., and R. B. Nehring. 2000. A simple technique used to filter and quantify the actinospore of *Myxobolus cerebralis* and determine its seasonal abundance in the Colorado River. Journal of Aquatic Animal Health 12:316–323.

Thompson, K. G., R. B. Nehring, D. C. Bowden, and T. Wygant. 1999. Field exposure of seven species or subspecies of salmonids to *Myxobolus cerebralis* in the Colorado River, Middle Park, Colorado. Journal of Aquatic Animal Health 11:312–329.

Vincent, E. R. 1996. Whirling disease and wild trout: the Montana experience. Fisheries 21(6):32–33.

Vincent, E. R. 2000. Whirling disease report. 1997–1998. Montana Fish, Wildlife and Parks, Helena, Montana.

Wavre, M., and R. O. Brinkhurst. 1971. Interactions between some tubificid oligochaetes and bacteria found in the sediment of Toronto Harbour, Ontario. Journal of the Fisheries Research Board of Canada 28:335–341.

Weiss, S., and S. Schmutz. 1999. Response of resident brown trout, *Salmo trutta* L., and rainbow trout, *Oncorhynchus mykiss* (Walbaum), to the stocking of hatchery-reared brown trout. Fisheries Management and Ecology 6:365–375.

Wipfli, M. S., J. Hudson, and J. Caouette. 1998. Influence of salmon carcasses on stream productivity: response of biofilm and benthic macroinvertebrates in southeastern Alaska, U.S.A. Canadian Journal of Fisheries and Aquatic Sciences 55:1503–1511.

Wolf, K., and M. E. Markiw. 1984. Biology contravenes taxonomy in the Myxozoa: new discoveries show alternation of invertebrate and vertebrate hosts. Science 225:1449–1452.

Wolf, K., M. E. Markiw, and J. K. Hiltunen. 1986. Salmonid whirling disease: *Tubifex tubifex* (Müller) identified as the essential oligochaete in the protozoan life cycle. Journal of Fish Diseases 9:83–85.

Xiao, C., and S. S. Desser. 1998a. Actinosporean stages of myxozoan parasites of oligochaetes from Lake Sasjewun, Algonquin Park, Ontario: new forms of triactinomyxon and raabeia. Journal of Parasitology 84:998–1009.

Xiao, C., and S. S. Desser. 1998b. Actinosporean stages of myxozoan parasites of oligochaetes from Lake Sasjewun, Algonquin Park, Ontario: new forms of echinactinomyxon, neoactinomyxum, aurantiactinomyxon, guyenotia, synactinomyxon and antonactinomyxon. Journal of Parasitology 84:1010–1019.

Xiao, C., and S. S. Desser. 1998c. The oligochaetes and their actinosporean parasites in Lake Sasjewun, Algonquin Park, Ontario. Journal of Parasitology 84:1020–1026.

Zendt, J. S., and E. P. Bergersen. 2000. Distribution, and abundance of the aquatic oligochaete host *Tubifex tubifex* for the salmonid whirling disease parasite *Myxobolus cerebralis* in the Upper Colorado River basin. North American Journal of Fisheries Management 20:502–512.

American Fisheries Society Symposium 29:167–179, 2002

Modeling *Myxobolus cerebralis* Infections in Trout: Associations with Habitat Variables

MONICA HINER[1] AND CHRISTINE M. MOFFITT[2]

Department of Fish and Wildlife Resources, University of Idaho, Moscow, Idaho 83844-1136, USA

208-885-7047 (phone); 208-885-9080 (fax); cmoffitt@uidaho.edu

ABSTRACT. We obtained data from several sources to explore selected habitat compartments of a more complex epizootic model of factors affecting *Myxobolus cerebralis* in free-ranging populations of fish. We examined histological sections from branchial and cranial tissues from Yellowstone cutthroat trout *Oncorhynchus clarki bouvieri* and rainbow trout *O. mykiss*, naturally exposed to *M. cerebralis* at nine locations within three river drainages in Idaho, for evidence of characteristic pathology of whirling disease. Mean water temperature during exposure, temperature coefficient of variation, and the density of two groups of benthic macroinvertebrates that can thrive in habitats of high organic content were significantly positively correlated with the mean lesion severity of infected fish. We used stepwise multiple linear regression models to test combinations of variables as predictors of disease severity. Models with average water temperature or average temperature variation during exposure and the density of oligochaetes and chironomids accounted for more than 80% of the variation.

Myxobolus cerebralis was first observed in natural populations of salmonids within Idaho in 1987. This parasite has since been confirmed in free-ranging salmonids from the Coeur d'Alene, St. Joe, East Fork of the Big Lost, Little Lost, South Fork of the Boise, Big Wood, South Fork of the Snake, Teton, East Fork of the Salmon, Pahsimeroi, and Lemhi River drainages (Keith Johnson, Idaho Department of Fish and Game, personal communication). In Montana and Colorado, whirling disease caused by *M. cerebralis* is suspected to be a contributing factor in the observed decline of free-ranging trout populations (Nehring and Walker 1996; Vincent 1996; Baldwin et al. 1998), and within Idaho, whirling disease likely contributes to populations declines of rainbow trout *Oncorhynchus mykiss* observed in the Big Lost drainage (Elle 1998). Interest in the biology and life cycle of *M. cerebralis* has increased because of its potential impact on free-ranging trout populations.

Disease prevalence and severity of *M. cerebralis* infections in natural populations has been shown to vary across large geographic areas, within drainages, and over short geographic reaches (Baldwin et al. 1998; Hiner and Moffitt 2001;

Sandell et al. 2001). In 1997 and 1998, field exposure studies conducted in the South Fork of the Boise River showed significantly different disease prevalence and severity within a 25 km river reach (Hiner and Moffitt 2001). The ability to identify factors that may play a role in the differential infection, observed within drainages and over short geographic reaches, may serve useful in the future when attempting to manage habitats where populations are severely affected by the parasite.

Many authors have acknowledged the importance of understanding the complexity of disease development and the numerous factors involved. Snieszko (1973) stressed the role of environment along with the host and pathogen in conceptual models of the disease process. More recently, specific host/pathogen/environment models have been proposed to understand disease and the disease process within fish populations (Hedrick 1998). Reno (1998) reviewed the importance of numerous factors in several host parasite infections, such as *M. cerebralis*, and identified variables affecting disease susceptibility, including fish density, contact rate with the pathogen, immunity, natural survival rates of both fish and worm populations, and parasite development within *Tubifex tubifex*.

Models of habitat suitability have been used extensively to quantify relationships among habitat parameters and dependent variables, such as fish

[1]Present address: Yurok Tribal Fisheries Program, 15900 Hwy 101 North, Klamath, California 95548, USA; Phone: (707) 482-2841; Fax: (707) 482-0384; 3-mail: monica@cc.northcoast.com

[2]Corresponding author: cmoffitt@uidaho.edu

abundance and species distribution (Lewis 1969; Orth and Maughan 1982; Layher and Maughan 1985; McClendon and Rabeni 1987; Wesche et al. 1987). Environmental variables used in habitat modeling have included water temperature, depth, velocity, substrate, elevation, and cover (Binns and Eiserman 1979; Layher et al. 1987; Hubert and Rahel 1989). While habitat suitability models are used extensively in predicting abundance and distribution of fish, little research has focused on quantifying the role of habitat parameters potentially affecting disease prevalence and severity, especially for multiple host parasites such as M. cerebralis.

Two variables that most likely affect the distribution and severity of M. cerebralis infections in natural environments are water temperature and abundance of T. tubifex. Temperature has been related to disease severity (Halliday 1973; Baldwin et al. 2000), and studies have also shown that the production, release, and survival of the infective triactinomyxon stage from T. tubifex are correlated with temperature (Markiw 1992; El-Matbouli et al. 1999). The abundance of T. tubifex likely has the largest potential influence on prevalence and extent of infection observed in natural fish populations, since infection prevalence within any worm population appears low and has been shown to range between 0.0% and 7.3% (Markiw 1986; Wolf et al. 1986; Rognlie and Knapp 1998; Zendt and Bergersen 2000). Areas with a high abundance of T. tubifex, where the parasite has been confirmed previously with water temperatures conducive to the increase the production, release, and survival of triactinomyxons, could serve as amplification areas for the parasite and result in increased disease prevalence and severity, compared with other areas.

T. tubifex are found in a wide variety of habitats but are often in high abundance in habitats with heavy loads of fine particulate matter and organic debris (Reynoldson et al. 1991). Identification of T. tubifex is difficult (Kathman and Brinkhurst 1998; Zendt and Bergersen 2000), as identification is based on microscopic sexual structures that are present in only a portion of the population. Predicting habitats where the worm would likely occur in high densities, by counting other invertebrate groups that share similar habitats, may be more promising. Both Chironomidae and Tubifidae are cosmopolitan in distribution but thrive in substrates with high accumulations of organic and silty deposits, where other benthic species cannot persist, and these invertebrate families are often associated with shallow impound-ments or degraded habitats (Hynes 1966; Hart and Fuller 1974; Reynoldson et al. 1991; Lampert and Sommer 1997; Zendt and Bergersen 2000).

Caged field exposures of strains of fish, known to be susceptible to M. cerebralis, that are similar in age (early fry) are a good tool to determine the extent of infection at specific locations within a river reach and have been previously used to document M. cerebralis infections (Hnath 1970; Horsch 1987; Baldwin et al. 2000, Hiner and Moffitt 2001). Sampling free-ranging fish populations for M. cerebralis prevalence and disease severity are most likely biased, since heavy infection may result in differential mortality (Baldwin et al. 1998), and migration of both infected and noninfected fish within free-ranging populations may also occur before sampling. By utilizing caged field exposures, relative comparisons for disease prevalence and severity can be made among study sites. Habitat variables can also be sampled at exposure sites to describe the precise habitat associated with a measured disease response in the exposed fish.

We used data from field exposures conducted by our laboratory and by the Idaho Department of Fish and Game in three M. cerebralis positive drainages to explore the relationships among water temperature, aquatic benthic macroinvertebrate communities, and resulting disease, as quantified by average lesion severity observed by histopathology. We quantified the range of disease response in the drainages and explored the relationships among selected habitat variables and average lesion severity as a tool to predict the risks of M. cerebralis infection in fish.

METHODS

Conceptual Model

Upon initiation of this project, we created a conceptual model of potential interactions among variables that likely influence the dynamics of M. cerebralis infections. The proposed model encompasses numerous factors and their potential interactions, including the distribution and abundance of T. tubifex, vulnerability of fish to infection, fish population dynamics, movement and migration, and stream habitat parameters (Figure 1). Due to limitations in data available for a retrospective analysis, we restricted our analysis to selected habitat elements of the Tubifex worm host that can affect the distribution and severity of disease within fish (Figure 1, dark shaded boxes).

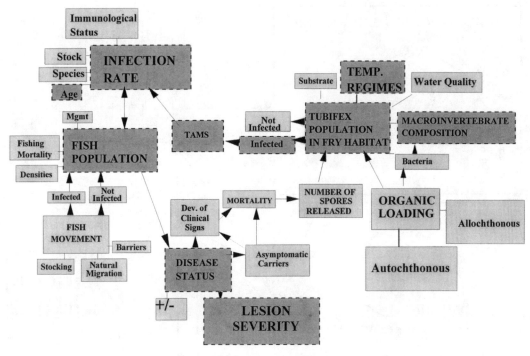

Figure 1. Conceptual epizootic model of *Myxobolus cerebralis* infections. Boxes with dark shading and dashed lines are explored in models and relationships in this study.

Field Exposures

Natural exposures of *M. cerebralis*-free Kamloops rainbow trout (Troutlodge strain, Sumner, Washington) and Yellowstone cutthroat trout *O. clarki bouvieri* (Henry's Lake, Idaho) were conducted in three drainages (South Fork Boise River, Teton River, and East Fork Big Lost River) between 1996 and 1998 (Figure 2). During all exposures, fry (30–40 mm in length) were placed in cages and exposed to natural river conditions for 10 d, during times of natural fry emergence. Circular exposure cages were made from 125-L heavy-duty plastic barrels with 22 cm × 41 cm openings on each side, which were covered with 1.0 mm mesh screen to prevent fish escapement. Cages were anchored to the river bottom using T-posts, so that the mesh openings were perpendicular to water flow and placed adjacent to the riverbank in areas with reduced flow, such as eddies.

Differences in sampling strategy between the University of Idaho and Idaho Department of Fish and Game resulted in variation among the number of replicate exposures conducted at each sample site. One replicate of each species was exposed at each site in Teton River and East Fork Big Lost River exposures, and two replicates of each species per site were used in South Fork Boise River exposures. After 10 d of exposure in the river, fry were collected and transported to pathogen-free water for continued rearing. Fry exposed in the East Fork of the Big Lost River were reared at Mackay Fish Hatchery (IDFG); fry from exposures in the Teton River were held at Montana State University Wild Trout Laboratory; and fry exposed in the South Fork of the Boise River were reared at the University of Idaho Fisheries Wet Laboratory. Time of rearing was variable among the three facilities, ranging between 1,148°C and 1,750°C cumulative temperature units (Table 1).

Analysis of Histopathology

At the end of rearing, all fish were euthanized (100 mg/L MS-222), and between 5 and 20 fish from each species exposure per site were removed for analysis of histopathology (Table 1). The heads of these fish were fixed in 10% neutral buffered formalin, prepared for paraffin histology, sectioned sagitally (5 μm thickness), and stained with hematoxylin and eosin (Luna 1968; Yasutake and Wales

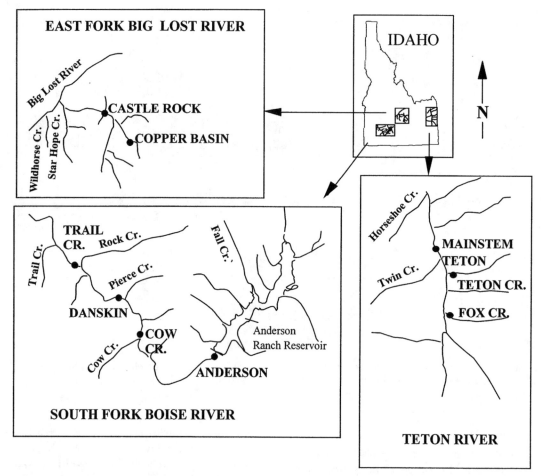

Figure 2. Location of study sites in Idaho indicated by a solid circle, in the East Fork of the Big Lost River (Copper Basin, Castle Rock), the South Fork of the Boise River (Anderson, Cow Creek, Danskin, Trail Creek), and the Teton River drainage (Fox Creek, Teton Creek, mainstem Teton).

1983). All sections were coded in our laboratory before interpretation to eliminate bias of reporting. Slides were examined by light microscopy for presence of spores and pathology typical of *M. cerebralis* infections and scored according to a severity index ranging from 0 to 4 (Baldwin et al. 2000). Two sections were mounted and stained for each fish, one section from the area where the anterior periorbital started to appear and a deeper section cut toward the midline. The higher score of the two sections examined was recorded for final reporting and analysis. Each fish was scored separately, and a mean lesion severity score was computed for each exposure site.

Habitat Variables

Water Temperature

We obtained records of water temperatures at exposure sites to examine the relationship between mean lesion severity and temperature parameters (mean temperature and coefficient of variation) during exposure. Electronic temperature records were taken at the exposure site during each field exposure, except for exposures in the East Fork of the Big Lost River. We obtained temperature records for Castle Rock for July 1997 (exposures were conducted in July 1996), and temperature records for Copper Basin were not available from

Table 1. Summary of location, date, fish species, temperature units post exposure, and lesion severity score of each fish examined from field exposures conducted in three drainages in Idaho (CTT = cutthroat trout; RBT = rainbow trout).

Location	Onset of exposure	Species	Number examined	Temp. units (°C) postexposure	0	1	2	3	4	Mean lesion severity
East Fork Big Lost River										
Copper Basin	8 July 1996	CTT	5	1,750	0	0	0	1	4	3.8
	8 July 1996	RBT	5	1,750	0	0	0	0	5	4.0
Castle Rock	8 July 1996	CTT	5	1,750	1	0	2	2	0	2.0
	8 July 1996	RBT	5	1,750	0	0	0	2	3	3.6
Teton Drainage										
Fox Creek	5 August 1997	CTT	20	1,690	2	1	1	1	15	3.3
	5 August 1997	RBT	20	1,690	0	1	0	1	18	3.8
Teton Creek	5 August 1997	CTT	20	1,690	1	1	1	1	16	3.5
	5 August 1997	RBT	20	1,690	1	0	1	2	16	3.6
Mainstem Teton	5 August 1997	CTT	20	1,690	3	2	4	1	10	2.65
	5 August 1997	RBT	20	1,690	6	1	0	1	12	2.6
South Fork Boise River										
Anderson	3 July 1998	CTT	20	1,148	20	0	0	0	0	0
	3 July 1998	RBT	20	1,148	20	0	0	0	0	0
Cow Creek	3 July 1998	CTT	20	1,148	17	1	0	0	2	0.45
	3 July 1998	RBT	20	1,148	15	1	2	1	1	0.6
Danskin	3 July 1998	CTT	20	1,148	9	0	3	5	3	1.65
	3 July 1998	RBT	20	1,148	13	1	2	2	2	0.95
Trail Creek	3 July 1998	CTT	20	1,148	0	0	0	0	20	4.0
	3 July 1998	RBT	20	1,148	0	0	0	0	20	4.0

the immediate study site but were obtained from a U.S. Forest Service monitoring station, 2.5 km downstream from the exposure site (Bart Gamett, USFS, personal communication). For all sites, temperature data were recorded in 15-min intervals, except at the Copper Basin site, which recorded in 1-h 15-min intervals. Average temperature during exposure and coefficient of variation (mean/SD) was calculated for all temperature records (Table 2).

Aquatic Benthic Macroinvertebrates

Within the East Fork of the Big Lost River, aquatic macroinvertebrate survey data were obtained from the Idaho Division of Environmental Quality (Thomas Herron, Idaho Division of Environmental Quality, personal communication). A Hess sampler was used to collect 3 replicate samples, approximately 1 km upstream from the Copper Basin exposure site in July 1995 (IDEQ 1999). No benthic macroinvertebrate samples were collected from the Castle Rock site.

For all sites in the Teton River drainage, benthic macroinvertebrate data were obtained from

the Stream Ecology Center at Idaho State University (Dr. Wayne Minshall, Idaho State University, personal communication). A Slack sampler (250 μm size mesh; Cuffney et al. 1993) was used to collect five replicate samples (each 0.52 m²) from the Fox Creek and mainstem Teton River exposure sites, between 24 April and 6 May 1998. No benthic macroinvertebrate samples were collected from the Teton Creek site.

For all sites in the South Fork of the Boise River, aquatic benthic macroinvertebrates were sampled by the University of Idaho in August 1998, after all field exposures of fry were completed. A D-frame kick net (mesh size approximately 500 μm) was used to collect 3 replicate samples from each site, except the Trail Creek site where only two replicate samples were taken, due to heavy loads of fine particulate matter. An area of substrate 1.0 m² and 0.1 m deep was sampled from the area directly surrounding the exposure cages. For all benthic macroinvertebrate samples (Copper Basin, Fox Creek, mainstem Teton River, Anderson, Cow

Table 2. Mean lesion severity, temperature parameters during exposure, and mean oligochaete and chironomid density for field all exposure sites in Idaho.

Location and date of exposure	Mean lesion severity	Temp. min/max (°C)	Mean daily temperature (°C)	Temp. coefficient of variation	Mean oligochaetes/m²	Mean chironomids/m²
East Fork Big Lost River (8–18 July 1996)						
Copper Basin	3.9	5.2 / 19.9	12.72	29.2867	30	1,116.7
Castle Rock	2.8	7.0 / 18.0	12.50	17.2457	No data	No data
Teton River Drainage (5–15 August 1997)						
Fox Creek	3.55	8.1 / 17.4	11.39	20.6831	163.1	1,400.0
Teton Creek	3.55	8.0 / 16.3	11.16	18.5844	No data	No data
Mainstem Teton	2.63	10.3 / 17.5	13.58	12.0774	29.2	655.8
South Fork Boise River (3–12 July 1998)						
Anderson	0	6.9 / 10.5	7.57	7.8929	257.0	3.0
Cow Creek	0.55	7.7 / 13.1	9.84	12.7815	297.8	98.3
Danskin	1.30	7.7 / 13.4	10.25	13.7795	64.2	90.2
Trail Creek	4	7.8 / 14.2	10.83	14.9825	2,394.8	371.3

Creek, Danskin, and Trail Creek), invertebrates were classified and enumerated by order, except for Diptera, which were further separated, and individuals in the family Chironomidae were identified and quantified. We calculated a mean abundance for oligochaetes and chironomids across all replicate samples for each site. Since sample area among the sites varied, depending on collecting agency and method used, mean sample abundance was extrapolated to mean density (number/m²), so that relative comparisons could be made across sites (Table 3).

Statistical Analyses and Modeling

We used a one-way analysis of variance (ANOVA) (SAS 1996) to compare differences in mean lesion severity scores between species across all sites with the GLM model y = species + error. The species effect was not significant (ANOVA $P = 0.786$), so lesion severity scores for both species were combined, and new mean lesion severity scores were computed for each of the nine exposure sites. A two-way nested ANOVA was used to examine differences in the mean lesion severity among drainages and within each drainage among study sites. We used the GLM model y = drainage + site (drainage) + error. Tukey's multiple comparisons were used to test whether differences were present among drainages, and we plotted least squared means values, for each site, to determine significant differences ($P < 0.05$). We used linear regression analysis and correlations to describe the relationship between mean lesion severity and each individual independent variable. Independent variables

used in our analysis were (1) mean temperature during exposure, (2) temperature coefficient of variation, (3) mean density of oligochaetes, (4) mean density chironomids, and (5) combined density of oligochaetes and chironomids. We used logarithmic transformations of each biotic parameter for statistical analysis. A stepwise multiple regression analysis (Timm and Mieczkowski 1997) was used to identify different combinations of the five independent variables that were significantly related to mean lesion severity and accounted for variation in lesion severity observed among the exposure sites. For the multiple linear regressions, only sites that had complete data sets for both temperature and invertebrates were included in the analysis (Table 2). As a precaution, we plotted the residuals and student residuals for each model and evaluated collinearity of variables.

RESULTS

Field Exposures

Significant differences were detected in lesion severity among the drainages (ANOVA $P = 0.0001$; Table 4). Across all sites within each drainage, fish exposed in the South Fork of the Boise River had lower mean lesion severity scores than fish exposed in the Teton River and East Fork of the Big Lost Rivers (ANOVA, Tukey's multiple comparison test; $P < 0.05$; Figure 3).

Mean lesion severity was significantly different among fish sampled from sites within the East Fork of

Table 3. Mean abundance of oligochaetes and chironomids per sample, sampler type, and extrapolated density for samples collected from exposure sites in Idaho.

East Fork Big Lost River (Hess sampler, 0.10 m²)							
	Composite of three samples					Mean abundance	Extrapolate number/m²
Copper Basin							
Chironomidae	335					111.67	1,116.7
Oligochaeta	9					3.0	30.0

Teton River Drainage (Slack sampler, area sampled 0.52 m²)							
	Sample					Mean abundance	Extrapolated number/m²
	1	2	3	4	5		
Fox Creek							
Chironomidae	104	120	102	84	3230	728.0	1,400.0
Oligochaeta	5	9	4	2	9	5.8	163.1
Mainstem							
Chironomidae	177	220	780	204	324	341.0	655.8
Oligochaeta	0	4	24	24	24	15.2	29.2

South Fork Boise River (D-frame kicknet, area sampled 1.0 m²)					
	Sample			Mean abundance	Extrapolated number/m²
	1	2	3		
Anderson					
Chironomidae	2	7	0	3.0	3.0
Oligochaeta	12	102	657	257.0	257.0
Cow Creek					
Chironomidae	175.5	81	38.3	98.3	98.3
Oligochaeta	589.5	153	150.8	297.8	297.8
Danskin					
Chironomidae	22.5	144	104	90.2	90.2
Oligochaeta	40.5	90	62	64.2	64.2
Trail Creek					
Chironomidae	144	598.5	—	371.3	371.3
Oligochaeta	3,246	1,543.5	—	2,394.8	2,394.8

the Big Lost and South Fork Boise Rivers; however, lesion severity did not significantly differ among exposure sites in the Teton River Drainage. Mean lesion severity among sites within the Teton River drainage ranged between 2.63 and 3.55. In the East Fork of the Big Lost River, fish sampled from the Castle Creek exposures had a significantly lower mean lesion severity score than fish exposed at Copper Basin ($P = 0.032$). Fish sampled from the Castle Creek exposures had a mean lesion severity score of 2.8, with one fish negative for the parasite. Fish sampled from the Copper Basin exposures were all positively infected and had severe infections (mean lesion severity 3.9). Infections ranged among the study sites within the 25-km study reach of the South Fork Boise River, with the uppermost site (Anderson) negative for the parasite, Cow Creek and Danskin sites showing mild lesion severity (mean lesion severity 0.55 and 1.3, respectively), and all fish sampled from Trail Creek exposures exhibiting severe infections with lesion severities of 4.

Modeling Habitat Variables

The correlations between mean lesion severity score and mean temperature or temperature variations during exposure were significant with the highest correlation between mean lesion severity scores and temperature coefficient of variation ($r^2 = 0.682$; $P = 0.022$; Figure 4).

We found no correlation between mean lesion severity score and oligochaete density ($r^2 = 0.003$; $P = 0.915$), likely due to the diversity of the species of oligochaetes. The oligochaete data were not detailed to identification of specific *Tubifex* worms. Of the study sites with data on aquatic benthic macroinvertebrates, all sites except Trail Creek had

Table 4. Summary of ANOVA models testing effects of drainage (East Fork Big Lost River, Teton River, South Fork Boise River) and site within drainage on severity of M. cerebralis infection, expressed by mean lesion score.

Source of variation	df	Sum of squares	F	P
Model: y = drainage + site (drainage) + error				
Drainage	2	14.8918	39.71	0.0001
Site (drainage)	6	21.3168	18.95	0.0001
Error	9	1.6875		
Total	17	37.8961		

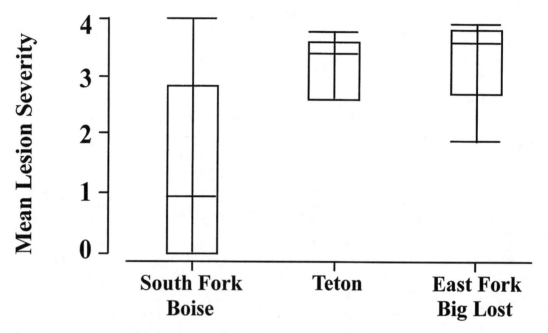

Figure 3. Boxplots of mean lesion scoresfor fish in exposure units in the South Fork of the Boise River (4 sites), Teton River drainage (3 sites), and East Fork Big Lost River (2 sites). Solid horizontal bars represent median values; boxes represent range of data between 25th and 75th percentiles; vertical bars indicate normal range of data.

mean densities of oligochaetes less than 300/m² (Table 2). Trail Creek had the highest density of oligochaetes (2,395/m²), and all fish were severely affected by whirling disease (mean lesion severity score = 4). We found a significant correlation between chironomid abundance and mean lesion severity score (r^2 = 0.758; P = 0.024) and a significant correlation between mean lesion severity and the mean density of oligochaetes and chironomids (r^2 = 0.761; P = 0.024; Figure 5).

Two multiple linear regression models provided significant relationships between mean lesion severity and habitat parameters. Both models included the density of oligochaetes and chirono-

mids. The first model included the mean density of oligochaetes and chironomids with mean temperature during exposure and accounted for 83% of the variation in lesion severity observed among the study sites (P = 0.013; Table 5). The second model replaced mean temperature during exposure with temperature coefficient of variation and accounted for 82% of the variation in mean lesion severity among the seven sites (P = 0.014; Table 5).

DISCUSSION

The use of data that was not collected simultaneously with field exposures of fry, or collected by

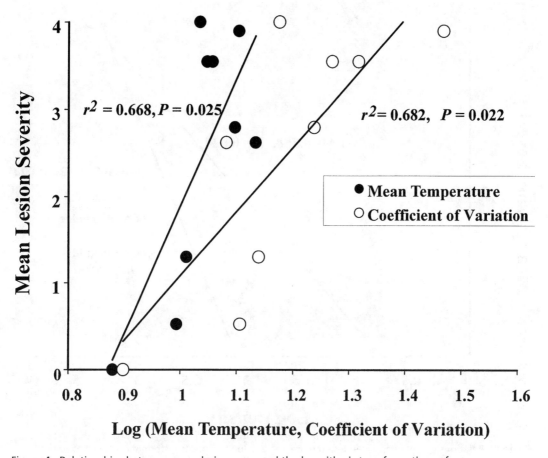

Figure 4. Relationships between mean lesion score and the logarithmic transformations of mean temperature during exposure (solid circles) and temperature coefficient of variation (open circles) for nine exposure sites.

other research teams, limits the extrapolation and likely increases the sources of error and variation. We have evidence that these trends remain similar over the years. In studies conducted in 1997 in the South Fork of the Boise River, the prevalence of *M. cerebralis* in exposed rainbow trout was nearly identical to that reported in this study at the same exposure sites (C. M. Moffitt, unpublished data). We did not have adequate samples to account for yearly differences in this study. Other published studies have used similar approaches to examine watershed trends (Baldwin et al. 1998; Sandell et al. 2001). Moreover, month of exposure (and associated water temperatures) may be more important than between-year variation (Baldwin et al. 2000).

Given these limitations, we observed variation in the extent of pathology caused by *M. cerebralis* among and within the three drainages, with higher mean lesion severity for all fish exposed in the Teton River and East Fork of the Big Lost River drainages than in three of the four exposures conducted in the South Fork of the Boise River (all Trail Creek exposures resulted in lesion severity scores of 4). Wild trout populations, sampled over several years throughout Montana, showed significant variation in prevalence and severity of *M. cerebralis* infections among drainages (Baldwin et al. 1998). Sandell et al. (2001) reported variable infection rates in rainbow trout fry exposed over two years at three sites in the Lostine River, Oregon.

In our study, mean water temperatures during exposure were correlated with higher mean lesion severity scores. The relationship was influenced by the Anderson site in the South Fork of the Boise River, which tested negative for the parasite and had the lowest mean temperature during exposure (7.57°C). Field exposure studies of rainbow trout in Montana showed that temperature during exposure

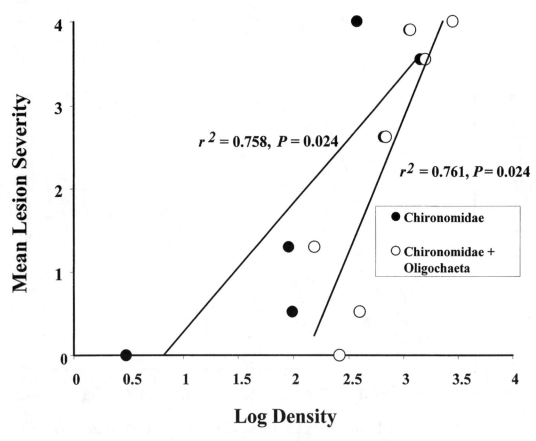

Figure 5. Relationship between mean lesion score and the logarithmic transformations of chironomid density (solid circles) and density of chironomids and oligochaetes (open circles) for seven exposure sites.

Table 5. Summary of multiple linear regression analyses for two models relating mean lesion score to temperature parameters and the density of oligochaetes and chironomids at seven exposure sites in the East Fork Big Lost River, Teton River, and South Fork Boise River drainages (AVTMP = mean temperature during exposure, CVTMP = temperature coefficient of variation, OLIGCHIR = density of oligochaetes and chironomids / m²).

Variable	Parameter estimate	Standard error	Parameter P	Model P	Model r²
Model: y = CVTMP + OLIGCHIR + error					
Intercept	−8.928	2.0795	0.013	0.014	0.82
CVTMP	1.707	0.8265	0.108		
OLIGCHIR	1.018	0.3349	0.039		
Model: y = AVTMP + OLIGCHIR + error					
Intercept	−13.276	2.0795	0.019	0.013	0.83
AVTMP	3.662	1.6951	0.097		
OLIGCHIR	1.057	0.3152	0.029		

was correlated with mean lesion severity scores (Baldwin et al. 2000). Halliday (1973) showed an increase in disease severity with increasing exposure temperature, with clinical signs of disease less evident in fish exposed and reared at 7°C compared with fish exposed and reared at 12°C and 17°C.

Temperature coefficient of variation, which measures the extent of diurnal temperature fluctuation, was also correlated with mean lesion severity score. This variable has largely been ignored in previous studies, and there is a lack of information on its effect on disease severity and prevalence. However, coefficients of variation of both stream depth and velocity have also been used to quantify habitat for predicting fish distribution and standing stock (Hermansen and Kreg 1984; Wesche et al. 1987; Hubert and Rahel 1989). Since calculation of temperature coefficient of variation includes both mean temperature during exposure and standard deviation of records, this variable may be more important than mean temperature to predict habitats likely to cause severe *M. cerebralis* infections. Diurnal temperature fluctuations are often associated with shallow areas that may have the opportunity for more thermal heating during the day and also may be more likely to have *Tubifex* habitat.

Assessing aquatic benthic macroinvertebrate communities and identifying areas with high densities of *T. tubifex* is likely a more precise way to measure infection risk. However, *T. tubifex* is difficult to identify and is rarely quantified in routine sample analysis (Kathman and Brinkhurst 1998; Zendt and Bergersen 2000). Our studies failed to demonstrate a positive relationship between the overall abundance of all oligochaete species and *M. cerebralis* lesion severity in fish. However, we observed positive correlations between mean lesion severity scores and both chironomid density and the density of chironomids and oligochaetes. The habitats with high density of both oligochaetes and chironomids were heavily silted and high in organic debris. Although chironomids are not linked to the life cycle of *M. cerebralis*, the group may be useful as a surrogate when predicting habitat quality, since they co-occur in high densities with *T. tubifex*, in degraded habitats where few other species persist (Hynes 1966; Hart and Fuller 1974; Lampert and Sommer 1997). Särrkä (1994) used abundance ratios between oligochaetes and chironomids to rank pollution status. We calculated these ratios, using our data, but did not observe significant relationships with mean lesion severity.

Two regression models showed significant relationships between environmental variables and the severity of lesions. The first model included temperature coefficient of variation and density of both oligochaetes and chironomids. Of the seven sites used for multiple linear regression analysis, the three sites with the highest mean severity of lesions (Copper Basin, Fox Creek, and Trail Creek) shared two profiles that the other sites with low to moderate lesion severity lacked: (1) all three sites had a temperature coefficient of variation greater than or equal to 15, and (2) the density of oligochaetes and chironomids combined was greater than 1,000/m^2.

A habitat parameter that was not quantified in our analysis but may be influential in controlling disease prevalence and severity is the presence of impoundments, both natural and man-made. Two sites with the highest mean lesion severity scores (Trail Creek and Copper Basin) were located directly below beaver dams in habitats with accumulations of fine particulate matter and organic debris. Beaver dams can affect the community structure of macroinvertebrates, likely from increased organic and fine particulate matter (Naiman et al. 1988). Impoundments serve as sediment traps, sometimes producing anoxic conditions in bottom sediments, creating optimal habitat for large densities of *T. tubifex* due to lack of competition from other species. Lotic invertebrates are replaced by lentic forms in impoundments, with chironomids and tubificids among the dominant taxon present (McDowell and Naiman 1986).

In addition to providing optimal habitat for *T. tubifex*, impounded reaches and reservoirs increase littoral habitat and diurnal warming of these areas during late spring and summer months and may support high production rates of the infective stage (triactinomyxon) of the parasite within *T. tubifex*. If large densities of *T. tubifex* are present in the littoral zones and deep sediments of impounded reaches, this could lead to "point source" production and release of triactinomyxons from infected worm populations. Zendt and Bergersen (2000) proposed that certain degraded and heavily organic habitats likely provided point source regions in the Upper Colorado River Basin. Such effects may also be observed downstream of a reservoir, due to reservoir eutrophication creating increased organic loading in tailwaters, which may potentially increasing the amount of habitat available for *T. tubifex* downstream as well (Thornton et al. 1990).

We acknowledge that numerous variables may have a potential impact on disease severity and prevalence of M. cerebralis infections. Differences in species and age susceptibility have been shown in M. cerebralis infections (Hoffman et al. 1962; Hoffman 1974; O'Grodnick 1979), and strain resistance has been reported in Ceratomyxa shasta infections (Ibarra et al. 1994; Bartholomew 1998). This study was conducted to encourage further investigations and models of epizootic interactions for comparative analysis with our results, to define characteristics of habitats associated with higher risk of infection.

ACKNOWLEDGMENTS

We thank Yasunari Kiryu for his work in preparation and analysis of histological samples. Data for modeling was provided by Christina Relyea and Wayne Minshall, Idaho State University; Steven Elle, Tony Lamansky, and Elizabeth Mamer, Idaho Department of Fish and Game; Thomas Herron, Idaho Division of Environmental Quality; and Bruce Smith and Bart Gamett, U.S. Forest Service. Christopher J. Williams and Kirk Steinhorst, University of Idaho provided advice on statistical analyses. We thank C. Michael Falter, University of Idaho; John Gay, Washington State University; Keith Johnson, Idaho Department of Fish and Game; and two anonymous reviewers for providing valuable suggestions for this manuscript. This is contribution 938 of the Forestry, Wildlife and Range Experiment Station. Funding was provided by the Idaho Department of Fish and Game.

REFERENCES

Baldwin, T. J., J. E. Peterson, G. C. McGhee, K. D. Staigmiller, E. S. Motteram, C. C. Downs, and D. R. Stanek. 1998. Distribution of Myxobolus cerebralis in salmonid fishes in Montana. Journal of Aquatic Animal Health 10:361–371.

Baldwin, T. J., Vincent, E. R., Silflow, R. M., and D. R. Stanek. 2000. Myxobolus cerebralis infection in rainbow trout (Oncorhynchus mykiss) and brown trout (Salmo trutta) exposed under natural stream conditions. Journal of Veterinary Diagnostic Investigation 12:312–321.

Bartholomew, J. L. 1998. Host resistance to infection by the Myxosporean parasite Ceratomyxa shasta: a review. Journal of Aquatic Animal Health 10:112–120.

Binns, N. A., and F. M. Eiserman. 1979. Quantification of fluvial trout habitat in Wyoming. Transactions of the American Fisheries Society 103:215–227.

Cuffney, T. F., M. E. Gurtz, and M. R. Meador. 1993. Methods for collecting benthic invertebrate samples as part of the national water-quality assessment program. U.S. Geological Survey. Open-file Report 93-406.

El-Matbouli, M., T. S. McDowell, and R. P. Hedrick. 1999. Effect of water temperature on the development, release and survival of the triactinomyxon stage of Myxobolus cerebralis in its oligochaete host. International Journal for Parasitology 29:627–636.

Elle, S. 1998. Wild trout investigations. Whirling disease studies. Annual Performance Report, Project F-73-R-18. Idaho Department of Fish and Game, Boise, Idaho.

Halliday, M. M. 1973. Studies of Myxosoma cerebralis, a parasite of salmonids. II. The development and pathology of Myxosoma cerebralis, in experimentally infected rainbow trout (Salmo gairdneri) fry reared at different water temperatures. Nordisk Veterinaermedicin 25:349–359.

Hart, Jr., C. W., and S. L. H. Fuller, editors. 1974. Pollution ecology of freshwater invertebrates. Academic Press, New York.

Hedrick, R. P. 1998. Relationships of the host, pathogen, and environment: implications for diseases of cultured and wild fish populations. Journal of Aquatic Animal Health 10:107–111.

Hermansen, H., and C. Kreg. 1984. Influences of physical factors on density of stocked brown trout (Salmo trutta fario L.) in a Danish lowland stream. Fisheries Management 15:107–110.

Hiner, M., and C. M. Moffitt. 2001. Variation in Myxobolus cerebralis infections in field exposed cutthroat, and rainbow trout in Idaho. Journal of Aquatic Animal Health 13:124–132. .

Hnath, J. 1970. Whirling disease in the state of Michigan. Journal of Parasitology 56(4) Section II, Part I, 273:149–150.

Hoffman, G. L. 1974. Fish age as related to Myxosoma cerebralis, cause of whirling disease. The Progressive Fish-Culturist 36:151.

Hoffman, G. L., C. E. Dunbar, and A. Bradford. 1962. Whirling disease of trout caused by Myxosoma cerebralis in the United States. United States Fish and Wildlife Service. Special Scientific Report-Fisheries. No. 427.

Horsch, C. 1987. A case history of whirling disease in a drainage system: Battle Creek drainage of the Upper Sacramento River Basin, California, USA. Journal of Fish Diseases 10:453–460.

Hubert, W. A., and R. J. Rahel. 1989. Relations of physical habitat to abundance of four nongame fishes in high-plains streams: a test of habitat suitability index models. North American Journal of Fisheries Management 9:332–340.

Hynes, H. B. N. 1966. The biology of polluted waters. Liverpool University Press, Liverpool.

Ibarra, A. M., R. P. Hedrick, and G. A. E. Gall. 1994. Genetic analysis of rainbow trout susceptibility to the myxosporean, Ceratomyxa shasta. Aquaculture 120:239–262.

IDEQ (Idaho Division of Environmental Quality). 1999. The 1999 beneficial use reconnaissance project work plan for wadable streams. Beneficial use reconnaissance project technical advisory committee, Boise, Idaho.

Kathman, R. D., and R. O. Brinkhurst. 1998. Guide to the freshwater oligochaetes of North America. Aquatic Resources Center, Tennessee.

Lampert, W., and U. Sommer. 1997. Limnoecology: the ecology of lakes and streams. Oxford University Press, New York.

Layher, W., and O. Maughan. 1985. Relations between habitat variables and channel catfish populations in prairie streams. Transactions of the American Fisheries Society 114:771–780.

Layher, W., O. Maughan, and W. Warde. 1987. Spotted bass habitat suitability related to fish occurrence and biomass and measurements of physicochemical variables. North American Journal of Fisheries Management 7:238–251.

Lewis, S. 1969. Physical factors influencing fish populations in pools of a trout stream. Transactions of the American Fisheries Society 98:14–19.

Luna, L.G, editor. 1968. Manual of histologic staining methods of the Armed Forces Institute of Pathology, 3rd Edition. McGraw-Hill Book Company, New York.

Markiw, M. E. 1986. Salmonid whirling disease: Dynamics of experimental production of the infective stage—the triactinomyxon spore. Canadian Journal of Fisheries and Aquatic Sciences 43:521–526.

Markiw, M. 1992. Experimentally induced whirling disease I. Dose response of fry and adults of rainbow trout exposed to the triactinomyxon stage of *Myxobolus cerebralis*. Journal of Aquatic Animal Health 4:40–43.

McClendon, D., and C. Rabeni. 1987. Physical and biological variables useful for predicting population characteristics of smallmouth bass and rock bass in an Ozark stream. North American Journal of Fisheries Management 7:46–56.

McDowell, D., and R. Naiman. 1986. Structure and function of a benthic invertebrate stream community as influenced by beaver (*Castor canadensis*). Oecologia 68:481–489.

Naiman, R., C. Johnston, and J. Kelley. 1988. Alteration of North American streams by beaver. BioScience 38(11):753–761.

Nehring, R. B., and P. G. Walker. 1996. Whirling disease in the wild: the new reality in the Intermountain West. Fisheries 21:28–32.

O'Grodnick, J. 1979. Susceptibility of various salmonids to whirling disease (*Myxosoma cerebralis*). Transactions of the American Fisheries Society 108:187–190.

Orth, D., and O. Maughan. 1982. Evaluation of the incremental methodology for recommending instream flows for fishes. Transactions of the American Fisheries Society 111:413–441.

Reno, P. W. 1998. Factors involved in the dissemination of disease in fish populations. Journal of Aquatic Animal Health 10:160–171.

Reynoldson, T. B., S. P. Thompson, and J. Bamsey. 1991. A sediment bioassay using the tubificid oligochaete worm *Tubifex tubifex*. Environmental Toxicology and Chemistry 10:818–819.

Rognlie, M. C., and S. E. Knapp. 1998. *Myxobolus cerebralis* in *Tubifex tubifex* from a whirling disease epizootic in Montana. Journal of Parasitology 84:711–713.

Särrkä, J. 1994. Lacustrine, profundal meiobenthic oligochaetes as indicators of trophy and organic loading. Hydrobiologia 278:231–241.

Sandell, T. A., A. Lorz, D. G. Stevens, and J. L. Bartholomew. 2001. Dynamics of *Myxobolus cerebralis* in the Lostine River, Oregon: implications for resident and anadromous salmonids. Journal of Aquatic Animal Health 13:142–150.

SAS. 1996. SAS/STAT user's guide, version 6.12. SAS Institute, Cary, North Carolina.

Snieszko, S. F. 1973. Recent advances in scientific knowledge and developments pertaining to diseases of fishes. Advances in Veterinary Medicine and Comparative Medicine 17:291–314.

Thornton, K. W., B. L. Kimmel, and F. E. Payne. 1990. Reservoir limnology: ecological perspectives. John Wiley and Sons, New York.

Timm, N. H., and T. A. Mieczkowski. 1997. Univariate and multivariate general linear models: theory and applications using SAS software. SAS Institute, Cary, North Carolina.

Vincent, E. R. 1996. Whirling disease and wild trout: the Montana experience. Fisheries 21:32–34.

Wesche, T. A., C. M. Goertler, and W. A. Hubert. 1987. Modified habitat suitability index model for brown trout in southeastern Wyoming. North American Journal of Fisheries Management 7:232–237.

Wolf, K., M. E. Markiw, and J. K. Hiltunen. 1986. Salmonid whirling disease: *Tubifex tubifex* (Müller) identified as the essential oligochaete in the protozoan life cycle. Journal of Fish Diseases 9:83–85.

Yasutake, W. T., and J. H. Wales. 1983. Microscopic anatomy of salmonids: an atlas. United States Department of the Interior, Fish and Wildlife Service. Resource Publication 150. Washington, D.C.

Zendt, J. S., and E. P. Bergersen. 2000. Distribution and abundance of the aquatic oligochaete host *T. tubifex* for the salmonid whirling disease parasite *M. cerebralis* in the Upper Colorado River Basin. North American Journal of Fisheries Management 20:502–512.

American Fisheries Society Symposium 29:181–193, 2002

Whirling Disease among Snake River Cutthroat Trout in Two Spring Streams in Wyoming

WAYNE A. HUBERT* AND MICHAEL P. JOYCE

U.S. Geological Survey, Wyoming Cooperative Fish and Wildlife Research Unit[1]
University of Wyoming, Laramie, Wyoming 82071-3166, USA

ROBERT GIPSON

Wyoming Game and Fish Department, Fish Division, Jackson, Wyoming 83001, USA

DAVID ZAFFT

Wyoming Game and Fish Department, Fish Division, Laramie, Wyoming 82070, USA

DAVID MONEY, DEEDRA HAWK, AND BRANDON TARO

Wyoming Game and Fish Department, Services Division, Laramie, Wyoming 82071-3312, USA

ABSTRACT. We assessed endemic age-0 cutthroat trout *Oncorhynchus clarki* for evidence of pathology associated with *Myxobolus cerebralis* in two streams formed by springs in western Wyoming. We hypothesized that the location of spawning sites in spring streams would affect the extent of exposure of cutthroat trout fry to M. *cerebralis* triactinomyxons (tams), occurrence of the parasite in their bodies, and clinical signs of whirling disease. The spring streams were warm relative to nearby streams flowing from the mountains or spawning and emergence of fry was early compared with fish in mountain streams. Tams were abundant early in the summer and clinical signs of whirling disease among age-0 fish were seen as early as mid-June in one stream. There were high densities of tams in one stream, and densities declined with upstream progression from May through July, whereas in the other stream, low densities of tams were observed in the downstream portion early in the summer, and they were not detected in July and August. Age-0 cutthroat trout were abundant; clinical signs of whirling disease were evident, and histological evidence of whirling disease was common in the stream where tams were abundant. Low densities of age-0 cutthroat trout and no clinical signs of whirling disease were observed in the stream where tams were not abundant. Among sentinel fish in the stream with abundant tams, we found extensive occurrence of M. *cerebralis*, with many fish showing clinical signs and histological evidence of pathology associated with M. *cerebralis*. The proportion of sentinel fish with clinical and histological signs of whirling disease decreased with upstream progression. In the stream with low tam, densities sentinel fish became infected with M. *cerebralis*, but there were essentially no clinical signs or histological indications of whirling disease.

Myxobolus cerebralis has been found in endemic and naturalized salmonid populations and has been associated with declines in recruitment and abundance of salmonids in Colorado and Montana (Nehring and Walker 1996; Vincent 1996). The parasite is widespread in the western United States and appears to be expanding its range (Bergersen and Anderson 1997). Research has shown differences in the effects of M. *cerebralis* on various salmonid species and subspecies endemic or introduced to the western United States (El-Matbouli

et al. 1992; Markiw 1992a; Thompson et al. 1999; Hedrick et al. 1999). Laboratory experiments have demonstrated that the magnitude of infection and the extent of signs of whirling disease are related to the size of fish upon exposure (Hedrick et al. 1999), exposure dose (Markiw 1991, 1992b), and water temperature (Halliday 1973).

Hedrick (1998) stated that disease of wild fish must be considered in the context of the complex interactions of numerous physical, chemical, biological, and ecological parameters (i.e., the aquatic ecosystem). Most research on the effects of M. *cerebralis* on salmonids has occurred in laboratory studies and among cultured populations (Thompson et al. 1999). Responses of wild salmonid populations to M. *cerebralis* are poorly understood

*Corresponding author: whubert@uwyo.edu
[1]The unit is jointly supported by the U.S. Geological Survey, University of Wyoming, Wyoming Game and Fish Department, and the Wildlife Management Institute.

(Walker 1997). Where declines in wild rainbow trout *Oncorhynchus mykiss* in rivers have been related to recruitment failures associated with *M. cerebralis*, there have been few descriptions of the dynamics of whirling disease among age-0 fish (Nehring and Thompson 2001).

Myxobolus cerebralis has been found in wild salmonids in the Salt River and its tributaries, a watershed that drains into the Snake River in western Wyoming. The Yellowstone cutthroat trout *Oncorhynchus clarki bouvieri* is endemic to this watershed. The fine-spotted form of this subspecies is common in the Snake River and some of its tributaries and is commonly referred to as the Snake River cutthroat trout (Behnke 1992). Recent experiments with captive Yellowstone cutthroat trout (Hedrick et al. 1999), and in particular the Snake River cutthroat trout (Thompson et al. 1999), are inconclusive regarding their susceptible to *M. cerebralis* relative to other species of salmonids. There is no evidence based on field studies as to the possible effects of *M. cerebralis* on endemic Snake River cutthroat trout within their natural range.

Some Snake River cutthroat trout in the Snake River and one of its major tributaries, the Salt River, carry out a unique life history involving migration from the rivers into tributaries formed by springs to spawn (Kiefling 1978). Snake River cutthroat trout fry emerge from redds and use the spring streams as nursery habitat. A relatively high proportion of Snake River cutthroat trout and other salmonid species in spring streams tributary to the Salt River have been found to be infected with *M. cerebralis*. However, the possible impacts of the parasite on age-0 Snake River cutthroat trout using spring streams are not known.

Our purpose was to evaluate the spring through summer dynamics of *M. cerebralis* triactinomyxon (tams) and age-0 cutthroat trout in streams formed by large springs relative to evidence of pathology associated with the parasite. We hypothesized that the location of spawning sites in spring streams used by Snake River cutthroat trout would affect the extent of exposure of their fry to tams, occurrence of *M. cerebralis* in their bodies, and clinical signs of whirling disease. Within spring streams, we hypothesized that the occurrence of tams would decrease with progression upstream. In such cases, fry produced in headwater areas would be exposed to fewer tams than fry in downstream reaches. The tendency of some

Snake River cutthroat trout to migrate to headwater areas to spawn might result in reduced impacts of *M. cerebralis* on their offspring.

STUDY AREA

Two spring streams, Christensen Creek and Perk Creek (Figures 1 and 2), on the floodplain of the Salt River in Lincoln County, Wyoming, were studied. Several headwater springs contribute to the overall discharge of each stream. The natural stream channels tend to be wide and shallow with sand substrate. Substantial reaches of both streams have been modified to enhance habitat for trout. Habitat modifications included the digging of pools (1.5–2.0 m deep) and construction of gravel-cobble riffles between pools. In addition to endemic cutthroat trout, naturalized rainbow trout *O. Nerka*, brown trout *Salmo trutta*, and brook trout *Salvelinus fontinalis* occur in the streams.

METHODS

Physical Parameters

Base maps of each spring stream were constructed from U.S. Geological Survey topographic maps (1:24,000 scale). The Christensen Creek study reach was 3.5 km long, and the Perk Creek study reach was 1.5 km long. The entire length of the study reach in each spring stream was surveyed to determine the spatial distribution of habitat features. A technique similar to the basin-wide inventory (Hankin and Reeves 1988) was used. Water temperature was monitored from March through August with recording thermometers placed at five sites over the length of each study reach. Discharge was measured at several locations over the length of each study reach, and staff gages were used to monitor variation in discharge (McMahon et al. 1996).

Spawning and Age-0 Salmonid Sampling

Snorkeling was conducted biweekly in 10 pools over the length of each study reach during March and April to determine fish species and length composition. Observed salmonids were visually identified to species and placed in three length classes (<31, 31–45, and > 45 cm total length, TL).

The entire length of each study reach was visually surveyed biweekly from the bank during the day from March through August to identify the locations of redds and age-0 salmonids. The

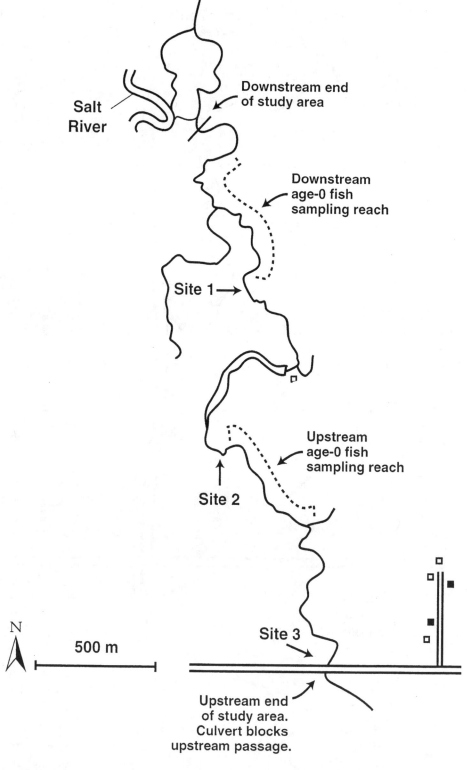

Figure 1. Map of the Christensen Creek study reach showing the locations of where tam sampling occurred and sentinel fish were placed (sites 1–3) and where age-0 fish were sampled.

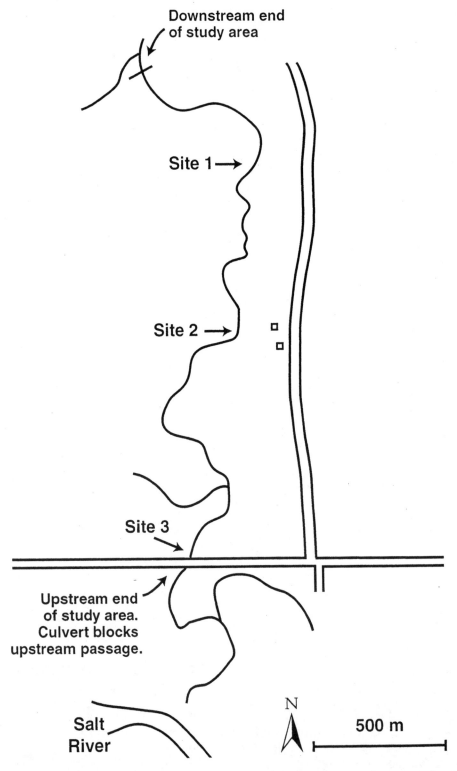

Figure 2. Map of the Perk Creek study reach showing the locations where tam sampling occurred and sentinel fish were placed (sites 1–3).

proportion of riffles with recently constructed (within the preceding 2 weeks) redds was used as an index of spawning activity because superimposition prevented identification of individual redds. During each survey, one bank of each stream was consistently walked, and the locations and numbers of age-0 salmonids observed along the stream margin were recorded. The number of age-0 salmonids observed per 100 m of stream bank during each survey was used as an index of their abundance. Clinical signs of whirling disease (black tail, spinal deformities, and erratic swimming) were noted when observed. A sample of age-0 fish was collected by dipnetting during each survey to assess fish species composition and lengths and to enable more detailed observations of clinical signs of whirling disease.

Tam Sampling

Tams were sampled biweekly at three sites (see Figures 1 and 2) over the length of each study reach from May through August, using the technique developed by Thompson and Nehring (2000). Three 423-L samples were filtered through a 20-micron filter at each site. The collected material was stained with crystal violet, and 10 sub-samples of 80 microliters each were drawn. Tams were identified and enumerated by examining the subsamples in a gridded Petri dish under a dissecting microscope. The mean number of tams observed among the subsamples was expanded to estimate the density of tams (number/m^3).

Pathology

We sampled age-0 cutthroat trout, from both an upstream and downstream reach of each spring stream, by electrofishing in early August. All captured age-0 salmonids were identified to species, measured, and observed for clinical signs of whirling disease. One hundred fish, from each reach in Christensen Creek, identified to be *Oncorhynchus*, were sacrificed for further analyses. Pelvic fins were removed and preserved in ethanol for determination of the presence of M. *cerebralis* DNA, using nested polymerase chain reaction (PCR) analysis (Andree et al. 1998). The caudal fin was removed and preserved in ethanol for analysis by the Fish Genetics Laboratory at Brigham Young University to determine if the fish were cutthroat trout, rainbow trout, or cutthroat trout × rainbow trout hybrids. The remainder of the body was preserved in 10% buffered formalin and sub-

mitted to the Washington Animal Disease Diagnostic Laboratory (WADDL) at Washington State University for standard histological analysis of cranial tissue (Hedrick et al. 1999). Each fish was graded using the MacConnell-Baldwin grading system: 0 = no infection; 1 = minimal infection, few areas of cartilage damage, and no inflammatory response; 2 = mild infection, several areas of cartilage damage, and focal inflammatory response; 3 = moderate infection, several areas of cartilage damage in each section, and diffuse inflammatory response; 4 = moderately severe infection, several to numerous areas of cartilage damage, and extensive inflammatory response; and 5 = severe infection, areas of cartilage damage throughout, and extensive inflammatory response.

Sentinel Fish

Sentinel fish experiments were conducted beginning 30 June and 16 July 1999 in Christensen Creek and 16 July 1999 in Perk Creek, to assess longitudinal variation in occurrence of M. *cerebralis* and clinical signs of whirling disease among age-0 Snake River cutthroat trout in each stream. Flooding prevented an experiment from being initiated in Perk Creek in late June. Snake River cutthroat trout were obtained from the Jackson National Fish Hatchery and were certified to be pathogen free. Each experiment consisted of 50 age-0 sentinel fish (mean TL = 40 mm) being placed in each of two enclosures at the three locations where tams were sampled in each stream (see Figures 1 and 2). They were exposed to flowing water for 7 d and then transported to the Wild Trout Research Laboratory at Montana State University in Bozeman for rearing. The fish were sacrificed 4 months postexposure when the fish averaged 82 mm TL. Individual fish were measured and observed for clinical signs of disease; pelvic fins were removed and preserved in ethanol; and bodies were preserved in 10% buffered formalin. Pelvic fins were assessed for the presence of M. *cerebralis* DNA, using PCR. Twenty-five randomly selected sentinel fish, from each site for the experiments that began 16 July 1999, were sent to WADDL for histological analysis and grading.

Chi-square analysis was used to assess differences in frequency data among study sites. Student's *t*-test and one-way analysis of variance (ANOVA) were used to compare means between study reaches or among study sites. Significance was determined at *P* less than 0.05 for all tests.

RESULTS

Habitat

Over the length of the study reaches, 55 constructed riffles and pools were identified in Christensen Creek, and 36 constructed riffles and pools were found in Perk Creek. Spawning habitat occurred almost exclusively in the constructed riffles, but a small amount of spawning gravel was found in association with a few spring upwellings near the headwaters of each stream.

Mean daily water temperatures were 7.5–8.0°C in March 1999 and increased to 8.0–8.5°C in August 1999 immediately downstream from the headwater springs of both study streams with little diel variation (less than 1°C). At the most downstream site in each study reach, mean daily water temperatures were 5.5–6.5°C in March and increased gradually to 10.0–11.5°C in August with substantial diel variation (up to 5°C).

Discharge in Christensen Creek was 0.45–0.55 m³/s at the downstream end of the study reach. About 50% of the discharge originated from a complex of headwater springs and the remainder from three tributaries formed by springs and springs within the main stem of the stream channel. Highest flows occurred during periods of snowmelt in May associated with surface runoff from the adjacent floodplain.

Discharge in Perk Creek was 0.35–0.45 m³/s at the downstream end of the study reach during most of the study (March through the middle of June and the middle of July through August). Approximately 40% of the discharge originated from a complex of headwater springs. The remainder came from two tributaries formed by springs and springs within the main-stem channel. However, during spring runoff (middle of June to the middle of July), the Salt River rose and inundated all study sites in Perk Creek. During this time, the water level was 1.5–2.0 m above normal; the stream flowed out of its banks; and the water was very turbid.

Salmonid Assemblages

During snorkeling observations in March and April 1999, 780 salmonids were observed in Christensen Creek, including 523 cutthroat trout, 130 brook trout, 71 brown trout, 52 rainbow trout, and 4 cutthroat trout × rainbow trout hybrids. Among the 523 cutthroat trout, 33% were less than 31 cm, 49% were 31–45 cm, and 19% were greater than 45 cm

TL. A total of 129 salmonids were observed in Perk Creek, including 55 cutthroat trout, 60 rainbow trout, 12 brown trout, and 2 brook trout. Among the 55 cutthroat trout, 7% were less than 31 cm, 80% were 31–45 cm, and 13% were greater than 45 cm TL. Cutthroat trout were observed over the length of both study reaches, with no evident of longitudinal patterns in abundance or length structure.

Snorkeling and visual observations from the shore did not identify any cutthroat trout, brown trout, or rainbow trout in the headwaters (upstream from tam-sampling site 3) of either Christensen Creek or Perk Creek. However, a few brook trout were observed in the headwaters of both streams. Tam-sampling site 3 was immediately downstream from a culvert under a road that appeared to block upstream movement of salmonids into the headwaters of both streams (Figures 1 and 2).

Spawning

Visual surveys of redds indicated that they were almost exclusively associated with constructed riffles in both spring streams. Redds were observed in both streams from March into June, but it could not be determined if the redds were constructed by cutthroat trout or rainbow trout. Recent redd construction in Christensen Creek was observed on 41 of 55 riffles over the length of the study reach, but no redds were observed in the 300-m segment immediately downstream from tam-sampling site 3 where there was no habitat modification (Figure 1). Thirty-six riffles were identified in Perk Creek in early March, but the seven most downstream riffles were inundated by a backwater when the Salt River rose in April, and they were no longer considered as potential spawning sites. Redd construction in Perk Creek was observed among 26 of the 29 riffles from late March into the middle of June (Figure 3).

Tams

Biweekly samples of tams indicated substantial differences in densities between the two spring streams (Table 1), with much higher densities in Christensen Creek. In Christensen Creek, densities declined with upstream progression over the three sampling sites from May through July, but the highest densities were observed at the middle-sampling site in August. Tams were collected at the most upstream site in Christensen Creek in only one sample (8 June, Table 1). In Perk Creek, tams were found only at the downstream and middle sites and only on 18 May and 14 June (Table 1).

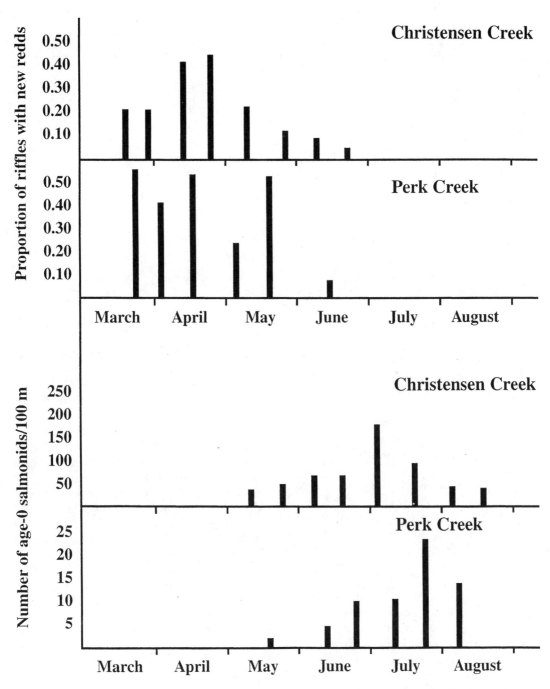

Figure 3. Proportions of riffles with new redds and number of age-0 salmonids/100 m visually observed over the Christensen Creek and Perk Creek study reaches on sampling dates from March through August 1999.

Table 1. Estimated densities (number/m³) of tams at downstream, middle and upstream sampling sites in the Christensen and Perk creeks study reaches during 1999.

	Sampling site		
Date	1 (Downstream)	2 (Middle)	3 (Upstream)
	Christensen Creek		
11 May	7.0	0	0
25 May	54.9	16.1	0
8 June	48.5	33.8	2.7
21 June	29.8	8.1	0
7 July	33.9	16.0	0
19 July	1.2	0	0
3 August	2.2	53.1	0
16 August	0	40.6	0
	Perk Creek		
18 May	0	4.6	0
14 June	2.2	2.3	0
28 June	0	0	0
12 July	0	0	0
26 July	0	0	0
10 August	0	0	0

Age-0 Salmonids

Age-0 salmonids were observed in both Christensen Creek and Perk Creek, beginning in May. The age-0 fish were *Oncorhynchus* and brown trout, but the *Oncorhynchus* could not be identified to species using morphological features. Age-0 salmonids were observed over the length of the Christensen Creek study reach, with the exception of the 300 m immediately downstream from tam-sampling site 3 (Figure 1). Among the age-0 fish in Christensen Creek that were visually observed from the bank, no clinical or behavioral signs of whirling disease were observed in May, but black tails and erratic swimming were seen from 7 June through August with their prevalence increasing through the summer. Among the dip-netted samples of age-0 salmonids collected from Christensen Creek, clinical signs of disease (predominantly black tails with some spinal deformities) were observed among both *Oncorhynchus* and brown trout beginning 20 June (length range = 24–50 mm TL) and continuing through the summer. Forty percent of the fish sampled by dip net in August (length range = 31–57 mm TL) showed clinical signs of disease.

The density of age-0 salmonids in Perk Creek was about one-tenth that observed in Christensen Creek (Figure 3). Age-0 salmonids were scattered over the entire length of the Perk Creek study reach, and no clinical signs of whirling disease were observed at any time during the summer.

On 2 and 3 August 1999, age-0 salmonids were collected by electrofishing near the downstream and middle sites where tams were sampled in Christensen Creek. Age-0 fish were rare near the upstream tam-sampling site (Figure 1), so a sample could not be collected. Totals of 268 age-0 fish were collected at the downstream site (119 *Oncorhynchus* and 149 brown trout) and 186 age-0 fish were collected at the middle site (154 *Oncorhynchus*, 30 brown trout, and 2 brook trout). Clinical signs of whirling disease (black tails and spinal deformities) were observed only among *Oncorhynchus* greater than 40 mm TL and among brown trout greater than 50 mm TL. The highest proportion of clinical signs occurred among *Oncorhynchus* of 51–60 mm TL at both the downstream and upstream sites.

The first 100 *Oncorhynchus* greater than 40 mm TL, collected from both the downstream and middle sites on Christensen Creek, were preserved for laboratory analyses. Genetic analysis indicated that all of the fish were cutthroat trout (R. P. Evans and D. K. Shiozawa, Brigham Young University, personal communication). Clinical signs of whirling disease were observed among fish from both the downstream and middle sites (Table 2).

Among cutthroat trout of 41–80 mm TL, 31% from the downstream site and 30% from the middle reach showed clinical signs of whirling disease. Similarly, among brown trout of 51–80 mm TL, 38% from the downstream site and 32% from the middle site showed clinical signs of whirling disease. The proportions were not significantly different between downstream and middle sites for either species. The PCR analysis indicated that 68% of the cutthroat trout from the downstream site and 89% from the middle site had M. cerebralis DNA in pelvic fin tissue. The proportions were not significantly different. Histological analysis of cutthroat trout indicated that 79% from the downstream site and 84% from the middle site were infected with M. cerebralis (Table 3), with no significant difference in the proportions. The mean histological grade was 2.2 among fish from the downstream site and 2.5 among fish from the middle site with no significant difference between sites or among length classes.

Identification of the presence of M. cerebralis among cutthroat trout, sampled from Christensen Creek, differed between PCR and histological analyses (Table 4). Of the 200 fish analyzed, 157 were identified as positive for M. cerebralis using PCR and 163 using histological analysis (grades 1–5). However, 27 of the 37 fish (73%) that graded 0 histologically were positive for M. cerebralis DNA using PCR. Conversely, 33 of the 156 fish (21%) that graded 1–5 were negative for M. cerebralis DNA using PCR. When either histological grades 1–5 or a positive PCR reading were used as a measure of the presence of M. cerebralis, 95% of the age-0 cutthroat trout were identified as being infected by the parasite.

On 3 August 1999, age-0 salmonids were sampled in Perk Creek. Approximately 200 m (15% of

Table 2. Frequency of clinical signs of whirling disease among cutthroat trout and brown trout sampled near the downstream and middle tam-sampling sites in Christensen Creek on 2 and 3 August 1999.

Length increment (mm)	Cutthroat trout		Brown trout	
	Percent with clinical signs	Number observed	Percent with clinical signs	Number observed
Site 1 (Downstream)				
≤ 40	0	18	0	0
41–50	22	55	0	3
51–60	46	9	52	29
61–70	20	5	38	50
71–80	0	2	31	42
81–90	0	0	17	12
91–100	0	0	1	8
Site 2 (Middle)				
≤ 40	0	37	0	0
41–50	18	39	0	1
51–60	42	45	80	5
61–70	32	25	43	7
71–80	14	7	8	12
81–90	0	0	0	5
91–100	0	0	0	0

Table 3. Histological analysis (using the MacConnell-Baldwin grading system) of age-0 cutthroat trout greater than 40 mm TL collected from near the downstream and middle tam-sampling sites in Christensen Creek on 2 and 3 August 1999.

Sampling site	Grade						mean grade
	0	1	2	3	4	5	
Downstream	21	15	18	21	23	2	2.2
Middle	16	15	17	11	23	5	2.5

the study reach) was sampled in the vicinity of the downstream tam-sampling site (Figure 2) yielding 92 age-0 fish (70 *Oncorhynchus*, 21 brown trout, and 1 brook trout). A similar length of stream was sampled near the upstream tam-sampling site, and 43 age-0 fish were collected (16 *Oncorhynchus*, 26 brown trout, and 1 brook trout). No fish collected at either site showed clinical signs of whirling disease. Due to the low densities of age-0 *Oncorhynchus* and the absence of clinical signs of whirling disease, no age-0 *Oncorhynchus* were sacrificed for laboratory analysis.

Sentinel Fish

Clinical signs (black tails and spinal deformities) of whirling disease occurred among sentinel fish from both study streams (Table 5), but their occurrence was significantly greater among fish from Christensen Creek (15%) than from Perk Creek (2%). Within Christensen Creek, the proportion of fish showing clinical signs of whirling disease declined with upstream progression in both experiments.

The PCR analysis identified M. *cerebralis* DNA in pelvic fin tissue of sentinel fish from Christensen Creek, but only 6 of 209 fish were positive for parasite DNA with no downstream to upstream trends. None of the sentinel fish from Perk Creek were positive for parasite DNA.

Histological analysis revealed 60% occurrence of M. *cerebralis* among sentinel cutthroat trout examined from Christensen Creek. Only 4% of the fish from Perk Creek showed histological evidence of infection by the parasite (Table 6). Downstream to upstream trends were observed in both study streams. Among the sentinel fish in Christensen Creek, 88% at the downstream site (mean grade = 3.1), 92% at the middle site (mean grade = 2.8), and none at the upstream site showed histological evidence of M. *cerebralis*. In Perk Creek, 12% of the fish at the downstream site (mean grade = 0.1) showed histological evidence of M. *cerebralis*, whereas no fish at the middle or upstream sites showed histological evidence of M. *cerebralis*.

DISCUSSION

Our purpose was to assess the dynamics of M. *cerebralis* tam occurrence and clinical signs of whirling

Table 4. Comparison of results obtained by nested polymerase chain reaction (PCR) analysis (Positive = *Myxobolus cerebralis* DNA detected in fin tissue, Negative = *M. cerebralis* DNA not detected in fin tissue) and histological analysis and grading using the MacConnell-Baldwin grading system among 200 age-0 cutthroat trout collected from Christensen Creek on 2 and 3 August 1999.

PCR	Grade						Total
	0	1	2	3	4	5	
Positive	27	22	28	24	49	7	157
Negative	10	8	7	8	10	0	43

Table 5. Occurrence of clinical signs (black tail or spinal deformities) of whirling disease among cutthroat trout at downstream (1), middle (2) and upstream (3) tam-sampling sites from sentinel fish experiments initiated on 30 June (Experiment 1) and 16 July (Experiment 2) 1999 in Christensen and Perk creeks.

Stream	Experiment	Site	Total number of survivors to termination	Number with clinical signs disease
Christensen	1	1	89	18
		2	95	14
		3	99	0
Christensen	2	1	84	23
		2	87	19
		3	95	8
Perk	2	1	98	1
		2	32	2
		3	90	2

Table 6. Frequencies of histological grades using the MacConnell-Baldwin grading system among 25 randomly selected sentinel cutthroat trout that survived to the termination of the experiment at the downstream (1), middle (2), and upstream (3) tam-sampling sites from the sentinel fish experiment initiated on 16 July 1999 in Christensen and Perk creeks.

Stream	Site	Grade						Mean
		0	1	2	3	4	5	
Christensen	1	3	4	3	0	8	7	3.1
	2	2	5	2	5	8	3	2.8
	3	25	0	0	0	0	0	0.0
Perk	1	22	2	1	0	0	0	0.1
	2	25	0	0	0	0	0	0.0
	3	25	0	0	0	0	0	0.0

disease among age-0 cutthroat trout in spring streams. We found that spawning occurred primarily on constructed riffles. Water temperatures of the springs (7.5°C) made the water warmer than streams flowing from the surrounding mountains during the spring. Consequently, spawning occurred 4–6 weeks earlier than in montane streams, and age-0 cutthroat trout emerged from redds earlier and grew to longer lengths by the end of the summer. Tams were observed in both streams in May. In Christensen Creek, they remained abundant through the summer at the downstream and middle sampling sites at mean daily water temperatures of 8–12°C. Clinical signs of whirling disease were first observed in June, shortly after cutthroat trout emerged from redds, and the frequency of clinical signs increased into August, in Christensen Creek.

In Christensen Creek, where tams were abundant and clinical signs of whirling disease were prevalent, we found that age-0 cutthroat trout of 41–80 mm TL in early August showed clinical signs of disease, with fish of 51–60 mm TL showing the highest proportions. However, histological grading did not vary significantly among fish from 41 to 80 mm TL. The occurrence of M. cerebralis in these fish was extensive, with 95% showing evidence of infection by PCR or histological analyses. Our observations suggest that relatively small fish can be used to monitor whirling disease within individual spring streams or to evaluate the relative magnitude of whirling disease among age-0 fish in different spring streams.

Seventy-nine percent of the age-0, free-ranging cutthroat trout collected from Christensen Creek in early August were positive for M. cerebralis DNA in their pelvic fin tissue, whereas only 3% of

the sentinel fish exposed for 7 d in July and reared for 4 months were positive for M. cerebralis DNA in pelvic fin tissue. Histological evidence (grades 1–5) of infection by M. cerebralis was extensive among the fish collected from the wild (82% of fish sampled) and the sentinel fish from the two downstream sites (90%). Extensive evaluation of our laboratory procedures indicated that false positives with PCR were not occurring among the sentinel fish. The discrepancy between PCR and histopathology may have been due to the tissue that was analyzed. Wild fish were exposed to tams up to the sampling date, but sentinel fish had not been exposed to tams for 4 months when they were sacrificed for analysis. We hypothesize that M. cerebralis DNA was rare in pelvic fins of sentinel fish 4 months after exposure because developmental stages had migrated from the fins to the central nervous system (El-Matbouli et al. 1992, 1999) or M. cerebralis DNA in the fins had been destroyed; consequently, their DNA was not detected by PCR analysis. However, this hypothesis needs to be tested through controlled experiments. Nevertheless, our findings indicate that pelvic fin tissue should not be used to assess the presence of M. cerebralis in sentinel fish experiments.

Clinical signs of whirling disease among age-0 cutthroat trout and brown trout were substantial in Christensen Creek (see Table 2). The relative histopathology and survival among age-0 cutthroat trout and brown trout in the wild, where clinical signs of disease are apparent, need to be assessed. When Thompson et al. (1999) compared rainbow trout, brook trout, brown trout, and different subspecies of cutthroat trout in controlled experiments, they found that one group of Snake River cutthroat trout survived better and developed fewer spores than any other group except brown trout, but the Snake River cutthroat trout were older than the brown trout when exposed. Another group of Snake River cutthroat trout, exposed at a younger age than brown trout, had lower survival than brown trout.

We hypothesized that the life history of endemic cutthroat trout in spring streams of the Salt River drainage may reduce susceptibility to whirling disease. Some of our results support the hypothesis that age-0 cutthroat trout in upstream areas of spring streams have reduced infection with M. cerebralis and clinical signs of whirling disease. The cutthroat trout in the sentinel fish experiments in Christensen Creek showed declines in occurrence of clinical signs and histological evidence of whirling disease with upstream progression. The hypothesis that tam densities decrease with upstream progression was supported by the results from both study streams. However, the facts that spawning sites were rare in upstream portions of the study areas and that upstream migration into the headwaters of both spring streams was blocked by culverts suggest that our findings should not be extrapolated to other spring streams.

We found a high degree of temporal concordance in the onset and prevalence of clinical signs of whirling disease and the relative abundance of tams at study sites. There was a high prevalence of clinical signs in Christensen Creek and little evidence of disease in Perk Creek. Similarly, the density of tams was substantially greater in Christensen Creek than in Perk Creek. In the vicinity of the upstream headwater springs of both streams, but upstream from culverts that blocked upstream fish movement, there was substantial organic sediment in both streams that could serve as habitat for Tubifex tubifex (Zendt and Bergerson 2000). In addition, the constructed pools within both streams had substantial sediment and organic matter in the bottoms that could serve as habitat for T. tubifex (Rognlie and Knapp 1998). The only substantial habitat difference between the two streams was the occurrence of flooding. Christensen Creek was not affected by flooding, whereas Perk Creek was inundated when the Salt River rose during spring runoff in June. The current velocities in Perk Creek when flooded were not sufficient to flush sediment from either the upstream headwater area or the bottom of the constructed pools. Communications with area residents and landowners suggest that flooding of Perk Creek, associated with spring runoff, is relatively common (perhaps occurring 1 out of 2 years), but it is extremely rare in Christensen Creek. Nevertheless, the density of tams was very low (prior to and after flooding), and there was essentially no evidence of whirling disease among age-0 salmonids in Perk Creek. However, this may indicate that M. cerebralis had only recently become enzootic in Perk Creek. Additional research, together with assessment of the hydrological, biological, and chemical characteristics of the two spring streams, may identify the mechanisms responsible for the differences in the occurrence of whirling disease among age-0 salmonids.

ACKNOWLEDGMENTS

We thank S. Sturlaugson and B. Barney for assistance in the field, K. Grande and personnel of the U.S. Fish and Wildlife Service at the Jackson National Fish Hatchery for providing fish and assisting in development of the sentinel fish experiments, C. Frasier for rearing the sentinel fish, E. MacConnell and several students at Montana State University for processing the sentinel fish when the experiments were terminated, T. Johnson for his consistent help and support, and L. Perkins for access to his property. The research was funded by the Whirling Disease Initiative of the National Partnership on Management of Wild and Native Cold Water Fisheries and the Wyoming Game and Fish Department.

REFERENCES

Andree, K. B., E. MacDonnell, and R. P. Hedrick. 1998. A nested polymerase chain reaction for the detection of genomic DNA of Myxobolus cerebralis in rainbow trout (Oncorhynchus mykiss). Diseases of Aquatic Organisms 34:145–154.

Behnke, R. J. 1992. Native trout of western North America. American Fisheries Society, Monograph 6, Bethesda, Maryland.

Bergersen, E. P., and D. E. Anderson. 1997. The distribution and spread of Myxobolus cerebralis in the United States. Fisheries 22(8):6–7.

El-Matbouli, M., T. Fischer-Scherl, and R. W. Hoffmann. 1992. Present knowledge on the life cycle, taxonomy, pathology, and therapy of some Myxosporea spp. important for freshwater fish. Annual Review of Fish Diseases 3:367–402.

El-Matbouli, M., R. W. Hoffmann, and C. Mandok. 1999. Light and electronic microscopic observations on the route of the triactinomyxon-sporoplasm of Myxobolus cerebralis from epidermis into rainbow trout cartilage. Journal of Fish Biology 46:919–935.

Halliday, M. M. 1973. Studies of *Myxobolus cerebralis*, a parasite of salmonids. II. The development and pathology of *Myxobolus cerebralis* in experimentally infected rainbow trout (*Salmo gairdneri*) fry reared at different temperatures. Nordisk Veterinaermedicin 25:349–358.

Hankin, D. G., and G. H. Reeves. 1988. Estimating total fish abundance and total habitat area in small streams based on visual estimation techniques. Canadian Journal of Fisheries and Aquatic Sciences 45:834–844.

Hedrick, R. P. 1998. Relationships of the host, pathogen, and environment: implications for diseases of cultured and wild fish populations. Journal of Aquatic Animal Health 10:107–111.

Hedrick, R. P., T. S. McDowell, K. Mukkatera, M. P. Georgiadis, and E. MacConnell. 1999. Susceptibility of selected inland salmonids to experimentally induced infections with *Myxobolus cerebralis*, the causative agent of whirling disease. Journal of Aquatic Animal Health 11:330–339.

Kiefling, J. W. 1978. Studies of the ecology of the Snake River cutthroat trout. Wyoming Game and Fish Department Fisheries Technical Bulletin 3, Cheyenne, Wyoming.

Markiw, M. E. 1991. Whirling disease: earliest susceptible age of rainbow trout to the triactinomyxid of *Myxobolus cerebralis*. Aquaculture 92:1–6.

Markiw, M. E. 1992a. Salmonid whirling disease. U.S. Fish and Wildlife Service Fish and Wildlife Leaflet 17.

Markiw, M. E. 1992.b. Experimentally induced whirling disease. I. Dose response of fry and adults of rainbow trout exposed to the triactinomyxon stage of *Myxobolus cerebralis*. Journal of Aquatic Animal Health 4:40–43.

McMahon, T. E., A. V. Zale, and D. J. Orth. 1996. Chapter 4. Aquatic habitat measurements. Pages 83–120 in B. R. Murphy and D. W. Willis, editors. Fisheries techniques, 2nd edition. American Fisheries Society, Bethesda, Maryland.

Nehring, R. B., and K. G. Thompson. 2001. Impact assessment of some physical, and biological factors in the whirling disease epizootic among wild trout in Colorado. Colorado Division of Wildlife Aquatic Research Special Report 76, DOW-R-S-76-01.

Nehring, R. B., and P. G. Walker. 1996. Whirling disease in the wild: the new reality in the Intermountain West. Fisheries 21(6):28–30.

Rognlie, M. C., and S. E. Knapp. 1998. *Myxobolus cerebralis* in *Tubifex tubifex* from a whirling disease epizootic in Montana. Journal of Parasitology 84:711–713.

Thompson, K. G., and R. B. Nehring. 2000. A simple technique used to filter, and quantify the actinospore of *Myxobolus cerebralis* and determine its seasonal abundance in the Colorado River. Journal of Aquatic Animal Health 12:316–323.

Thompson, K. G., R. B. Nehring, D. C. Bowden, and T. Wygant. 1999. Field exposure of seven species or subspecies of salmonids to *Myxobolus cerebralis* in the Colorado River, Middle Park, Colorado. Journal of Aquatic Animal Health 11:312–329.

Vincent, E. R. 1996. Whirling disease and wild trout: the Montana experience. Fisheries 21(6):32–33.

Walker, P. G. 1997. Whirling disease problem reveals need to redirect agency fish health programs. Fisheries 22(8):4.

Zendt, J. S., and E. Pl Bergersen. 2000. Distribution, and abundance of the aquatic oligochaete host *Tubifex tubifex* for the salmonid whirling disease parasite *Myxobolus cerebralis* in the Upper Colorado River basin. North American Journal of Fisheries Management 20:502–512.

Section 6

Diagnostic Methods

American Fisheries Society Symposium 29:197–211, 2002

A Review of the Approaches to Detect *Myxobolus cerebralis*, the Cause of Salmonid Whirling Disease

KARL B. ANDREE* AND RONALD P. HEDRICK

Department of Medicine and Epidemiology, School of Veterinary Medicine
University of California at Davis, Davis, California 95616, USA

ELIZABETH MACCONNELL
Montana Fish Wildlife and Parks, 1400 19th St, Bozeman, Montana 59715, USA

ABSTRACT. The impacts of whirling disease on the aquaculture of trout over the past 80 years has resulted in the development of simple to more complex methods to detect the causative agent *Myxobolus cerebralis*. With the absence of effective treatments, efficient detection of this pathogen is critical. It is our goal to give a historical perspective to these developments while discussing the relative merits of the different techniques.

The current widespread distribution of *M. cerebralis* is in part due to the movements of hatchery-raised trout for stocking for sport fishing that were often infected but showing no signs of disease. Employing more sensitive diagnostic procedures to detect the pathogen in lightly infected fish is one important means to prevent this inadvertent spread of *M. cerebralis*.

The detection of spores of the correct size and shape and in the known target tissues is the basis of the currently accepted diagnosis of *M. cerebralis* infection. Unfortunately, many myxosporean spores share similar morphology, and improved approaches to discriminating *M. cerebralis* from other *Myxobolus* spp. found in salmonid tissues are needed. There have been progressive improvements in the detection and discrimination of *M. cerebralis* spores in fish tissues, including the introduction of mechanical and enzymatic means to release spores from skeletal elements, analyses of pooled or batch samples, and the use of labeled antibodies for the specific identification of spores. More sensitive and specific detection limits for the parasite in fish tissue and oligochaetes have been obtained using polymerase chain reaction (PCR) and *in situ* hybridization. The benefits of these new molecular approaches are already evident as research tools, and their validation for use as more broadly applied diagnostic reagents should be an immediate goal.

Myxobolus cerebralis, the etiological agent of whirling disease, is the most intensively studied of the myxosporean parasites of fish yet no effective treatments for this disease have been developed. Experimental oral administrations of fumagillin have been shown to reduce the severity of infection, but the parasite is not eliminated from the host (El-Matbouli and Hoffmann 1991). Avoidance of the parasite, to prevent infections in captive fish, is therefore a primary management approach. Avoidance relies on effective disinfection of water or sediments by the use of various chemicals (Hoffmann and Hoffmann 1972), ultraviolet (UV) light (Hoffmann 1975; Hedrick et al. 2000), and ozone (M. El-Matbouli, University of Munich, unpublished data). All of these methods can be effective in preventing contact between the parasite and susceptible trout.

Accurate detection of the parasite provides a measure of the success of management procedures aimed at controlling whirling disease. Due to the impacts of whirling disease on wild rainbow trout in certain rivers of the Intermountain West of the United States, testing to detect the pathogen is routinely conducted both on wild and captive-reared stocks of trout to 1) understand the current distribution of *M. cerebralis*, 2) identify point sources and monitor the spread of infection, and 3) determine prevalence and severity of infection in wild trout populations. Surveillance programs can also target other hosts and locations where the parasite might be present, including worms, water, and bird feces.

---

*Corresponding author: kbandree@ucdavis.edu

Whirling disease was first described by Hofer (1903) and came to be regarded as a condition found primarily among hatchery fish (Halliday 1976). The disease spread to the United States from Europe in the 1950s with shipments of trout for human consumption and is now found in 23 states (Hoffmann 1990; Andree et al. 1999). Since the early 1990s, the parasite has become a serious concern to wild salmonid populations in the states of the Intermountain West (Nehring and Walker 1996; Vincent 1996). The importance of detecting the parasite has therefore grown with commercial, state, and federal rearing of trout (Hoffmann 1990) and with the potential detrimental impacts the parasite can have on wild and native trout populations (Nehring and Walker 1996).

The name "whirling disease" is derived from the tail-chasing swimming behavior exhibited during the acute phase of the disease. Concurrent with the aberrant swimming behavior, the fish usually exhibits caudal melanosis due to pressure upon nerves in this region by the host inflammatory response to the parasite. The inflammation impinges on root ganglia along the spine preventing control of the melanocytes of the skin, resulting in melanosis in the caudal third of the body (El-Matbouli et al. 1995). The aberrant swimming behavior has been recently attributed to impaired motor control due to the parasite and the host inflammatory response constricting the upper spinal cord and lower brain stem (Rose et al. 2000) and not due to the disruption of the auditory organs, as was previously thought. Developmental stages of M. cerebralis reach the central nervous system via peripheral nerves after penetration of the skin. The skin, and the mucous pores especially, is the main site of attachment for the infectious triactinomyxon (TAM) stage (El-Matbouli et al. 1995). However, the skin and central nervous tissue are sites for proliferation of the parasite and not sporogenesis (El-Matbouli et al. 1995). After reaching skeletal elements, the parasite destroys cartilage by the production of lytic enzymes. Destruction of the cartilage, if extensive, can lead to permanent skeletal deformities of various severity that may include caudal lordosis and/or scoliosis, misaligned jaws, foreshortening of the snout, exophthalmia, and foreshortening of the operculum. It is the destruction of the cartilaginous elements that leads to development of permanent skeletal deformities, which characterize the chronic phase of the disease (Hedrick et al. 1998). These skeletal abnormalities will persist even after caudal melanosis sub-

sides. Sporogenesis begins after the parasite has reached the cartilage. Through sequential divisions, microscopic multicellular spores with protective valves are formed, which then can lie dormant in the skeleton. In cases where skeletal aberrations are minimal, the fish can be considered asymptomatic carriers, harboring spore aggregates in the bones and cartilage that are released after death of the fish.

The development in the fish host represents only half of the life cycle of M. cerebralis. The parasite also requires an oligochaete worm Tubifex tubifex, in which a second sequential proliferative and sporogonic development takes place. The triactinomyxon or actinosporean stage that forms during the sporogenesis in the worm is the infective stage that attaches to the fish (Figure 1A; Wolf and Markiw 1984; Wolf et al. 1986, Markiw and Wolf 1983). The degree to which the disease develops in the fish host is dependent on the species and age of the salmonid (Hedrick et al. 1998) and the dose of the infective stage. The number of infective stages that might be found in the water is dependent on the type, strain, and abundance of the oligochaete host (Beauchamp et al. 2001) and environmental variables, many that remain to be more fully understood.

The initial management approaches to control whirling disease in Europe and the United States were rapid and severe, with destruction of fish stocks followed by hatchery disinfection (Horsch 1987; Hoffmann 1990). Fish were also destroyed in rivers and streams by employing toxic chemicals in an effort to contain the parasite (Hnath 1970). As the life cycle of the parasite became better understood, additional control measures were possible. Reducing the oligochaete habitat by replacing earthen-bottom ponds with concrete raceways and rearing young fish in well water prior to exposure to open-water supplies often effectively prevented the onset of whirling disease. Trout that were raised in these conditions, however, were often infected. The stocking of these trout for sport fishing is one means by which the pathogen has spread to new geographic regions. Treating incoming water with UV light or using strictly well water for all stages of rearing were measures that completely prevented M. cerebralis infection in trout raised in certain facilities (Hoffmann 1975; Hedrick et al. 2000).

The current management of whirling disease depends upon sensitive and specific detection methods for M. cerebralis. Several methods have been described for this purpose, including relative-

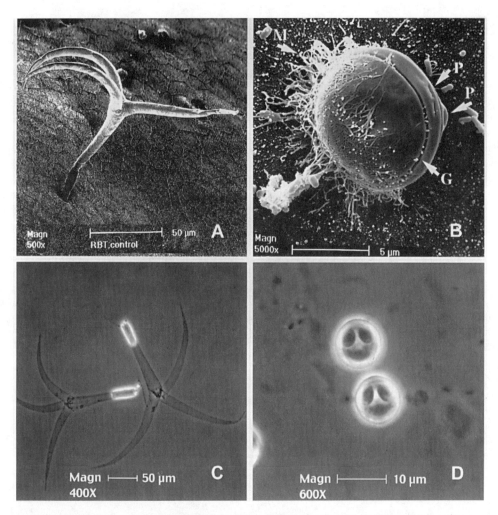

Figure 1. Scanning electronmicrographs and light photomicrographs showing actinosporean and myxosporean stage spores of *M. cerebralis*. 1A and 1C show the actinosporean stage of *M. cerebralis*, which develops in the oligochaete worm host. 1A shows the actinosporean spore stage immediately after attachment, via the polar filaments, to the skin of a rainbow trout. 1B and 1D show the myxosporean stage spore, which develops in the salmonid fish host. Using scanning electronmicroscopy, the mucous coating on the posterior of the spore (M), polar capsule pores for extrusion of polar filaments (P), and parasutural grooves (G) are evident.

ly new molecular approaches. In this review, we examine the historical development of these detection methods for M. *cerebralis* and discuss their strengths and weaknesses, to address particular management concerns.

DIAGNOSTIC METHODS

A positive diagnosis of whirling disease depends on observation of the characteristic gross clinical signs and demonstration of the presence of M. *cerebralis* spores associated with cartilage. These signs include a tail-chasing swimming behavior and cau-

dal melanosis during the acute phase. Later in infection, these signs may be accompanied by deformities present throughout the skeleton but usually most pronounced in the cranium and spine. These deformities can vary considerably in severity, making light infections difficult to identify. Also, bacterial cold-water disease, electroshock injuries, and dietary insufficiencies can cause skeletal deformities and/or melanosis, which resembles whirling disease in salmonids (Margolis et al. 1996). Diagnoses of whirling disease require demonstrating the presence of M. *cerebralis* spores (provided ample time has passed for them to develop) by microscopic examination of extracts from cranial skeleton and stained tissue sections from the same region. Although presporogonic stages can be observed in stained tissue sections (Hermanns and Korting 1985), the unique spore stages that form as the end point of myxosporean sporogonic development are the accepted standard to identify species in the phylum Myxozoa. Due to the tissue tropism of M. cerebralis, early procedures developed for diagnosis of whirling disease involved excision and then crushing of skeletal elements prior to microscopic observation (Plehn 1924; Lucky 1970). However, several Myxobolus spp. exhibit similar morphology (Figure 2) and can be found in cranial tissues (Hedrick et al. 1991). It is, therefore, insufficient to base diagnoses of whirling disease on gross clinical signs or the presence of spores in tissue extracts alone. Ultimately, confirmation of whirling disease depends on observation of characteristic gross clinical signs and detection of spores of the required size and shape associated with lesions in cartilage in stained tissue sections from suspect salmonid fish (Figure 3).

The current and standard procedures recommended by the Fish Health Section (FHS) of the American Fisheries Society (AFS) require that heads of the fish be processed by either of two approved extraction techniques, followed by microscopic examination to demonstrate presence of spores (see below). The skeletal elements of the head are subjected to enzymatic digestion and centrifugation, or the head is ground directly and the homogenate is passed through a plankton centrifuge. In both procedures, the goal is to isolate spores for microscopic observation. Additionally, fish heads are placed in fixative and, following standard paraffin embedding, sections are prepared and stained with hematoxylin and eosin and examined to confirm that the spores or developmental stages of the parasite are present in cartilage (Lorz and Amandi 1994). A modification of these guidelines, used by some laboratories, is to cut fish heads in half in a midline saggital plane and process each half separately. One half is used for spore isolation and enumeration, and the other half is used for histological analysis. A weakness of this approach is the asymmetric distribution of spores and developmental stages between the two halves of the head, which, particularly at low parasite concentrations, can lead to disparate results between histological and spore enumeration analyses (Andree et al. 1998).

Spore Description

Taxonomic classification of members of the genus Myxobolus to the species level can be difficult, and the aid of an experienced parasitologist for descriptions of the size and morphologic criteria associated with the spore may be needed (Lom 1987; Wolf and Markiw 1976). When spores or gross signs of the disease are misinterpreted, misdiagnoses of whirling disease can occur that may have expen-

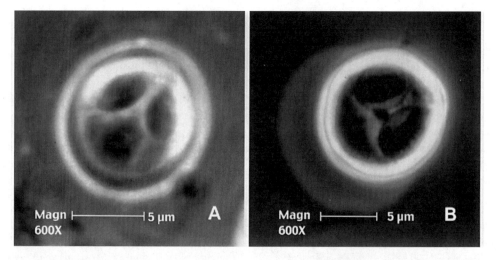

Figure 2. Light photomicrographs of spores isolated from rainbow trout cranial tissue. In Figure 2A is *Myxobolus cerebralis* shown after purification using the pepsin trypsin digest method. In Figure 2B is an unnamed *Myxobolus* species found in PTD extracts and localized by histology in the brain stem. This specimen is shown after purification on a Percoll gradient.

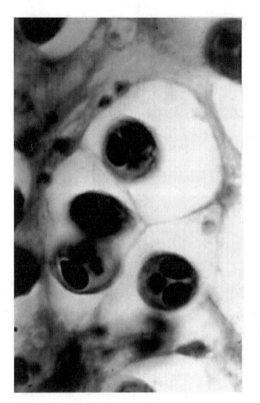

Figure 3. A section of an inclusion of *M. cerebralis* spores in cranial bones stained with methylene blue.

sive consequences for the aquaculturist (Margolis et al. 1996). The myxosporean stage spore of M. *cerebralis* has a lenticular shape measuring 9.7 μm in length and 8.5 μm in width, with two equally-sized polar capsules of 4.2 × 3.1 μm (Lom and Hoffmann 1970). The polar filaments make 5–6 turns within the polar capsules and are expelled through pores at the anterior end of the spore (Figure 1B). The posterior aspect of the M. *cerebralis* spore is covered by a mucous envelope that can be visualized by negative staining with dyes such as India ink (Lom and Hoffmann 1971). This character has been described as variable by some authors, and its appearance may be dependent on spore maturity (Mitchell 1989). There is a parasutural groove visible by scanning electron microscopy along each side of the suture line where the two protective valves join (Figure 1B) (Lom and Hoffmann 1971).

The myxosporean spore stage has been the historical focal point for the diagnosis of whirling disease. Detection of the characteristic size and shape of the spores by simple microscopic proce-

dures made this stage of the parasite most recognizable, particularly since other stages found in the fish (e.g., in the skin and nerves) were unknown or more difficult to observe, or their morphology could not be attributed to a single species. In addition, antigen and DNA-based diagnostic approaches that resolved certain of these concerns had yet to be developed or applied to M. *cerebralis* detection.

Spore Staining Techniques

The presence of spores in heavily infected fish can be demonstrated by simply crushing dissected skeletal elements and observing them by light microscopy. The detection of spores in lightly infected fish, including fish with no clinical signs of whirling disease, is considerably more difficult. Various methods have therefore been developed to isolate and concentrate spores and to stain them using a variety of dyes such as methylene blue (Figure 3), malachite green, or silver nitrate to facilitate their detection. Although the original methods of Plehn (1924) did not advocate staining, MacLean (1971) found that the use of malachite green aided in the identification of spores, especially for detection of spores in crude preparations of macerated bones. Wolf and Markiw (1979) compared a variety of stains to find one suitable for use in conjunction with the pepsin-trypsin digestion method and plankton centrifugation. Such a stain would need to impregnate the surface of the spores and remain stable through the changes of pH and mechanical shear forces that spores experience during these procedures. Among the stains used, many were lost during the spore purification process or damaged spores by altering their morphology. Silver nitrate was retained in the spores and left the spores intact. However, the silver nitrate staining protocol added 3 h to the purification procedure, so it was not advocated for standard use but as a tool to aid in measuring the efficiency of spore recovery from the selected skeletal elements (Wolf and Markiw 1979).

Some stains, though not species-specific, have been used for the identification of the actinosporean stages of M. *cerebralis*. Crystal violet has been used successfully for staining actinosporean stages to aid in their identification in water filtrates (Thompson and Nehring 2000). More specialized staining methods, such as fluorescein diacetate (FDA) and propidium iodide (PI), have been used to indicate the viability of germ cells in the sporoplasm of the actinosporean stage of M. *cerebralis* and other myxosporeans (Yokoyama et al. 1997).

Mechanical and Enzymatic Isolation of Spores

Early diagnostic procedures, such as those used by Plehn (1924), provided turbid preparations of crushed bone, making spore identification difficult. A more complex method involving sequentially pulverizing and filtering samples, followed by centrifugation to remove bony matrix, was proposed by Prasher et al. (1971). A clear outline for quantifying spores was also provided in this protocol. Realizing the extensive nature of infestation of spores throughout the axial skeleton and the skull, this protocol was modified further to include the bones of the lower jaw and gills (Tidd et al. 1971). Yet, a more thorough separation of spores from bony material was achieved using enzymatic digestion of the bones (Landolt 1973; Markiw and Wolf 1974b; Markiw and Wolf 1974c). A comparison of spore purification methods found that spore yields were significantly increased when the mechanical and enzymatic treatments of bone were followed by stepped gradient centrifugation (Markiw and Wolf 1974a).

Based largely upon the accuracy and ability to batch sample, the AFS/FHS Blue Book (Lorz and Amandi 1994) adopted the enzymatic pepsin and trypsin digestion (or PTD), as developed by Markiw and Wolf (1974b). This method involves the sequential digestion of the bony elements of the skeleton after removal of soft tissue. The digest of the bones is filtered and concentrated prior to purification by dextrose gradient centrifugation. The resultant material is resuspended in a small volume for examination by light microscopy for the presence of spores. This procedure is coupled to the examination of stained tissue sections from the same cranial tissues used in the digest. Recent modifications to the PTD include replacing the pepsin solution if the initial pH of 1.8 rises above 3.8, as above this pH the pepsin activity decays. A second modification reduces the concentration of trypsin for frozen samples from a final concentration of 0.5% to 0.05%, thereby preventing unwanted degradation of the spores (Linda Chittum, Utah Division of Wildlife Resources, personal communication). In our laboratory, an additional modification is the replacement of dextrose with percoll (Sigma Chemical Company) for the gradient centrifugation.

The plankton centrifugation method was developed as an alternate, time-saving method for spore isolation from fish tissues (O'Grodnick 1975). In this procedure, fish heads are homogenized, and the homogenate is filtered through gauze. This crude filtrate is then passed through a plankton centrifuge and the resulting pellet examined microscopically. While the preparation for microscopic observation is more turbid, this alternative has been accepted by the AFS/FHS as an approved method for M. cerebralis detection (Lorz and Amandi 1994). Observation of spores from the plankton centrifugation method can be improved by the addition of a trypsin treatment (0.25%) to clarify the material collected from the plankton centrifuge (Markiw and Wolf 1974c).

The PTD and plankton centrifuge methods are well suited to batch sampling and are effective means for extracting and concentrating the spores from fish tissues. Both methods provide the ability to quantify the numbers of spores per sample. Since these methods do provide quantitative results, special care should be taken to clean the equipment between batch samples, to remove all spores that may adhere to the apparatus and contaminate subsequent samples. The PTD method is often preferred for quantitative analysis, since this method has been demonstrated to allow detection of a greater proportion of spores than simple physical separatory methods, due to release of the spores from bony inclusions after the enzymatic digestion (Markiw and Wolf 1974a, 1974c). By comparison, the plankton centrifuge spore preparations are considerably more turbid than those from the PTD method, making spore identification more difficult. Weaknesses of both procedures include a significant investment in time and labor and the ability to detect only the hardened spore stage of the parasite. The length of time from initial exposure to the parasite until these spores are fully formed depends on numerous factors, including water temperature (Halliday 1973). Therefore, periods from 3 to 11 months post exposure to the parasite may be required before fish are suitable for sampling using these methods. Lastly, these procedures are not designed for, nor do they detect, presporogonic stages in the fish host or any of the developmental or sporogonic stages in the oligochaete host.

Histological Assessment of Disease in Fish

Examination of mounted stained tissue sections is required to demonstrate the presence of spores, first identified in pepsin-trypsin digest or plankton centrifuge extracts, which are associated with cartilage lesions, in order to confirm a diagnosis of whirling disease. While histological assessment is more costly and labor intensive, making it less conducive to batch testing, it is still required to prevent misdiagnoses when *Myxobolus* spp. that are not M. *cerebralis* are found in cranial tissues other than cartilage. Tissues are fixed and slides prepared following the methods described by the AFS/FHS Blue Book (Lorz and Amandi 1994). Until recently, there have been no guidelines for assessing the degree or severity of infection associated with the parasite. A more objective, quantitative measure of lesions, developed by one of the authors (E. MacConnell) and Tom Baldwin (Washington State University), is now available and has been widely used to evaluate the impact of M. *cerebralis* on exposed fish (Baldwin et al. 2000). The histological assessment uses a graded scale based on the number and size of cartilage lesions, associated inflammation, and involvement of surrounding tissues. All cartilaginous tissue in each section should be examined. Best results are obtained when the grade is based on two head or whole body sections per fish (Table 1 and Figure 4). This scale provides a numerical standard for more objective comparisons between and within diagnostic laboratories. The grading scale has also been directly beneficial to research that requires evaluation of the responses to infection among fish in field and laboratory trials.

Immunologic Methods

Labeled antibodies could, in theory, provide a more species-specific, objective detection method. The successful use of serologic methods, such as direct (DFAT) or indirect fluorescent antibody tests (IFAT), requires that the initial antigens used for antibody production be well purified (Goldman 1968). Unfortunately, preparations of spores purified by mechanical and/or enzymatic means usually contain considerable amounts of fish tissue. These spore preparations can be further cleared of bony debris by two-phase separations or gradient centrifugation (Contos and Rothenbacher 1974; Hamilton and Canning 1988; Kozel et al. 1980; Markiw and Wolf 1974a). Initial difficulties experienced with antibody-based detection methods included auto-fluorescence of spores (Putz and McElwain 1969) and the inability to detect circulating antibodies by DFAT in serum from infected rainbow trout (Halliday 1974). Later reports by Griffin and Davis (1978), however, demonstrated the presence of antibodies in the serum of infected fish, using an IFAT. Subsequent studies with polyclonal antisera prepared in rabbits injected with myxosporean spores of M. *cerebralis* demonstrated conserved antigens with the actinosporean stages by DFAT (Markiw 1989). Hamilton and Canning (1988) used polyclonal mouse anti-M. *cerebralis* antibodies to confirm conservation of antigens among presporogonic and spore stages of M. *cerebralis*. However, cross-reactivity between spores of different *Myxobolus* spp. became apparent, as demonstrated in the DFAT with M. *cartilaginis* (Markiw and Wolf 1978). Despite these problems, the IFAT and DFAT can still serve as useful investigative tools for the study of particular antigens of the parasite, some of which may represent important immunogens in the fish.

Table 1. MacConnell-Baldwin numerical scale for scoring lesion severity.

Grade	Description
0	No abnormalities noted. *M. cerebralis* is not present.
1	Small, discrete, focus or foci of cartilage degeneration. No or few leukocytes associated with lesion.
2	Single, locally extensive focus or several smaller foci of cartilage degeneration and necrosis. Inflammation is localized. Few to moderate numbers of leukocytes infiltrate lytic cartilage.
3	Multiple foci (usually 3 to 4*) of cartilage degeneration and necrosis. Moderate numbers of leukocytes associated with lytic cartilage. Inflammation has minimal or mild impact on surrounding tissues.
4	Multifocal (usually 4 or more sites*) to coalescing areas of cartilage degeneration and necrosis. Moderate to large numbers of leukocytes border and infiltrate lytic cartilage. Locally extensive areas of granulomatous inflammation found, involving surrounding tissues.
5	Multifocal (usually 6 or more*) to coalescing areas of cartilage necrosis. Moderate to large numbers of leukocytes border and infiltrate necrotic cartilage. Granulomatous inflammation is extensive with severe impact on surrounding tissues. This classification is characterized by loss of normal architecture (as shown in this section of the nares) and reserved for the most severely affected fish. If in doubt, classify as a grade 4.

* Numbers of lesions typical for sections of head, not whole body sections.

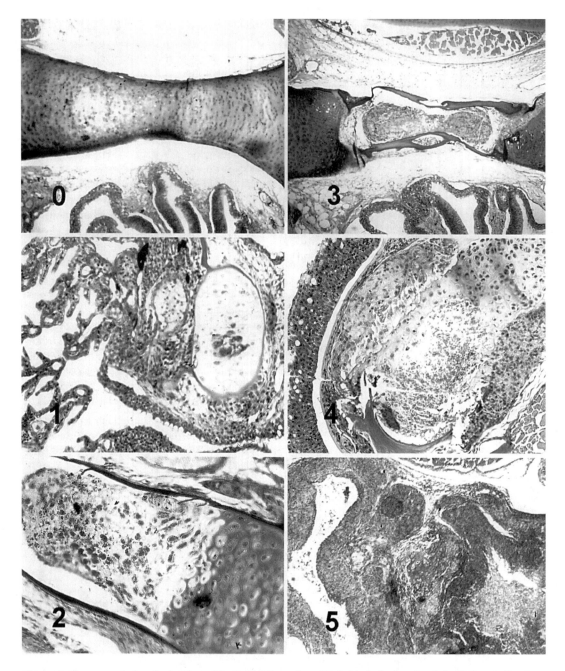

Figure 4. The numerical scale developed by E. MacConnell and T. Baldwin for scoring infection severity of salmonid fish by *M. cerebralis*. Grades 0, 3, and 5 are sections through the nares. Grades 1, 2, and 4 are sections through gill arches. Grades 0, 1, 3, and 5 are shown at a magnification of 250X. Grades 2 and 4 are shown at a magnification of 630X. See Table 1 for description of grades.

Polymerase Chain Reaction Tests for the Detection of *Myxobolus* cerebralis

While the approved diagnostic methods for M. *cerebralis* have distinct advantages, they require significant expenditures of time and money and ultimately rely on the proper identification of the myxosporean spore stage. The problems with the current immunological methods have also been discussed. An important shortcoming of the currently approved diagnostic methods is the inability to detect all stages of the parasite throughout the life cycle. An alternative approach employing DNA-based detection of the organism has therefore been developed in our laboratory (Andree et al. 1998). Genomic DNA provides a stable target to detect the parasite throughout its life cycle. We have used the small subunit ribosomal DNA (SSU rDNA or 18S rDNA) of M. *cerebralis* as the target for the PCR, as these genes occur in multiple copies in the genome and their sequences can be compared with a growing database of related organisms.

Smothers et al. (1994) published the first rDNA sequences for myxosporeans. Since that time, additional rDNA sequences have been obtained and DNA-based tests have been developed for a number of myxosporeans, including *Ceratomyxa shasta*, *Kudoa thyrsites*, *Myxobolus cerebralis*, and PKX (Palenzuela et al. 1999; Hervio et al. 1997; Andree et al. 1998; Saulnier and De Kinkelin 1997). An important criterion for the design of these PCR tests is their ability to detect only the parasite of concern and not closely related species. The particular primer sequences used for detection of M. *cerebralis* were chosen after comparison to related myxosporean species and host 18S rDNA to find sights most unique to M. *cerebralis* (Table 2). Most importantly, it was demonstrated that the PCR test for M. *cerebralis* did not amplify DNA from *Myxobolus* spp. that may be found in the salmonid fish hosts for the parasite (Andree et al. 1998). Detection of genomic DNA of M. *cerebralis* is possible by either a nested test or by employing one set of primers in a single

amplification to save time and money in reagents (Table 2). Detection of infection with M. *cerebralis*, using the nested PCR assay, is demonstrated by the specific amplification of a 415 bp DNA fragment visible in ethidium bromide stained agarose gels. A second amplicon of 508 bp may appear in concert with the 415 bp fragment in a dose-dependent fashion, under the conditions used by Andree et al. (1998). The second 508 bp band appears due to carryover of Round One primers, when primer and specific target DNA concentration is high (Figure 5). Only one band of 508 bp appears when using the single-round amplification assay. The program used in the thermal cycler for the nested test generally uses the following parameters: denature samples at 95°C for 5 min; amplify using 35 cycles of 1 min at 95°C, followed by 2.5 min at 65°C, followed by 1.5 min at 72°C; hold samples at 72°C for 10 min for a final elongation step. When testing for parasite DNA using a single round of amplification, the annealing and polymerization times can be shortened to 30 s each. The single-round PCR has been shown to detect fewer fish with light infections than the nested test (Andree et al. 1998); however, both PCR assays have a threshold of detection that is 10-fold lower than either of the AFS/FHS-approved spore enumeration methods. Additionally, using the currently approved AFS/FHS methods, specificity and sensitivity are achieved through two separate tests, while the PCR assay, in principal, can provide both in a single assay. For the above stated reasons, by January 2004, a single-round version of this test as modified by Pisces Molecular (Boulder, Colorado) (Schisler et al. 2001) will be authorized by the American Fisheries Society Fish Health Section, for use as the sole confirmatory test for detection of M. *cerebralis*.

Although the PCR assay for detection of M. *cerebralis* is not currently approved by the AFS/FHS as a substitute to the traditional procedures, it has already shown its utility as a powerful research tool to detect light and or early infections in fish and oligochaetes. In a recent comparison of

Table 2. Probes for *Myxobolus cerebralis* rDNA, their sequence (5′–3′) and the diagnostic applications for which they are suitable.

Probe	Sequence	Application
Tr 5-16	GCATTGGTTTACGCTGATGTAGCGA	PCR (single and nested) / ISH
Tr 3-16	GAATCGCCGAAACAATCATCGAGCTA	PCR (nested) / ISH
Tr 5-17	GCCCTATTAACTAGTTGGTAGTATAGAAGC	PCR (nested)
Tr 3-17	GGCACACTACTCCAACACTGAATTTG	PCR (single and nested) / ISH

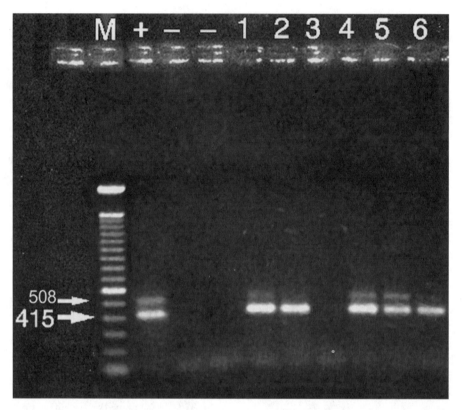

Figure 5. The nested PCR assay for *Myxobolus cerebralis* will generate a diagnostic band of 415 bp. Under circumstances where primer and target DNA concentrations are high, carryover of primers from round one can generate a 508 bp band as well. Using the primer combination for the single round assay generates only the 508 bp band. Lanes are marked as follows: M is 1 μg of 100 bp ladder; (+) is plasmid p18Tr29 bearing the 18S rDNA of *M. cerebralis*; (−) are uninfected fish DNA; 1–6 are wild rainbow trout samples.

PTD and the single-round PCR for detection of M. *cerebralis* in free-ranging and hatchery-raised salmonid fish, the PCR identified M. *cerebralis* DNA more frequently than did PTD. Among the free-ranging fish tested, 31.3% tested positive by PCR, while only 14.5% were found to harbor spores when analyzed by PTD (Schisler et al. 2001). Other examples of where the PCR test has been useful include detecting parasite DNA and therefore demonstrating (1) the inability of some subspecies of T. *tubifex* to act as a host for M. *cerebralis*, (2) the effects of thermal shifts on the parasite burden of T. *tubifex* (El-Matbouli et al. 1999), (3) the effectiveness of UV irradiation or ozone for killing actinosporean stages in water sources (Hedrick et al. 2000), (4) the presence or absence in river drainages of oligochaetes infected with M. *cerebralis*, (5) the presence of the parasite in water filtrates, and (6) differences or distinguishing between different actinosporeans in water and oligochaete samples. The PCR is acutely sensitive to detecting very small amounts of target DNA. In laboratory tests, the nested M. *cerebralis* PCR assay could detect the DNA of a single germ cell within the sporoplasm of the actinosporean stage in a background of 300 ng of host genomic DNA (Andree et al. 1998). A disadvantage of the current PCR assay is the inability to effectively quantify the severity of infection in the host.

As with all PCR assays, start up involves a considerable financial investment in equipment and reagents. A properly trained technician is also needed to run the test. All DNA diagnostic assays provide results that are directly related to the quality of the sample obtained. Quality in reference to a DNA sample pertains to purity of the DNA and lack of degradation. Clean and careful technique is paramount for the collection of the fish biopsy/necropsy and throughout the DNA extraction and amplification processes. Also, cross-contamination between samples is always a concern when running highly sensitive assays, such as the PCR. Therefore, extra care should be taken to clean reusable items in 10% bleach for 10 min and/or to use disposable items, such as collection tubes, collection bags, razors, gloves, etc., as much as possible. Once the DNA extraction has commenced, pulse centrifugation is recommended prior to opening the DNA extraction tubes for addition of reagents, to ensure no liquid is adherent to the inside of the tube cap. This can help to prevent sample material getting spread to other samples from the outer surface of the tubes or on the surface of gloves. Methods for DNA extraction can vary from phenol emulsion type organic extractions to the spin column chromatography used in many commercial kits. An advantage of the commercial kits is the assurance that all reagents have been prepared in a facility lacking the DNA your laboratory is working with, which precludes the possibility of contamination of these reagents.

Collecting samples for the PCR assay (Andree et al. 1998) requires obtaining a lethal sample from the cranial region. The testing of worms is simplified, since they can be collected in pools of 10–20, or individually, and require no homogenization prior to DNA extraction (authors' unpublished data). For field collection of samples, or in situations where DNA extraction will proceed at a later time, tissues can be placed and stored in 70–90% ethanol or, alternatively, kept on ice for some hours prior to freezing. The ethanol can be aspirated away just before addition of the lysis buffer. Nonlethal sampling has been used for PCR assay detection in sentinel fish for early detection of M. cerebralis. This method involves retrieving fin clips or skin scrapings from the exposed fish after 24–48 h exposure to suspect waters. Nonlethal sampling of sentinel fish could be especially useful for fisheries managers in an effort to identify those river systems where the parasite is present without the need to sacrifice endangered or threatened wild salmonids.

Collection of water samples can be done directly or after concentration from large volumes by filtration. Water or water filtrates can be fixed with 95% ethanol to a final concentration of 70%. Later, the solid material in the water is collected by centrifugation, followed by removal of the supernatant prior to DNA extraction. The results from PCR assays from nonfish samples should be carefully considered. In ecosystems where M. cerebralis occurs, only myxosporeans from the salmonid fish host have been studied carefully, to date. Thus, samples may contain other parasites or organisms that have not specifically been examined for their reaction in the PCR assay. As a confirmation of the PCR results from such samples, random samples of positive PCR products might be sequenced to confirm the identity of the sequence, by comparison with the 18S rDNA sequence deposited in GenBank (Andree et al. 1997). A commercial laboratory (Pisces Molecular) has reported good results conducting PCR assays on water filtrates and worms in addition to typical fish samples (Schisler et al. 2001).

In Situ Hybridization Procedure for the Detection of *Myxobolus* cerebralis

An in situ hybridization (ISH) protocol has been developed, using digoxigenin labeling of the primers used for the PCR assay for the detection of *Myxobolus cerebralis* in infected tissues of either the worm or the fish host (Antonio et al. 1998). The labeled primers (probes) for this protocol hybridize specifically to the parasite rDNA present in sections of infected tissues. Either temperature of the hybridization, or wash steps, or altering the salt concentration in the wash buffers controls the stringency of probe binding. Digoxigenin-labeled probes, binding to parasite DNA, are detected by standard immunohistochemical protocols, using a monoclonal anti-digoxigenin antibody conjugated to horseradish peroxidase. Positive staining appears as tight foci of purple precipitate against a brown background from the Bismarck brown counterstain. This procedure allows the identification of very low numbers of parasites within tissues that might otherwise be unrecognizable with conventional stains. While this procedure is less sensitive than PCR and limited by the number of samples that can be processed concurrently, the major strength of this method is the ability to obtain spatial and temporal distribution information about the parasite in either host. Particularly useful applications of this technique include investigations of portals of entry, routes of spreading through tissues, and discrimination of different actinosporean types and stages in the oligochaete host.

Resource Management

Controlling whirling disease in wild trout populations has become a major concern for resource managers. Complete depopulation has traditionally been an approach employed with captive and, in a few cases, wild trout populations. After the first detection in the United States, aggressive steps were taken to eliminate M. cerebralis infections that resulted in destruction of numerous fish, and disinfection of hatchery facilities (Hoffmann 1990). For example, during a four-year period (1970–1974), the Michigan Department of Natural Resources, in an effort to eradicate M. cerebralis, repeatedly used chlorine and rotenone on 7 mi of the Tobacco River after M. cerebralis was found in wild trout. This effort ultimately failed to remove the parasite when it reappeared in a 1975 survey (Hnath 1970). In California, M. cerebralis was first found in Monterey County in 1965 and subsequently spread to several areas in the state (Modin 1998). Many infections were not accompanied by gross clinical signs of whirling disease until a severe epizootic occurred in 1995 among rainbow trout at the Mt. Whitney State Fish Hatchery. After the removal of the source of infection in at least three sites in California, spores of the parasite can no longer be detected in wild populations in the same area (Modin 1998). Earlier in California in 1985, myxosporean stage spores 12–15 mm in diameter that resembled those of M. cerebralis were found in head tissues of rainbow trout from the Darrah Springs State Fish Hatchery. Stock destruction and hatchery disinfection was ordered, but the spores were later found to be from nervous tissue and were identified as M. kisutchi rather than M. cerebralis (Horsch 1987).

Management strategies for hatchery rearing of salmonid fish arrived at empirically, before the life cycle of the parasite was described in 1984, are still commonly employed. These include (1) the use of concrete raceways to reduce worm habitat, (2) stocking of fish into waters with the parasite after they are at least 15 cm in length, (3) rearing fish on well water or treatment of incoming water with UV irradiation, and (4) limiting transport of M. cerebralis positive fish.

In 1988, an emergency meeting was convened after the appearance of whirling disease in the Colorado River, Colorado, USA. By 1994, the 1988 outbreak had led to a dramatic decline in salmonid populations in several areas of the intermountain west of the United States (Nehring and Walker 1996; Vincent 1996). This initiated a more intense surveillance of wild salmonid populations across the United States for whirling disease, which involved the use of both AFS/FHS approved diagnostic procedures and the PCR assay. Has PCR made an impact on resource management decisions? The answer would seem to be yes. The Colorado Division of Wildlife now utilizes the single-round PCR assay to certify captive-reared, endangered fish species prior to stocking "triple A" waters. The fish are grown in isolation units and screened by examination of stained tissue sections and PCR before stocking. Also, in Colorado, the PCR has been used to screen sentinel fish in an effort to eliminate M. cerebralis from at least one hatchery water supply. The PCR provides managers with a sensitive and rapid assay that yields results unachievable through other means (Linda Chittum, Utah Division of Wildlife Resources, personal communication). This is an example of how the PCR can be useful in making decisions regarding fishery management, although it does not directly involve a certification of fish. A PCR-positive result cannot take precedence over negative PTD or histological examination results until the PCR assay has been accepted as a validated diagnostic test. However, with the advantages of the PCR and the limits of the current methods in mind, it would be prudent to pursue validation of the PCR assay for the detection of M. cerebralis.

CONCLUDING REMARKS

Prior to recognizing that the actinosporean and myxosporean stages were the same organism, diagnostic developments focused on the myxosporean stage spores. Later, spore isolation methods were perfected to the level that antigen preparations injected into mice or rabbits could be used to develop antibody-based detection techniques. However, the changing antigenic compositions of the life stages make these detection procedures for myxosporeans more difficult. DNA-based detection procedures offer many advantages over conventional methods, and these are evident from significant research advances that have utilized these new techniques (identifying susceptible and non-susceptible hosts, more sensitive means of analyzing hosts in a given watershed, etc.). Now, even newer technologies, such as real-time quantitative PCR that allows sensitive detection of concentrations of the parasite DNA without the need to run gels, are available. This would improve batch sampling further and avoid some of the concerns regarding contamination that are inherent to any typical PCR assay where tubes of amplified DNA are opened for loading and running gels.

Although the PCR and ISH represent improved detection methods for the fish and especially the oligochaete hosts, further test development is needed. Possible areas of improvement include more research to conclusively demonstrate their effectiveness in detecting M. *cerebralis* in piscivorous bird feces (Taylor and Lott 1978; El-Matbouli et al. 1992a) and water filtrates. Currently, all accepted detection methods for M. *cerebralis* are lethal. A more thorough examination and standardization of the nonlethal sampling methodology for the PCR are also needed. Research in these areas, combined with the eventual reduced costs for real-time quantitative PCR, will help to fully develop the rDNA probes now used to detect M. *cerebralis*. The research needed to improve and validate these new diagnostic procedures should be an immediate goal.

ACKNOWLEDGMENTS

Thanks are given to Mansour El-Matbouli and Beth MacConnell for contributing some of the photographic images for this work. Funding was provided in part by the U.S. Fish and Wildlife Service, the California Department of Fish and Game, and the Whirling Disease Foundation.

REFERENCES

Andree, K. B., S. J. Gresoviac, and R. P. Hedrick. 1997. Small subunit ribosomal RNA sequences unite alternate actinosporean and myxosporean stages of *Myxobolus cerebralis* the causative agent of whirling disease in salmonid fish. Journal of Eukaryotic Microbiology 44(3):208–215.

Andree, K. B., E. MacConnell, and R. P. Hedrick. 1998. A polymerase chain reaction test for detection of *Myxobolus cerebralis*, the causative agent of salmonid whirling disease in fish, and a comparison to existing detection techniques. Diseases of Aquatic Organisms 34(2):145–154.

Andree, K. B., M. El-Matbouli, and R. P. Hedrick. 1999. Comparison of 18S and ITS-1 rDNA sequences of specific geographic isolates of *Myxobolus cerebralis*. International Journal of Parasitology 29:771–775.

Antonio, D. B., K. B. Andree, T. S. McDowell, and R. P. Hedrick. 1998. Detection of *Myxobolus cerebralis* in rainbow trout (*Oncorhynchus mykiss*) and oligochaete tissues using a nonradioactive *in situ* hybridization protocol. Journal of Aquatic Animal Health 10:338–347.

Baldwin, T. J., E. R. Vincent, R. M. Siflow, and D. Stanek. 2000. *Myxobolus cerebralis* infection in rainbow trout (*Oncorhynchus mykiss*) and brown trout (*Salmo trutta*) exposed under natural stream conditions. Journal of Veterinary Diagnostic Investigations 12:312–321.

Beauchamp, K. A., R. D. Kathman, T. S. McDowell, and R. P. Hedrick. 2001. Molecular Phylogeny of tubificid oligochaetes with special emphasis on *Tubifex tubifex* (Tubificidae). Molecular Phylogenetics and Evolution 19(2):2126–2124.

Contos, N., and H. Rothenbacher. 1974. An efficient concentration and purification method for spores of *Myxosoma cerebralis*. Progressive Fish-Culturist 36:101–102.

El-Matbouli, M., and R. W. Hoffmann. 1991. Prevention of experimentally induced whirling disease in rainbow trout (*Oncorhynchus mykiss*) by fumagillin. Diseases of Aquatic Organisms 10:109–113.

El-Matbouli, M., and R. W. Hoffmann. 1992a. Effect of freezing, aging and passage through the alimentary canal of predatory animals on the viability of *Myxobolus cerebralis* spores. Journal of Aquatic Animal Health 3:260–262.

El-Matbouli, M., T. Fischer-Scherl, and R. W. Hoffmann. 1992b. Present knowledge of the life cycle, taxonomy, pathology, and therapy of some myxosporea spp. important for freshwater fish. Annual Review of Fish Diseases 3:367–402.

El-Matbouli, M., R., W. Hoffmann, and C. Mandok. 1995. Light and electron microscopic observations on the route of the triactinomyxon-sporoplasm of *Myxobolus cerebralis* from epidermis into rainbow trout (*Oncorhynchus mykiss*) cartilage. Journal of Fish Biology 46:919–935.

El-Matbouli, M., T. S. McDowell, D. B. Antonio, K. B. Andree, and R. P. Hedrick. 1999. Effect of water temperature on the development of the development, release and survival of the triactinomyxon stage of *Myxobolus cerebralis* in its oligochaete host. International Journal of Parasitology 29:627–641.

Goldman, M. 1968. Fluorescent antibody methods. Academic Press, New York and London.

Griffin, B., and E. Davis. 1978. *Myxosoma cerebralis*: detection of circulating antibodies in infected rainbow trout (*Salmo gairdneri*). Journal of the Fish Research Board of Canada 35:1186–1190.

Halliday, M. M. 1973. Studies on *Myxosoma cerebralis*, a parasite of salmonids. II. The development and pathology of *Myxosoma cerebralis*, in experimentally infected rainbow trout (*Salmo gardneri*) fry reared at different temperatures. Nordic Veterinary Medicine 25:349–358.

Halliday, M. M. 1974. Studies on *Myxosoma cerebralis*, a parasite of salomids. IV. A preliminary immunofluorescent investigation of the spores of *Myxosoma cerebralis*. Nordic Veterinary Medicine 26:173–179.

Halliday, M. M. 1976. The biology of *Myxosoma cerebralis*: the causative agent of whirling disease of salmonids. Journal of Fish Biology 9:339–357.

Hamilton, A. J., and E. U. Canning. 1988. The production of mouse anti-*Myxosoma cerebralis* antiserum from Percoll-purified spores and its use in immunofluorescent labeling of Historesin-embedded cartilage derived from infected rainbow trout. Journal of Fish Diseases 11:185–190.

Hedrick, R. P., M. El-Matbouli, M. A. Adkison, and E. Mac-
Connell. 1998. Whirling disease: re-emergence among
wild trout. Immunological Reviews 166:365–376.

Hedrick, R. P., T. S. McDowell, G. D. Marty, K. Mukkatira, D. B. Antonio, K. B. Andree, Z. Bukhari, and T.
Clancy. 2000. Ultraviolet irradiation inactivates the
waterborne infective stages of Myxobolus cerebralis: a
treatment for hatchery water supplies. Diseases of
Aquatic Organisms 42:53–59.

Hedrick, R. P., A. Wishkovsky, J. C. Modin, and R. J. Toth.
1991. Three myxosporeans found in the cranial and
branchial tissues of rainbow trout in California. Journal of Aquatic Animal Health 3(1):55–62.

Hermanns, W., and W. Korting. 1985. Improved method
for the detection of non-sporulated Myxosoma cerebralis, Hofer 1903, causative agent of whirling disease (WD) of salmonids. Berlin, Munich. Tierarztl
Wschr 98:126–129.

Hervio, D. M. L., M. L. Kent, K. Khattra, J. Sakanari, H.
Yokoyama, and R. H. Devlin. 1997. Taxonomy of
Kudoa species Myxosporea), using small-subunit
ribosomal DNA sequence. Canadian Journal of
Zoology 75:2112–2119.

Hnath, J. G., 1970. Whirling disease (Myxosoma cerebralis) in the State of Michigan. Journal of Parasitology 56(4) section 2, part 1, resume 273.

Hofer, B. 1903. Uber die drehkrankheit der regenbogenforelle. Allgemeinen Fisch 28:7–8.

Hoffmann, G. L. 1975. Whirling disease (Myxosoma cerebralis): control with ultraviolet irradiation and effect
on fish. Journal of Wildlife Diseases 11:505–507.

Hoffmann, G. L. 1990. Myxobolus cerebralis, a worldwide
cause of salmonid whirling disease. Journal of
Aquatic Animal Health 2:30–37.

Hoffmann, Sr., G. L., and G. L. Hoffmann, Jr. 1972. The
effects of chemicals on spores in vitro, and of calcium oxide as a disinfectant in simulated ponds. Journal of Wildlife Diseases 8:49–53.

Horsch, C. M. 1987. A case history of whirling disease in
a drainage system: Battle Creek drainage of the
Upper Sacramento River Basin, California, USA.
Journal of Fish Diseases 10:453–460.

Kozel, T. R., M. Lott, and M. Taylor. 1980. Isolation of
Myxobolus cerebralis (Whirling Disease) spores from
infected fish by use of a physical separation technique. Canadian Journal of Fisheries and Aquatic
Sciences 37:171–174.

Landolt, M. L. 1973. Myxosoma cerebralis: isolation and
concentration from fish skeletal elements—trypsin
digestion method. Journal of the Fisheries Research
Board of Canada 30:1713–1716.

Lom, J. 1987. Myxosporea: a new look at long-known parasites of fish. Parasitology Today 3(11):327–332.

Lom, J., and G. L. Hoffmann. 1970. Redescription of the
spores of Myxosoma cerebralis. Second International
Congress of Parasitology, 6–12 September 1970,
Washington.

Lom, J., and G. L. Hoffmann. 1971. Morphology of the
spores of Myxosoma cerebralis (Hofer, 1903) and M.

cartilaginis (Hoffmann, Putz and Dunbar, 1965).
Journal of Parasitology 57:1303–1308.

Lorz, H. V., and A. Amandi. 1994. VI. Whirling disease of
salmonids. In J. C. Thoesen, editor. Suggested procedures for the detection and identification of certain finfish and shellfish pathogens, 4th edition.
American Fisheries Society, Fish Health Section,
Bethesda, Maryland.

Lucky, Z. 1970. Pathological changes and diagnostics of
myxosomosis of the rainbow trout (Salmo gairdneri
irideus) Acta Vet Brno, Supplement 1:19–29.

MacLean, D. G. 1971. A simplified procedure for detecting
Myxosoma cerebralis (Whirling disease) spores in large
lots of fish. The Progressive Fish-Culturist 33:203.

Margolis, M. L., M. L. Kent, and P. Bustos. 1996. Diseases of
salmonids resembling myxosporean Whirling Disease,
and the absence of Myxosoma cerebralis, in South
America. Diseases of Aquatic Organisms 25:33–37.

Markiw, M. E. 1989. Salmonid whirling disease: myxosporean and actinosporean stages cross-react in
direct fluorescent antibody test. Journal of Fish Diseases 12:137–141.

Markiw, M. E., and K. Wolf. 1974a. Myxosoma cerebralis:
Comparative sensitivity of spore detection methods.
Journal of the Fisheries Research Board of Canada
31:1597–1600.

Markiw, M. E., and K. Wolf. 1974b. Myxosoma cerebralis:
isolation and concentration from fish skeletal elements—sequential enzymatic digestions and purification by differential centrifugation. Journal of the
Fisheries Research Board of Canada 32:15–20.

Markiw, M. E., and K. Wolf. 1974c. Myxosoma cerebralis:
Trypsinization of plankton centrifuge harvests
increases optical clarity and spore concentration.
Canadian Journal of Fisheries and Aquatic Sciences
37:2225–2227.

Markiw, M. E., and K. Wolf. 1978. Myxosoma cerebralis:
fluorescent antibody techniques for antigen recognition. Journal of the Fisheries Research Board of
Canada 35:828–832.

Markiw, M. E., and K. Wolf. 1983. Myxosoma cerebralis
(Myxozoa: Myxosporea) etiologic agent of salmonid
whirling disease requires tubificid worm (Annelida:
Oligochaetes) in its life cycle. Journal of Protozoology 30:561–564.

Mitchell, L. G. 1989. Myxobolid parasites (Myxozoa:
Myxobolidae) infecting fishes in western Montana,
with notes on histopathology, seasonality, and
intraspecific variation. Canadian Journal of Zoology
67:1915–1922.

Modin, J. 1998. Whirling disease in California: A review
of its history, distribution, and impacts, 1965–1997.
Journal of Aquatic Animal Health 10(2):132–142.

Nehring, R. B., and Walker, P. G. 1996. Whirling disease
in the wild: the new reality in the intermountain
west. Fisheries 21:28–32.

O'Grodnick, J. J. 1975. Whirling disease Myxosoma cerebralis spore concentration using the continuous plankton centrifuge. Journal of Wildlife Diseases 11:54–57.

Palenzuela, O., G. Trobridge, and J. L. Bartholomew. 1999. Development of a polymerase chain reaction diagnostic assay for *Ceratomyxa shasta*, a myxosporean parasite of salmonid fish. Diseases of Aquatic Organisms 36:45–51.

Plehn, M. 1924. Praktikum der Fischkrankheiten Pages 301–474 *in* Handbuch der Binnenfischerei Mitteleuropas I. Stuttgart.

Prasher J. B., W. M., Tidd, and R. A. Tubb. 1971. Techniques for extracting and quantitatively studying the spore stage of the protozoan parasite *Myxosoma cerebralis*. The Progressive Fish-Culturist 33(4):193–196.

Putz, R. E., and I. McElwain. 1969. Serodiagnosis of whirling disease. Progressive Sport Fisheries Research.

Rose, J. D., G. S. Marrs, C. Lewis, and G. Schisler. 2000. Whirling disease behavior, and its relation to pathology of brain stem, and spinal cord in rainbow trout. Journal of Aquatic Animal Health 12(2):107–118.

Saulnier, D., and P. De Kinkelin. 1997. Polymerase chain reaction primers for investigations on the causative agent of proliferative kidney disease of salmonids. Journal of Fish Diseases 20:467–470.

Schisler, G. J., E. P. Bergersen, P. G. Walker, J. Wood, and J. Epp. 2001. Comparison of single-round polymerase chain reaction (PCR) and pepsin-trypsin digest (PTD) methods for detection of *Myxobolus cerebralis*. Diseases of Aquatic Organisms 45:109–114.

Smothers, J. F., C. D. von Dohlen, L. H. Smith, and R. D. Spall. 1994. Molecular evidence that the myxozoan protists are metazoans. Science 265:1719–1721.

Taylor, R., and M. Lott. 1978. Transmission of salmonid whirling disease by birds fed trout infected with *Myxosoma cerebralis*. Journal of Protozoology 25:105–106.

Thompson, K. G., and R. B. Nehring. 2000. A simple technique used to filter, and quantify the actinospore of *Myxobolus cerebralis* and determine its seasonal abundance in the Colorado River. J. Aquat. Animal Health 12(4):316–323.

Tidd, W. M., R. A. Tubb, and V. Wright. 1971. Modification of the *Myxosoma cerebralis* spore extraction technique. The Progressive Fish-Culturist 35(4):227–228.

Vincent, E. R. 1996. Whirling disease and wild trout: the Montana experience. Fisheries 21:32–34.

Wolf, K., and M. E. Markiw. 1976. *Myxosoma cerebralis*: In vitro sporulation of the myxosporidan of salmonid whirling disease. Journal of Protozoology 23:425–427.

Wolf, K., and M. E. Markiw. 1979. *Myxosoma cerebralis*: A method for staining spores and other stages with silver nitrate. Journal of Fisheries Research Board of Canada 36:88–89.

Wolf, K., and M. E. Markiw. 1984. Biology contravenes taxonomy in the myxozoa: new discoveries show alternation of invertebrate and vertebrate hosts. Science 225:1449–1452.

Wolf, K., M. E. Markiw, and J. K. Hiltunen. 1986. Salmonid whirling disease: *Tubifex tubifex* (Muller) identified as the essential oligochaete in the protozoan life cycle. Journal of Fish Diseases 9:83–85.

Yokoyama, H., T. Danjo, K. Ogawa, and H. Wakabayashi. 1997. A vital staining technique with fluorescein diacetate (FDA) and propidium iodide (PI) for the determination of viability of myxosporean and actinosporean spores. Journal of Fish Diseases 20(4):281–286.

American Fisheries Society Symposium 29:213–214, 2002

A Brief Critique of Methods of Sampling and Reporting Pathogens in Populations of Fish

CHRISTOPHER J. WILLIAMS

Division of Statistics, University of Idaho, Moscow, Idaho 83844-1104, USA

208-885-2802; 208-885-7959; chrisw@uidaho.edu

CHRISTINE M. MOFFITT

Department of Fish and Wildlife Resources, University of Idaho, Moscow, Idaho 83844-1136, USA

208-885-7047; 208-885-9080; cmoffitt@uidaho.edu

Studies of the disease status and distribution of pathogens in natural and captive populations of fish and of the implications of fish health in fisheries management have increased in recent years. With this new focus is a desire to use retrospective analyses from a variety of sources in order to examine trends on a larger geographic scale. To compare data and to model various outcomes with accuracy, estimates of the prevalence and of the confidence limits are needed. For years, pathologists have considered the implications of sample size on estimations of pathogen prevalence in fish populations, and sampling protocols used for screening have been developed and adapted from recommendations by Ossiander and Wedemeyer (1973) and Simon and Schill (1984). However, these methods addressed samples in which individual fish were examined for the agent of interest.

Confounding the interpretation of the outcome of tests is the use of pooled samples from several fish, processed in a single assay, and low numbers of pooled samples. The American Fisheries Society Fish Health Bluebook (Thoesen 1994) recommends a pooling of samples for analysis of several specific pathogens, including most viral screenings, and for many analyses of parasites, such as *Myxobolus cerebralis* (the causative agent of salmonid whirling disease), in which whole heads or half heads from like size fish are combined and digested with pepsin-trypsin to free spores for quantification.

Likelihood-based methods can be used to estimate apparent prevalence and to construct confidence intervals that can be used for samples with any combination of group sizes. This method of estimation extends the work of Worlund and Taylor (1983), whose method was only applicable for equal group sample sizes. Our analysis includes calculation of sampling distribution to assess variance and mean-squared error, to show how these parameters change, as a function of the number and size of pooled groups and the disease prevalence.

RESULTS AND DISCUSSION

We examined the statistics surrounding the outcome of sample analysis, based on samples ranging from low prevalence to 90% prevalence and several pooling strategies for samples of 30 and 60 fish. We calculated estimates of apparent prevalence for samples collected and screened for M. *cerebralis* infections by Utah Division of Wildlife scientists. Our calculations assume perfect specificity and sensitivity of assays, but we provide one method for adjusting these values when specificity and sensitivity are known. When all pools were positive, the apparent prevalence was 100%, but the bounds of the confidence interval ranged from 31% to 100%. A positive pool indicates that at least one fish in the pool was positive; however, when a pool is negative, all fish in the pool are presumed negative. Interpretations of data sets that are based only on the outcome of positive pools may be misleading, as the percentage of pools positive, when any single pool is scored negative, is higher than the maximum likelihood estimates of apparent prevalence. The confidence interval bounding estimates are generally lower when larger numbers of groups are used in analysis and when samples have fewer fish per pool. In populations with higher prevalence, the use of pooled samples significantly reduces the confidence of estimates.

In planning studies, one should avoid a sampling plan that is likely to yield all positive pools. When prevalence is high, one should not use pools, or else should use more pools of fewer fish or a combination of strategies. The variability of apparent prevalence estimates from the pooled samples can be quite large, particularly as disease prevalence increases. This method for estimating apparent prevalence from grouped fish samples, with any combination of sample sizes, will assist those planning sampling programs in determining the most effective way to get information about fish populations. We hope that these results will remind investigators to consider the desirability of estimating apparent prevalence from pooled samples and to account for the sampling variability of these estimates. We believe that future studies of disease prevalence should incorporate methods such as ours and should account for sampling variability when drawing inferences.

A SAS computer program for calculating disease prevalence estimates and confidence intervals, using these methods, is available (Williams and Moffitt, 2001).

ACKNOWLEDGMENTS

We are grateful to Steven Intelmann, University of Idaho, for gathering and organizing the pathogen database and for assisting with earlier interpretations of maximum likelihood estimates. We thank Ron Goede and Chris Wilson, Utah Division of Wildlife, for providing data on M. *cerebralis* monitoring. The Western Regional Aquaculture Center, U.S. Department of Agriculture, funded this study.

REFERENCES

Ossiander, F. J., and G. Wedemeyer. 1973. Computer program for sample sizes required to determine disease incidence in fish populations. Journal of Fisheries Research Board of Canada 30:1383–1384.

Simon, R. C., and W. B. Schill. 1984. Tables of sample size requirements for detection of fish infected by pathogens: three confidence levels for different infection prevalence and various population sizes. Journal of Fish Diseases 7:515–520.

Thoesen, J. C. Editor. 1994. Suggested procedures for the detection and identification of certain finfish and shellfish pathogens. 4th edition, Version 1. Fish Health Section, American Fisheries Society.

Williams C. J., and C. M. Moffitt. 2001. A critique of methods of sampling and reporting pathogens in populations of fish. Journal of Aquatic Animal Health 13:300–309.

Worlund, D. D., and G. Taylor. 1983. Estimation of disease incidence in fish populations. Canadian Journal of Fisheries and Aquatic Sciences 40:2194–2197.

Section 7

Management and Control

American Fisheries Society Symposium 29:217–225, 2002
© 2002 by the American Fisheries Society

Whirling Disease Prevention, Control, and Management: A Review

ERIC J. WAGNER

Fisheries Experiment Station, 1465 West 200 North, Logan, Utah 84321, USA
ewagner@sisna.com

ABSTRACT. Whirling disease occurs in salmonid fishes infected by the myxosporean parasite *Myxobolus cerebralis*. This review summarizes the literature regarding control and management of the disease. Suggested methods for killing the myxospore stage include thorough drying, heating for 10 min at 90°C, calcium hydroxide at more than 0.5% for 24 h, and calcium oxide or KOH at more than 0.25% for 24 h. Chlorine was also effective at 1600 ppm in 24-h exposure or 5,000 ppm in a 10-min exposure. Roccal (alkyl dimethylbenzylammonium chloride) at more than 200 ppm active ingredient was also effective. Calcium cyanide at 4,000 kg/ha has been used effectively for control in infected ponds. Treating incoming water with 2537 Å of ultraviolet (UV) light at dosages greater than or equal to 35,000 microwatt-sec/cm² was effective in preventing infection of rainbow trout fry. Filtration of water through a 25 μm commercial filter cartridge did not reduce or eliminate the disease, but sand-charcoal filters have been used successfully in France. Prophylactic treatment of incoming water with 0.5 ppm chlorine for 2 h once a week over a 4 month period reduced infection by 63–73%. Electrical charges of 1–3 kV pulsed for 1–25 times at 99 μsec per pulse have killed a high proportion of the triactinomyxon stage. Triactinomyxons were also inactivated by 260 ppm chlorine as sodium hypochlorite, 10% hydrogen peroxide for 10 min, or 50% povidone-iodine for 10 min. Various drugs have been tested, some of which reduced the level of infection but failed to eliminate the parasite. Management strategies to control the disease include control of the worm host and its habitat, stocking larger fish into infected waters, not stocking infected fish, education, enforcement of disease regulations, and stocking less susceptible species.

Whirling disease, caused by the myxosporean *Myxobolus cerebralis*, has been associated with significant declines in wild rainbow trout *Oncorhynchus mykiss* populations in Montana and Colorado. In both states, yearling rainbow trout densities indicated a 90% drop in young-of-the-year abundance in infected reaches of important fisheries, such as the Madison, Gunnison, and Colorado Rivers (Vincent 1996; Nehring and Walker 1996). The parasite has a widespread distribution in the United States in regions where trout are reared, and it is recognized as a prohibitive pathogen by many states and countries. Clearly, whirling disease is a significant problem.

In its life cycle, the parasite infects two hosts, alternating between a salmonid host and the oligochaete worm *Tubifex tubifex* (Markiw and Wolf 1983). The triactinomyxon is the developmental stage of the parasite released by the worm and is the stage that infects the fish host. Consumption of infected worms may also result in infection (Wolf and Markiw 1984). The triactinomyxon is a type of actinosporean, a group of organisms once thought to be separate species that parasitized worms (Janiszewska 1955) but now are considered alternate stages of myxosporean parasites (Wolf and Markiw 1984; Kent et al.1994; Brinkhurst 1996). After entering the fish host, the sporoplasm cells of the triactinomyxon replicate and become trophozoites, feeding on cartilage tissue (Hedrick et al. 1998). This presporogonic phase then shifts to the sporogonic phase, which culminates in the formation of the mature myxospore (Hedrick et al. 1998). Upon death of the fish host, the myxospore is released into the environment, where its consumption by the worm host renews the cycle. Control of the parasite is aimed at these various stages, which differ greatly in susceptibility to control measures.

Prior to understanding the complete life cycle, research was primarily aimed at the resistant myxospore stage. The myxospores can tolerate freezing at −20°C for at least 3 months and are still viable after passage through the guts of predators such as northern pike *Esox lucius*, black-crested night herons *Nycticorax nycticorax*, or mallard ducks *Anas platyrhynchos* (Taylor and Lott 1978; El-Matbouli and Hoffmann 1991a). There have been reports from Europe of myxospores remaining viable in dry pond beds for twelve years (Bauer 1962).

This article summarizes whirling disease control measures attempted to date. These measures include methods for the control of the worm host and its habitat, physical and chemical control of both the myxospore and triactinomyxon stages, and oral drugs aimed at the trophozoite stage. These measures are largely applicable to aquaculture. Also discussed are some general guidelines for management of the disease in the wild and the scientific rationale behind them.

CONTROL STRATEGIES

Control of Worm Host

One control strategy is to break the life cycle by removal of the worm host that produces the triactinomyxon stage. This approach is more applicable to aquaculture, but recent experimentation with electricity suggested that treatment of 'hot spots' in the wild may soon be practical (R. Ingraham and T. Claxton, Hamilton High School, Montana, personal communication). Using an electrophoresis chamber and worms from a commercial supplier, these students found that after 1,000-s exposure to low-level DC voltage, almost all the worms were dead within 48 h.

Thorough drying of ponds may be effective in certain climates. However, *Tubifex tubifex* has the ability to form resistant cysts, living for up to 14 d in dried mud (Kaster and Bushnell 1981). If the encysted worms were wetted periodically, they survived up to 70 d. Worms may also burrow down to moist depths that never dry. No work has been conducted, to date, that evaluates the ability of an infected worm to produce actinospores after encysting.

The lampricide TFM (3-trifluoromethyl-4-nitrophenol) has been used to control sea lamprey *Petromyzon marinus* in the Great Lakes, without harming fish. In an evaluation of the chemical's effect on nontarget organisms (Lieffers 1990), a reduction in the abundance of oligochaetes and other invertebrates was observed after treatment (4.2–14.0 mg/L). The widespread killing of nontarget organisms makes this chemical and others inappropriate for treatment of field ecosystems. However, it may be applicable for aquaculture ponds. In ponds, treatment without fish present would be advisable, since TFM can reduce oxygen and increase ammonia, apparently through its effect on photosynthesis (Dawson et al. 1992). Toxicity of TFM to fish is pH-dependent, increasing at pH 7.25 or lower (Bills and Johnson 1992). Further testing may provide appropriate dosages and application techniques for controlling oligochaete worms with fish present, as well as without.

A number of researchers have examined the effects of water quality variables and pollutants on the survival of *T. tubifex* (Chapman et al. 1982; Whitley 1968; Whitten and Goodnight 1966). For aquaculture applications, manipulation of water quality variables would be more practical for control of the worms than chemicals potentially toxic to fish. For pH, Whitley (1968) reported 72-h LC_{50} values of 5.8 and 9.7 at 20°C for a mix of *T. tubifex* and *Limnodrilus hoffmeisteri* without sediment. Chapman et al. (1982) reported that 96-h LC_{50} values for the acid and basic ends of the pH spectrum were pH 3.5 and 10.5, respectively, for *T. tubifex* held at 10°C without sediment. In tests on the effects of salinity, *T. tubifex* was more tolerant than many other freshwater species, with a 96-h LC_{50} of 14‰ in sediment (Chapman et al. 1982). A mixed culture of *T. tubifex* and *L. hoffmeisteri* was tolerant of shorter exposures in 2 cm of fine silt; salinities of 95‰ for 100 min or 24 h resulted in mortalities of only 18% and 60.6%, respectively (E. Wagner, unpublished data).

High temperature may be more effective in eliminating worms than salinity or application of other chemicals that cannot penetrate the sediment, but achieving the high temperatures presents challenges as well. The 96-h lethal temperature (LT_{50}) was 34°C in a study by Birtwell and Arthur (1980) and 35°C in another (Chapman et al. 1982). Shorter exposures to high temperatures on a mix of *T. tubifex* and *Limnodrilus hoffmeisteri* have recently been tested; LT_{50} values ranged from 38.1°C to 39.2°C in 10-min exposures and from 33.8°C to 36.6°C in 100-min exposures of worms acclimated to 6°C, 9°C, 13°C, or 22°C (Wagner and coauthors, Utah Division of Wildlife, unpublished data). Sublethal high temperature may be effective in controlling triactinomyxon release (El-Matbouli et al. 1999). These authors noted that infected worms transferred from 15°C to 25°C or 30°C ceased triactinomyxon production after 4 d, and developmental stages had degenerated, effectively purging the worm of infection.

Chemical and Physical Control Methods for Myxospore and Actinospore Stages

A variety of physical and chemical control measures have been tested. Many of these studies were conducted before the role of *T. tubifex* in the life cycle was known, so the measured result was a decrease in infection without a determination of where the life cycle was broken. For example, Hoffman and O'Grodnick (1977) noted that contaminated mud dried for 13–19 months was incapable of inducing infection, possibly indicating that the myxospore stage is susceptible to desiccation. Alternatively, the lack of the worm host in the test tank may have prevented infection of the fish. Similar observations are noted for the study of Hoffman and Hoffman (1972), in which quicklime (CaO) applied to simulated ponds at concentrations of 380 g/m² or greater, for two weeks, prevented infection of rainbow trout (Table 1). Treatment of simulated earthen ponds (2 cm infected mud in aquaria) with either 4,550 g/m² hydrated lime (CaOH) or up to 1,200 ppm chlorine, for 18–24 h, did not destroy all the myxospores (Hoffman and O'Grodnick 1977).

Various chemicals applied to aquaculture ponds in field tests have been reported to reduce infection. Calcium cyanide was effective in disinfecting ponds, whereas CaO was less effective (Table 1; Bauer 1962). Calcium cyanamide (488 g/m²) applied to dirt ponds, combined with chlorine gas (300 ppm) disinfection of incoming spring water, prevented recurrence of whirling disease the following year in a Pennsylvania trout hatchery (Hoffman and Dunbar 1961). However, some German culturists prefer CaO because it does not have the undesirable nutrient-fertilizing effect of calcium cyanamide (Hoffman and Hoffman 1972).

Chlorine has been one of the most effective chemicals for disinfection. For supernatant from infected mud, 10 ppm chlorine for 30 min was sufficient to prevent infection of rainbow trout fry (Hoffman and O'Grodnick 1977). Likely, the chlorine killed triactinomyxons in the supernatant water, rather than the myxospore, which would have needed the worm host to complete the parasite's life cycle. Research by Wagner and his coauthors (Utah Division of Wildlife, unpublished data) also indicated that triactinomyxons are sensitive to chlorine, succumbing to concentrations as low as 13 ppm in 10-min treatments. However, because low temperatures reduced the efficacy, 130–260 ppm chlorine for 10 min was recommended for 100% kill of triactinomyxons. For treatment of the myxospore, 200 ppm chlorine, for 18–24 h, gave variable results (Hoffman and Putz 1969); 400 ppm killed 36–90% of myxospores (Hoffman and Hoffman 1972). Recent data from E. MacConnell and her coauthors (U.S. Fish and Wildlife Service, Bozeman, Montana, personal communication), based on methylene blue vital staining (dead spores take the stain, live spores do not), suggested that 5,000 ppm chlorine, for 10 min, was sufficient for destruction of myxospores.

Another control strategy in hatcheries is to disinfect incoming water. Ozonation has also shown promise in preliminary tests for treatment of the actinospore stage (Hedrick et al. 1998). Markiw (1992a) noted that a hatchery water supply treated for four months with 0.5 ppm chlorine, for 2 h once a week, reduced infection by 63% and

Table 1. A list of chemicals causing distortion and probable death of *Myxobolus cerebralis* spores.

Chemical	Concentration	Duration	Citation
Calcium hydroxide	0.5 and 2.0%	24 h	Hoffman and Putz 1969
Calcium oxide (quicklime)	0.25, 0.5, and 1.0%	6, 2, & 2 d	Hoffman and Hoffman 1972
	380 g/m2 (3360 lb/acre)	2 weeks	
Potassium hydroxide	0.25, 0.5, and 1.0%	2 d	Hoffman and Hoffman 1972
Sodium hydroxide	0.5%	24 h	Hoffman et al. 1962
Available chlorine as	1,600 ppm	24 h	Hoffman and Putz 1969
sodium hypochlorite	5,000 ppm	10 min	MacConnell et al., personal
communication			
Alkyl dimethylbenzylammonium chloride (Roccal®)	200 and 800 ppm	24 h	Hoffman and Putz 1969
Calcium cyanide	4,000 kg/ha	no data	Bauer 1962

73% in two lots of trout. This prophylactic treatment did not harm the fish. However, the use of chlorine may be prohibited by the U.S. Food and Drug Administration (FDA) and state water quality regulations, as it is not an approved compound for discharging from hatcheries.

Filtration has not been effective when using membranes to prevent passage of triactinomyxons, but three-dimensional material has been a better filter. For example, filtration of water through a 25 μm commercial filter cartridge did not reduce or eliminate the disease (Hoffman 1974). Research by the author has shown that triactinomyxon recovery from 10 or 20 μm filters decreased with increasing water volume filtered. In these tests, water spiked with a known number of M. cerebralis triactinomyxons was filtered through either 10 or 20 μm mesh Nitex cloth. Recovery rates of 80 ± 26 (SD) and 62 ± 6% for 10 and 20 μm filters, respectively, were recorded when 2 L were filtered through, but this rate dropped to 54 ± 17 (10 μm) and 40 ± 32% (20 μm) when 8 L was filtered. The filtrate was not examined for triactinomyxons due to the large volume, so it is not clear if the triactinomyxons that were not recovered had passed through or were somehow destroyed in the filtration process. Hoffman et al. (1962) noted that sand-charcoal filters had been used successfully in France.

Hoffman (1974) found that treating water with 2537 Å of ultraviolet (UV) light, at dosages of more than 35 milliwatt-sec/cm^2 (mWs/cm^2), was effective in preventing infection of rainbow trout fry. At lower doses, 18 and 27.65 mWs/cm^2, infectivity was reduced by 31–86% and 86–100%, respectively, in comparison to control fish in untreated water (Hoffman 1975). Hoffman (1975) noted that soviet researchers have also had some

success using UV treatment for whirling disease control (Ivanov et al. 1968). O'Grodnick and Gustafson (1974) successfully treated a flow of 31 L/min with 35 mWs/cm^2 of UV after filtration through a 25 μm mesh filter. Hedrick et al. (2000) noted that a dose of 1,300 mWs/cm^2 was required to inactivate 100% of triactinomyxons held under a static collimated beam of UV. Ultraviolet treatment systems have been developed recently for decontaminating water infected with another pernicious parasite, Cryptosporidium parvum (Clancy et al. 1998). The Cryptosporidium Inactivation Device (CID), tested by Clancy et al. (1998), relies on trapping pathogens on 2 mm sintered stainless steel filters and irradiating them with UV light. Hedrick et al. (2000) tested the device at UV dosages of 4,000–8,000 mWs/cm^2 and found no evidence of infection among juvenile rainbow trout dosed with 7,539 or 64,200 irradiated triactinomyxons/fish.

In work with the triactinomyxon stage, using vital stains, Wagner and coauthors (Utah Division of Wildlife, unpublished data) evaluated the effects of various physical and chemical treatments (Table 2). Thorough drying (>1 h) was effective in killing the actinospore stage. They also found that hydrostatic pressure up to 6.2×10^7 Pa (9,000 psi) for 5 min significantly reduced viability to 43–60%, compared with 70–74% viability in controls. Recent research (E. Wagner et al., this volume) has indicated that electrical charges of 1–3 kV, pulsed for 1–25 times at 99 μsec/pulse, killed a high proportion of triactinomyxons. Drying or freezing triactinomyxons for at least an hour was effective in killing 100%. Lethal concentrations to triactinomyxons were high when using either hydrogen peroxide (10% H_2O_2 for 10 min) or povidone-iodine

Table 2. Summary of effective physical and chemical treatments for killing the triactinomyxon stage of *Myxobolus cerebralis*. CID = *Cryptosporidium* inactivation device.

Treatment	Concentration	Duration	Citation
Chlorine as sodium Hypochlorite	260 ppm	10 min	Wagner et al., personal communication
	10 ppm	30 min	Hoffman and O'Grodnick 1977
Povidone-iodine	50%	60 min	Wagner et al., personal communication
Hydrogen peroxide	10%	10 min	Wagner et al., personal communication
UV light-static	1300 mWs/cm^2		Hedrick et al. 2000
UV light-CID	4,000–8,000 mWs/cm^2		Hedrick et al. 2000
Temperature	75°C	5 min	Wagner et al., personal communication
Drying		>1 h	Wagner et al., personal communication
Freezing		>1 h	Wagner et al., personal communication
Electricity	25 99-μsec pulses of 3 kV each		Wagner et al., this volume

(50% for 60 min). Temperatures above 75°C, for at least 5 min, were lethal. Markiw (1992b) noted that triactinomyxon survival was temperature dependent; only 2 d at 23–24°C, but 7–8 d at 7°C.

Heat has been effective in causing the distortion and probable death of myxospores. Hoffman and Putz (1969) examined spores after heating in 0.85% saline to 60°C, 80°C, and 100°C. These temperatures were effective in killing myxospores, whereas temperatures of 40°C or room temperature were not. Later tests by Hoffman and Markiw (1977) indicated that heating spores for 10 min at 90°C was effective in killing the spores, as determined by methylene blue staining. Heating at lower temperatures progressively reduced the mortality percentage (80°C, 98%; 70°C, 60%; 60°C, 34%; 50°C, 24%) in five trials. Heating for longer periods (up to 100 min) at 70°C increased the percentage of dead spores but still did not reach 100% (Hoffman and Markiw 1977). Despite the lower temperature, smoking fish at 66°C for 40 min was effective in killing spores (Wolf and Markiw 1982).

Drug Treatment

Various drugs have been tested for their ability to reduce the severity of clinical whirling disease, with limited success (Table 3). Medicated pellets, containing 0.1% fumagillin (dicyclohexylamine), fed to rainbow trout reduced clinical infections in two tests (El-Matbouli and Hoffmann 1991b). In the first test, fish were fed from day 30 to day 160 post infection and, in the second, from day 14 to day 64 post infection; 73–100% of nonmedicated fish had severe infections, whereas only 10–20% of medicated fish harbored spores (El-Matbouli and Hoffmann 1991b). Staton et al. (this volume) also tested fumagillin at several hatcheries, using either natural exposure or controlled exposure to triactinomyxons. At doses of 3.7 or 7.5 mg/kg body weight per day, for 10 d, fumagillin was not effective in controlling the disease. Furthermore, higher mortality in some treatment groups indicated possible toxicity from the drug, suggesting that higher doses hold little promise. For another myxosporean *Thelohanellus hovorkai*, a 0.1% fumagillin diet (10 mg/kg fish) fed to koi carp *Cyprinus carpio*, for 3–7 weeks, significantly reduced prevalence rates and eliminated mortality and morbidity (Yokoyama et al. 1999). Molnar et al. (1987) also noted significantly reduced infections of *Sphaerospora renicola* after feeding common carp *Cyprinus carpio* 0.1% fumagillin in the feed. However, control of *Myxobolus cyprini* or *Thelohanellus nikolskii* in the same fish was not achieved with fumagillin treatment (Molnar et al. 1987), indicating differences among myxosporeans in susceptibility to the drug.

In drug efficacy tests with rainbow × cutthroat trout hybrids fed medicated feed, Taylor et al. (1973) found that furazolidone inhibited spore formation. However, the drug affected feed palatability, and growth in this group was half that of con-

Table 3. A list of drugs tested for use against *Myxobolus cerebralis*.

Drug	Dosage	Reduction of incidence (%) Lot 1	Lot 2	Citation
Acetarsone (Stovarsol)	10–1000 mg/kg fish/d (3 d/week for 6 months)	suppression[1]		Hoffman et al. 1962
Amprolium	13–18 mg/kg fish	17	0	Taylor et al. 1973
	24–44 mg/kg fish	50	0	Taylor et al. 1973
Fumagillin	1 g/kg feed fed at 1% of body weight	73		El-Matbouli and Hoffman 1991
	3.7 or 7.5 mg/kg fish for 10 d	0		Erdahl and Staton, personal communication
Furazolidone	152–194 mg/kg fish	100	39	Taylor et al. 1973
Merck 930	8–15 mg/kg fish	0	—	Taylor et al. 1973
	33–64 mg/kg fish	0	0	Taylor et al. 1973
Nicarbazine	6–14 mg/kg fish	17	0	Taylor et al. 1973
	30–60 mg/kg fish	22	0	Taylor et al. 1973
Oxytetracycline	68–152 mg/kg fish	39	—	Taylor et al. 1973
Sulfamerazin	15–36 mg/kg fish	0	0	Taylor et al. 1973

[1]No percentages given. This data was cited from Scolari (1954). Additional trials by Hoffman et al. (1962), using concentrations up to 100 times that recommended by Scolari, (1954) indicated this was not a promising drug.

trols. Also, some fish on furazolidone still had trophozoites and granulomas. Russian literature (Bauer 1962) suggested that osarsol added to feed was effective in controlling the disease, but no concentrations or treatment regimen was given. Acetarsone (Stovarsol) used at 10 mg/kg body weight for 3 d suppressed the disease but did not eliminate it (Hoffman et al. 1962). Alderman (1986) tested the same drug and found no reduction in pathology, spore frequency, and severity. However, the drug proguanil reduced the frequency of pathology and spores, as well as severity of the disease (Alderman 1986). Alderman (1986) also found that clamoxyquin reduced the frequency of clinical signs, though to a lesser extent than proguanil, but was more effective at reducing the frequency and severity of sporulation. Similarly, O'Grodnick and Gustafson (1974) noted that furoxone and benomyl reduced losses and infection of young salmonids, but none prevented or eliminated the disease. Even if these drugs were effective, registration of these drugs through the FDA for use with fish would require many years of testing and millions of dollars of funding.

MANAGEMENT STRATEGIES

Reduce or Eliminate Habitat of Alternate Host

Areas of high sediment loading have been observed to be 'hot spots' of parasite infection, producing higher numbers of triactinomyxons, resulting in higher prevalence rates and more severe infections (Thompson and Nehring 2000; Allen 1999). For wild fish, efforts to reduce introduction of sediment into the streams should help reduce the amount of habitat for the parasite's alternate host. This means controlling excess sediment from sources such as road construction, logging, grazing, mining, and recreation. Benefits would not only include reduced worm habitat, but improved egg and fish survival, less gill damage, better growth rates, and reduced stress as well (Newcombe and MacDonald 1991; Lake and Hinch 1999). For aquaculturists, avoiding earthen ponds for culturing fish and keeping concrete systems free of organic waste and sediment are good management strategies (Markiw 1992a). Infected hatcheries in Colorado have applied this strategy, in concert with disinfection of incoming water, and have been able to avoid reinfection (Nickum 2000). The Bellevue-Watson Fish Hatch-

ery of Colorado has implemented a strategy of lining earthen ponds with 50-mil white plastic sheets, then alternating culture between ponds every 2 months (Nickum 2000). The pond not in use is drained and cleaned, removing any habitat for worms as well as any worms that may have been infected. Since incubation of the spore in the worm host takes about 3 months before triactinomyxons are produced (Markiw 1992a), infection of fish is avoided.

Stocking Strategies

Stocking larger fish may be a practical alternative for aquaculturists and state wildlife agencies. Fish are less susceptible to the disease as they grow older (Markiw 1992c), since the cartilage attacked by the trophozoite is largely converted to bone in older fish. Rasmussen (1965) reported significant improvements in survival of rainbow trout in Danish trout farms, by rearing fry to 5 cm total length in concrete tanks before stocking into infected dirt ponds. Hoffman (1990) recommended using fish that are at least 6 cm long. E. Wagner and coauthors (unpublished data, Utah Division of Wildlife Resources) noted that fingerling rainbow trout stocked at a size of 10 g, or about 100 mm, into an irrigation reservoir, survived well and provided an excellent fishery, despite infection levels reaching 100% after stocking. E. K. N. Ryce and her coauthors at the Montana Cooperative Fisheries Research Unit (Montana State University, personal communication) have noted in laboratory tests that mortality dropped significantly if rainbow trout were exposed after 9 weeks post hatching. Thompson et al. (1999) also noted that rainbow and cutthroat trout naturally exposed to M. cerebralis infection at larger mean weights survived better than smaller counterparts.

Stocking of infected fish into parasite-endemic areas, as well as disease-free areas, is not recommended. This practice may exacerbate problems by increasing the dose of triactinomyxons. Markiw (1992c) demonstrated that rainbow trout exposed to low numbers (1 or 10) of triactinomyxons did not develop spores. Higher doses of triactinomyxons resulted in proportionately more spores being recovered from infected fish, presumably overwhelming the immune system. Nehring and Thompson (2001) observed a similar relationship between the triactinomyxon numbers and infection severity in the wild. Fish should not be transferred from positive sites (Hoffman et al. 1962). The disease is not considered egg-transmissible

(O'Grodnick 1975), so expansion programs for sensitive species, such as cutthroat trout, may benefit from egg transfer, if disease-free sources are not available.

For management of naturally reproducing populations in positive waters, selection of resistant species or strains for stocking is one of the few management options currently available. Salmonids vary in susceptibility (see MacConnell and Vincent, this volume; O'Grodnick 1979; Markiw 1992a). Selection for resistance may take many generations, as Ryce et al. (2001) noted no difference in susceptibility to M. *cerebralis* infection between rainbow trout recruited during the early years of infestation and those recruited before. Grayling appeared to be resistant to infection at doses of 1,000–2,000 triactinomyxons/fish, whereas this dose caused infection in bull trout *Salvelinus confluentus* (Hedrick et al. 1999). Cutthroat trout are more resistant to the disease than rainbow trout and produce fewer myxospores per fish (Walker and Nehring 1995; Hedrick et al. 1999; Markiw 1992a). In contrast, Thompson et al. (1999) found that three subspecies of cutthroat trout native to Colorado suffered higher mortality than rainbow trout exposed to the same ambient levels of triactinomyxons within the Colorado River over a period of 9 months. In that study, chronic stress, resulting from the use of cages (Strange et al. 1978), may have been a factor in the difference. With these susceptibility differences and lower spore load in mind, cutthroat trout may be better candidates than rainbow trout for stocking or wild-fish management in infected waters, especially in the West. The lower spore load should help reduce the number of triactinomyxons being produced in an area and, in turn, reduce disease severity.

Among the geographic variants of some species, there may be slight differences in susceptibility. For example, E. Wagner and coworkers (unpublished data, Utah Division of Wildlife) found that after exposure to 1,000 triactinomyxons/fish, the Bear Lake Bonneville cutthroat trout O. *clarki utah* had significantly lower prevalence rates than a southern Bonneville form, Snake River cutthroat trout or Yellowstone cutthroat trout O. *clarki bouvieri*.

Other Strategies and Future Research Needs

The best management is to avoid infecting negative waters, containing the infection through enforcement of disease regulations, public education, and disinfection. Another management option is the use of different life history strategies of salmonids that reduce the risk of infection. For example, rainbow trout in the Madison River in Montana spawn either in mainstem sites or in the tributaries (Downing 2000). Fry emerging from tributary redds may experience lower triactinomyxon doses or not be exposed at all until entry into the mainstem portion of the river. Differences in spawning times may also be a factor in infection. Fry of early spawners may emerge sooner, possibly reducing exposure to infection until a later age at which they are less vulnerable. These life history differences will likely be selected naturally by the populations in the river, where other variables that affect mortality, such as food availability, flow, and predation risk, will guide selection for the most appropriate survival strategy. Monitoring of these natural selection variables, in relation to whirling disease, is a daunting task. Nonetheless, it should provide insight into the role of M. *cerebralis* and other parasites in the natural selection process.

Future research into control of the disease is needed. Further exploration of electrical charge effects and designing these into affordable and practical devices for disinfection is needed. Also, research into the mechanisms of polar filament discharge may provide a means of controlling the parasite. Premature discharge, induced artificially, could prevent later infection of fish.

Enhancement of the immune response may be one avenue of research. Immunological studies have indicated that rainbow trout produce antibodies against M. *cerebralis*, but protection against infection has not been demonstrated (Griffin and Davis 1978; Markiw 1992a). Low levels of actinospore exposure over time need to be tested to see if this results in increased antibody titers and reduced susceptibility to subsequent infection. Benign species of *Myxobolus* may induce immunity if used as vaccines, either used as a whole or by focusing on certain immunogenic antigens. A greater understanding of the environmental determinants influencing the severity of the disease should lend greater insight into control measures and management strategies that minimize mortality. Until future research provides additional approaches to controlling whirling disease, the data summarized above should be helpful in efforts directed at control and eradication of the parasite.

REFERENCES

Alderman, D. J. 1986. Whirling disease chemotherapy. Bulletin of the European Association of Fish Pathologists 6(2):38–40.

Allen, M. B. 1999. Factors influencing the distribution of *Myxobolus cerebralis*, the causative agent of whirling disease, in the Cache LaPoudre River, Colorado. Master's thesis, Colorado State University, Fort Collins, Colorado.

Bauer, O. N. 1962. The ecology of parasites of freshwater fish. Bulletin of the State Scientific Research Institute of Lake and River Fisheries 49:3–189. (Translated from Russian for the National Science Foundation, Washington, DC by the Israel Program for Scientific Translations, Jerusalem).

Bills, T. D., and D. A. Johnson. 1992. Effects of pH on the toxicity of TFM to sea lamprey larvae and nontarget species during a stream treatment. Great Lakes Fishery Commission Technical Report 57:7–19.

Birtwell, I. K., and D. R. Arthur. 1980. The ecology of tubificids in the Thames estuary with particular reference to *Tubifex costatus* (Claparéde) and *Tubifex tubifex* (Müller). Pages 331–381 *in* R. O. Brinkhurst and D. G. Cook, editors. Aquatic oligochaete biology. Plenum, New York.

Brinkhurst, R. O. 1996. On the role of tubificid oligochaetes in relation to fish disease with special reference to the Myxozoa. Annual Review of Fish Diseases 6:29–40.

Chapman, P. M., M. A. Farrell, and R. O. Brinkhurst. 1982. Relative tolerances of selected aquatic oligochaetes to individual pollutants and environmental factors. Aquatic Toxicology 2:47–67.

Clancy, J. L., T. M. Hargy, M. M. Marshall, and J. E. Dyksen. 1998. Inactivation of *Cryptosporidium parvum* oocysts in water using ultraviolet light. Journal of the American Water Works Association 90:92–102.

Dawson, V. K., D. A. Johnson, and J. F. Sullivan. 1992. Effects of the lampricide 3-trifluoromethyl-4-nitrophenol on dissolved oxygen in aquatic systems. Great Lakes Fishery Commission Technical Report 57:21–33.

Downing, D. C. 2000. Spawning, and rearing ecology of Madison River rainbow trout in relation to whirling disease infection risk. Master's thesis, Montana State University, Bozeman, Montana.

El-Matbouli, M., and R. W. Hoffmann. 1991a. Effects of freezing, aging, and passage through the alimentary canal of predatory animals on the viability of *Myxobolus cerebralis* spores. Journal of Aquatic Animal Health 3:260–262.

El-Matbouli, M., and R. W. Hoffmann. 1991b. Prevention of experimentally induced whirling disease in rainbow trout *Oncorhynchus mykiss* by Fumagillin. Diseases of Aquatic Organisms 10:109–113.

El-Matbouli, M., T. S. McDowell, D. B. Antonio, K. B. Andree, and R. P. Hedrick. 1999. Effect of water temperature on the development, release and survival of the triactinomyxon stage of *Myxobolus cerebralis* in its oligochaete host. International Journal for Parasitology 29:627–641.

Griffin, B. R., and E. M. Davis. 1978. *Myxosoma cerebralis*: Detection of circulating antibodies in infected rainbow trout (*Salmo gairdneri*). Journal of the Fisheries Research Board of Canada 35:1186–1190.

Hedrick, R. P., M. El-Matbouli, M. A. Adkison, and E. MacConnell. 1998. Whirling disease: re-emergence among wild trout. Immunological Reviews 166:365–376.

Hedrick, R. P., T. S. McDowell, K. Mukkatira, M. P. Georgiadis, and E. MacConnell. 1999. Susceptibility of selected inland salmonids to experimentally induced infections with *Myxobolus cerebralis*, the causative agent of whirling disease. Journal of Aquatic Animal Health 11:330–339.

Hedrick, R. P., and seven coauthors. 2000. Ultraviolet irradiation inactivates the waterborne infective stages of *Myxobolus cerebralis*: a treatment for hatchery water supplies. Diseases of Aquatic Organisms 42:53–59.

Hoffman, G. L. 1974. Disinfection of contaminated water by ultraviolet irradiation, with emphasis on whirling disease (*Myxosoma cerebralis*) and its effect on fish. Transactions of the American Fisheries Society 103:541–550.

Hoffman, G. L. 1975. Whirling disease (*Myxosoma cerebralis*): control with ultraviolet irradiation and effect on fish. Journal of Wildlife Diseases 11:505–507.

Hoffman, G. L. 1990. *Myxobolus cerebralis*, a worldwide cause of salmonid whirling disease. Journal of Aquatic Animal Health 2:30–37.

Hoffman, G. L., and C. E. Dunbar. 1961. Studies on *Myxosoma cerebralis* (Hofer) Plehn (Protozoa: Myxosporidea) the cause of whirling disease of trout. Annual meeting of the American Society of Parasitologists, Abstract 53, August 27–31, Lafayette, Indiana. Journal of Parasitology 47 (4, section II):29.

Hoffman, G. L., C. E. Dunbar, and A. Bradford. 1962. Whirling disease of trout caused by *Myxosoma cerebralis* in the United States. Special Scientific Report No. 427. U.S. Fish and Wildlife Service, Washington, D.C.

Hoffman, G. L. Sr., and G. L. Hoffman, Jr. 1972. Studies on the control of whirling disease (*Myxosoma cerebralis*). Journal of Wildlife Diseases 8:49–53.

Hoffman, G. L., and M. E. Markiw. 1977. Control of whirling disease (*Myxosoma cerebralis*): use of methylene blue staining as a possible indicator of effect of heat on spores. Journal of Fish Biology 10:181–183.

Hoffman, G. L., and J. J. O'Grodnick. 1977. Control of whirling disease (*Myxosoma cerebralis*): effects of drying, and disinfection with hydrated lime or chlorine. Journal of Fish Biology 10:175–179.

Hoffman, G. L., and R. E. Putz. 1969. Host susceptibility and the effect of aging, freezing, heat, and chemicals on spores of *Myxosoma cerebralis*. Progressive Fish-Culturist 31:35–37.

Ivanov, A. P., R. Y. Kosyreva, and N. L. Nechaeva. 1968. Implementation for health of salmonids in fish culture in southern U.S.S.R. (in Russian). 1972. Food and Agriculture Organization of the United Nations, Symposium on the major communicable fish diseases in Europe and their control. FI:EIFAC 72/SC II-Symp 18.

Janiszewska, J. 1955. Actinomyxidia: morphology, ecology, history of investigations, systematics, development. Acta Parasitologica Polonica 2:405–443.

Kaster, J. L., and J. H. Bushnell. 1981. Cyst formation by *Tubifex tubifex* (Tubificidae). Transactions of the American Microscopy Society 100:34–41.

Kent, M. L., L. Margolis, and J. O. Corliss. 1994. The demise of a class of protists: taxonomic and nomenclatural revisions proposed for the protist phylum Myxozoa Grassé, 1870. Canadian Journal of Zoology 72:932–937.

Lake, R. L., and S. G. Hinch. 1999. Acute effects of suspended sediment angularity on juvenile coho salmon (*Oncorhynchus kisutch*). Canadian Journal of Fisheries and Aquatic Sciences 56:862–867.

Lieffers, H. J. 1990. Effects of the lampricide 3-trifluoromethyl-4-nitrophenol on macroinvertebrate populations in a small stream. Great Lakes Fisheries Commission Technical Report 55, Ann Arbor, Michigan.

Markiw, M. E. 1992a. Salmonid whirling disease. Fish and Wildlife Service Leaflet 17. Washington, D.C.

Markiw, M. E. 1992b. Experimentally induced whirling disease II. Determination of longevity of the infective triactinomyxon stage of *Myxobolus cerebralis* by vital staining. Journal of Aquatic Animal Health 4:44–47.

Markiw, M. E. 1992.c. Experimentally induced whirling disease I. Dose response of fry and adults of rainbow trout exposed to the triactinomyxon stage of *Myxobolus cerebralis*. Journal of Aquatic Animal Health 4:40–43.

Markiw, M. E., and K. Wolf. 1983. *Myxosoma cerebralis* (Myxozoa: Myxosporea) etiological agent of salmonid whirling disease requires tubificid worm (Annelida: Oligocheata) in its life cycle. Journal of Protozoology 30:561–564.

Molnar, K., F. Baska, and C. Székely. 1987. Fumagillin, an efficacious drug against renal sphaerosporosis of the common carp *Cyprinus carpio*. Diseases of Aquatic Organisms 2:187–190.

Nehring, R. B., and K. G. Thompson. 2001. Impact assessment of some physical, and biological factors in the whirling disease epizootic among wild trout in Colorado. Colorado Division of Wildlife, Special Report No. 76, Fort Collins, Colorado.

Nehring, R. B., and P. G. Walker. 1996. Whirling disease in the wild: the new reality in the Intermountain West. Fisheries 21(6):28–30.

Newcombe, C. P., and D. D. MacDonald. 1991. Effects of suspended sediments on aquatic ecosystems. North American Journal of Fisheries Management 11:72–82.

Nickum, M. J. 2000. Whirling disease forces hatchery modernization in Colorado. Hatchery Magazine 1(2):10–12,25.

O'Grodnick, J. J. 1975. Egg transmission of whirling disease. Progressive Fish-Culturist 37:153–154.

O'Grodnick, J. J. 1979. Susceptibility of various salmonids to whirling disease (*Myxosoma cerebralis*). Transactions of the American Fisheries Society 108:187–190.

O'Grodnick, J. J., and C. C. Gustafson. 1974. A study of the transmission, life history, and control of whirling disease of trout. Federal Aid to Fish Restoration Progress Report F-35-R-6. U.S. Fish and Wildlife Service, Washington, D.C.

Rasmussen, C. J. 1965. Control of whirling disease in Danish trout farms. European Inland Fisheries Advisory Commission Technical Paper No. 2:14–15.

Ryce, E. K. N., A. V. Zale, and R. B. Nehring. 2001. Lack of selection for resistance to whirling disease among progeny of Colorado River rainbow trout. Journal of Aquatic Animal Health 12:63–68.

Scolari, C. 1954. Sull'impiego dello Stovarsolo nella profilassi del "capostorno" O "lentosporiasi" delle trote d'allevamento. Clinica Veterinaria 77:50–53.

Strange, R. J., C. B. Schreck, and R. D. Ewing. 1978. Cortisol concentrations in confined juvenile chinook salmon (*Oncorhynchus tshawytscha*). Transactions of the American Fisheries Society 107:812–819.

Taylor, R. E. L., S. J. Coli, and D. R. Junell. 1973. Attempts to control whirling disease by continuous drug feeding. Journal of Wildlife Diseases 9:302–305.

Taylor, R. L., and M. Lott. 1978. Transmission of salmonid whirling disease by birds fed trout infected with *Myxosoma cerebralis*. Journal of Protozoology 25:105–106.

Thompson, K. G., R. B. Nehring, D. C. Bowden, and T. Wygant. 1999. Field studies of seven species or subspecies of salmonids to *Myxobolus cerebralis* in the Colorado River, Middle Park, Colorado. Journal of Aquatic Animal Health 11:312–329.

Thompson, K. G., and R. B. Nehring. 2000. A simple technique used to filter and quantify the actinospore of *Myxobolus cerebralis* and determine its seasonal abundance in the Colorado River. Journal of Aquatic Animal Health 12:316–323.

Vincent, E. R. 1996. Whirling disease and wild trout: the Montana experience. Fisheries 21(6):32–33.

Walker, P. G., and R. B. Nehring. 1995. An investigation to determine the cause(s) of the disappearance of young wild rainbow trout in the upper Colorado River, in Middle Park, Colorado. Colorado Division of Wildlife Report, Denver, Colorado.

Whitley, L. S. 1968. The resistance of tubificid worms to three common pollutants. Hydrobiologia 32:193–205.

Whitten, B. K., and C. J. Goodnight. 1966. Toxicity of some common insecticides to tubificids. Journal of the Water Pollution Control Federation 38:227–235.

Wolf, K., and M. E. Markiw. 1982. *Myxosoma cerebralis*: inactivation of spores by hot smoking of infected trout. Canadian Journal of Fisheries and Aquatic Sciences 39:926–928.

Wolf, K., and M. E. Markiw. 1984. Biology contravenes taxonomy in the Myxozoa: new discoveries show alternation of invertebrate and vertebrate hosts. Science 225:1449–1452.

Yokoyama, H., Y. S. Liyanage, A. Sugai, and H. Wakabayashi. 1999. Efficacy of fumagillin against haemorrhagic thelohanellosis caused by *Thelohanellus horvorkai* (Myxosporea:Myxozoa) in colored carp, *Cyprinus carpio* L. Journal of Fish Diseases 22:243–245.

American Fisheries Society Symposium 29:227–238, 2002
© 2002 by the American Fisheries Society

The Effect of Water Characteristics on Viability of the *Myxobolus cerebralis* Actinospore

MARK A. SMITH[1], ERIC J. WAGNER* AND AMY HOWA
Utah Division of Wildlife Resources, Fisheries Experiment Station
1465 W. 200 North, Logan, Utah 84321, USA

ABSTRACT The effect of pH, total hardness, salinity, and dissolved oxygen on viability of the tri-actinomyxon (TAM) stage of the Myxozoan parasite *Myxobolus cerebralis* was evaluated. The vital stains propidium iodide and fluorescein diacetate were used to determine if TAMs were alive or dead. Tests were conducted at 9°C and 14°C. Experiments were conducted at salinities of 5, 10, 15, 20, and 30 ‰, at pH from 4 to 10, at total hardness of 10 or 650 mg/L, and dissolved oxygen at 0–0.5 mg/L. Viability was not significantly affected by low dissolved oxygen or the total hardness levels tested. Few differences were observed related to temperature. Viability decreased over time in controls to roughly half after 72 h at either temperature. Salinity significantly affected mortality, even after 1 h. For disinfection, 20–30 ‰ salinity for at least 6 h would be required to effectively kill all TAMs. The further pH moved from circum-neutral, the more TAM viability decreased. Differences in water quality, especially pH, could partially explain differences observed in severity of whirling disease among geographic regions.

Myxobolus cerebralis is responsible for whirling disease in salmonid fishes. Native to Europe, the parasite was inadvertently introduced to the eastern United States in the mid-1950s (Markiw 1992a). The parasite has continued its western movement and to date can be found in states from coast to coast. However, the impacts of M. *cerebralis* on salmonid populations have varied widely with geographic location. In the northeastern United States and certain western waters, impacts of the disease on wild fish can be described as minimal (Hulbert 1996). However, many western trout waters have seen severe declines in wild trout populations. In Colorado, Nehring and Walker (1996) reported a nearly complete loss of four successive year classes (1991–1994) of rainbow trout fry in the upper Colorado River following the discovery of whirling disease. Similarly, on Montana's famed Madison River, discovery of whirling disease in 1991 is believed to be the chief cause of a 90% reduction in rainbow trout in two sections of the river by 1994 (Vincent 1996). To date there are no clear explanations for these regional variations in the impact of this parasite.

One possibility is that the characteristics of the water itself may provide the parasite better opportunities to infect its fish host. A more favorable water quality could extend the life of the infective stage, increasing the probability of finding a host. Many characteristics of surface water can be traced to the geology and climate of the region in which it lies. In the northeastern United States, trout streams are in general lower in pH and have softer water when compared with the majority of western streams (Schofield 1982). Exceptions to these generalizations can be found of course, but it is possible that geographic differences in water quality could partially help explain variations observed with whirling disease outbreaks.

The parasite M. *cerebralis* has a complex life cycle involving both an oligochaete worm and fish as hosts (Markiw and Wolf 1983). The infective stage for fish is the actinospore referred to as the triactinomyxon (TAM). After excretion from the oligochaete host (*Tubifex tubifex*), the TAMs are suspended in the water column and must encounter a fish in order to continue the life cycle of the organism. Once excreted, TAMs are short lived. Markiw (1992b) found that depending upon temperature, TAMs were viable for less than 8 d. El-Matbouli et al. (1999) found that more than 60% of TAMs were still viable after 15 d at water temperatures up to 15°C. In contrast, the resistant

[1]Present address: Wyoming Cooperative Fish and Wildlife Unit, University of Wyoming, Laramie, WY 82070, USA
*Corresponding author: nrdwr.ewagner@state.ut.us

myxospore stage of the parasite may last for up to 12 years in wet mud (Bauer 1962). Because of a shorter window of viability, the TAM may be the weakest link in the life cycle of the organism.

In this study, the possibility that water quality may impact the life span of the actinospore, and thus the success of the parasite in a given water, is explored. Because of the TAM's fragile nature and the short period of opportunity it has to find a fish, interactions may be limited in environments that are not conducive to the actinospore. If water quality variables do negatively impact the actinospore, this may help explain the variability in infection rates seen across space and time. Additionally, such knowledge could be helpful in developing techniques aimed at disrupting the life cycle of the parasite.

The objective of this study was to examine the effects of salinity, pH, dissolved oxygen, and total hardness on TAM viability.

METHODS

Experimental Design

The effects of salinity, pH, dissolved oxygen (DO), and total hardness on TAM viability were tested at two temperatures, 9°C and 14°C. Duration of exposures were 1, 6, 24, 48, and 72 h. Live TAMs were obtained from laboratory oligochaete cultures 3 times per week (Monday, Wednesday, Friday) by siphoning water from the cultures through a 20 μm Nitex filter (Aquaculture Research/Environmental Associates, Homestead, Florida). TAMs were kept cold in a cooler with ice packs and used for experiments within 24 h.

For each experiment there were three treatments with three replicates per treatment. The three treatments were a baseline, control and the target group. Determiniation of the baseline variability was made at the beginning of each experiment (time = 0) from the TAM stock solution using staining methods described below. Target groups were exposed to the experimental water quality characteristic, and control groups were exposed to fresh well water.

Assessment of TAM Viability

After TAMs were exposed to their respective treatments, they were concentrated by filtering through a small 10 μm mesh filter, which was then thoroughly rinsed with 20 mL of fresh well water. As the last aliquot of water was filtered, 100 μL of

retentate from the filter was transferred to each of three slides for staining.

Staining was based on the techniques of Jones and Senft (1985) and Markiw (1992b), where fluorescein diacetate (FDA) and propidium iodide (PI) are used simultaneously. An aqueous stock solution of PI (52 mg/mL) was made and kept frozen (−40°C) in 1 mL aliquots. Vials were then thawed on the test day. A concentrated stock solution of FDA (5 mg/mL acetone) was kept at −40°C. On the day of an experiment, 80–100 μL of the concentrated FDA solution was added to 8 mL of well water to make a working stock of FDA stain. This amount varied slightly to adjust for minor changes over time in the strength of the stain fluorescence and for the amount of organic matter in the sample. For each microscope slide, 50 μL of both FDA and PI were added to the 100 μL sample of recovered TAMs. Slides were cover slipped and incubated at 9–12°C in a dark, humid chamber for 45 min, then examined.

Using an epifluorescence microscope, live TAM sporozoites appear green with the FDA staining and dead sporozooites are red from the PI staining. On occasion sporozoites stained both red and green; these were recorded as possibly viable, while those stained green or red were recorded as viable and dead, respectively. All classifications of the possibly viable TAMs were included in the total TAM count. The total number of TAMs varied according to their availability and averaged 89, 99, 100, and 100 per slide in the salinity, pH, hardness, and DO tests, respectively.

Salinity Tests

Salinity experiments were conducted at concentrations of 5, 10, 15, 20, and 30‰. Stock solutions of salt water were made using n_iodized rock salt dissolved in well water. For each salinity test, equal parts of TAM stock solution and salt water were combined in test tubes to achieve the required salinity. The test tube contents were thoroughly mixed, covered, and stored at experiment temperature. After the exposure time had elapsed, the TAMs were collected and enumerated as described.

pH Tests

The effects of pH on TAM viability were tested at pH 4.0, 5.0, 6.0, 8.5, 9.0, 9.5, and 10. The well water used for controls naturally varied from pH 7.5 to 8.2. The pH for treatments was maintained by addition of either 0.1 N NaOH or 5% H_2SO_4 to hatchery well water. The pH was monitored throughout the exposure using a digital meter calibrated daily using standard pH buffers. Differences between initial and final pH levels of controls were minor, varying from 0.00 units in 1 h treatments to 0.1–0.25 units in 72 h treatments.

Small 10 μm filters were used for exposing TAMs to differing pH. These were constructed from plastic 50 mL centrifuge tubes. For each filter, the top 50 mm of the tube was cut off. Another 10 mm section of the tube was cut off and a small section removed to decrease the ring diameter to just fit within the 50 mm high section. The cut ring stretched and secured a small circle of 10-μm mesh Nitex snugly within the cylinder. The edges of the mesh were sealed with silicone to prevent loss of TAMs at the edges. The completed filters measured approximately 25 mm deep and 25 mm in diameter. Filters were weighted using small steel bolts taped to the outside to keep them neutrally buoyant. Lids from the centrifuge tubes were used to keep samples from leaking out.

Experiments were conducted by first filtering the TAM stock solution through the 10 μm filter and then refilling with water at the test pH. The filters were capped and placed into a 40 L tank containing water at the pH to be tested. Water from the tank was able to readily pass back and forth across the filter membrane. After the exposure time elapsed, filters were removed and the contents were rinsed with 20 mL fresh well water. Slides were made, read, and recorded as above.

Total Hardness Tests

The effects of total hardness on TAM viability was tested for water considered very hard and soft as measured using standard titration techniques (APHA 1989). Well water used for controls had a total hardness of approximately 260 mg/L. To test hard water, a stock solution of 1000 mg/L hard water was made using powdered $CaCl_2$, The stock solution was mixed in equal parts with TAM stock solution in test tubes, producing a 650 mg/L solution. The test tubes were covered and stored at 9°C and 14°C throughout exposure. After appropriate time had elapsed, TAMs were collected and enumerated.

The effects of soft water on TAM viability was tested using de-ionized well water mixed with regular well water to achieve 10 mg/L stock solution of soft water. TAMs were put onto small 10 μm filters and the water was allowed to drain off, as in pH experiments. The filters were then filled with soft

water at the test temperature, covered and placed in 1-L containers filled with soft water. Following exposures, samples were stained, incubated, and enumerated as noted above.

Dissolved Oxygen Tests

The effect of low DO on TAM viability was tested at 0–0.5 mg/L oxygen. The well water used for control groups ranged from 6 to 8 mg/L oxygen. Low DO was accomplished by bubbling nitrogen (N_2) gas through an air stone in a small vial of TAM stock solution. After the oxygen was depleted, the vial contents were carefully transferred into small test tubes. Test tubes were filled to overflowing and sealed carefully to avoid trapping any air within them. Dissolved oxygen was measured with a digital meter (calibrated for elevation and temperature) at the beginning and end of each exposure. Dissolved oxygen was 0.0 mg/L at the end of the 72 h period for both temperature tests. After the exposure time had elapsed, test tube contents were poured onto filters and rinsed. Slides were prepared as noted above.

Statistical analysis

Data were analyzed using SPSS Advanced Statistics 7.0 (SPSS 1996). Data were arc-sine transformed and tested for normality using the Shapiro-Wilk test. The Mann–Whitney U-test was used to compare between freshwater controls at the two temperatures, separately for each time period. The same test was used to compare DO and total hardness treatments with their respective controls. The effects of salinity and pH were analyzed using a one-way analysis of variance (ANOVA) separately for each time period. Subsequent mean comparisons were conducted using the Least Significant Difference test. An α-level of 0.05 was used for all analyses.

RESULTS

Effect of Temperature on Controls

When the effect of temperature (9°C versus 14°C) was compared between controls for each exposure time (Figure 1), the only significant difference ($P \leq 0.05$) was at 24 h. Time did prove to negatively affect TAM viability at both temperatures with nearly 50% of TAMs from control groups dead after 72 h.

Salinity

All concentrations of salt water (5–30‰) had negative effects on TAM viability at both 9°C (Figure 2a) and 14°C (Figure 2b). Even at 1 h there were

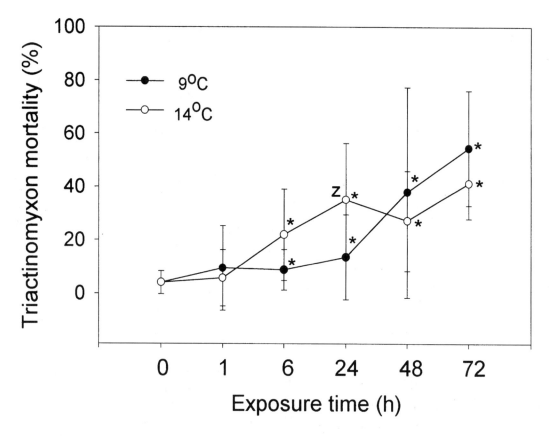

Figure 1. The effect of time and temperature on TAM viability for control groups from salinity tests. Reported as mean % of TAMs dead ± SD. Significant differences ($P \leq 0.05$) between temperature groups within a sampling time are noted by a letter. Asterisks denote significant differences from the baseline (Time = 0).

significant differences ($P \leq 0.05$) between all salt and freshwater treatments. Increasing the concentration of salt decreased the time required to kill TAMs. Salt concentrations of 20 and 30‰ had a rapid negative effect on TAM viability, killing more than 80% of TAMs within the first hour and nearly all TAMs after 6 h. Lower concentrations of salt (5 and 100/00) required 48 h to kill more than 80% of TAMs.

pH

TAM viability was impacted by pH. The further pH moved from circum-neutral, TAM viability decreased. On the basic end of the scale, pH of 8.5 and greater had significant ($P \leq 0.05$) negative effects on the viability of TAMs at 9°C (Figure 3) and 14°C (Figure 4). At 9°C, TAM viability was significantly lower than controls after 6 h of exposure to pH 9, 9.5, or 10. Viability of TAMs at pH 8.5 was slightly, but significantly, different from the controls

(pH 7.95 and 8.1, respectively) at 24 and 72 h.

At 14°C, pH 9.5 and 10 treatments had significantly more dead TAMs than the control after 6 h. Viability at pH 9 was significantly reduced after 24 h; mortality increased to 79–88% compared with controls that had 16–30% mortality. Survival at pH 8.5 was only significantly reduced from the controls at 24 and 72 h (pH 7.5 and 8.0, respectively).

Acidity was also found to negatively affect TAM viability. At 9°C (Figure 5), pH 4 induced significantly more dead TAMs than the controls after only 1 h. Reduction in viability at pH 5 was significant after 6 h and at pH 6 was only significantly reduced from the controls (pH 8.2) at 72 h. At 14°C (Figure 6), there were significantly more dead TAMs in the pH 4 treatment than the control after only 1 h. Reduction in viability at pH 5 was significant after 6 h and at pH 6 viability was significantly lower than controls (pH 7.9–8.1) after 24 h.

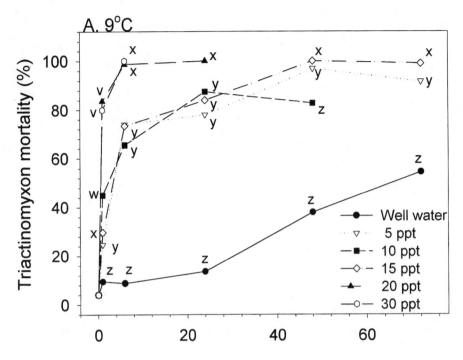

Figure 2a. Mean percentage of dead TAMs at reported saline concentrations over time, at 9°C. Significant differences ($P \leq 0.05$) between saline concentrations within a sampling time are noted by a different letter.

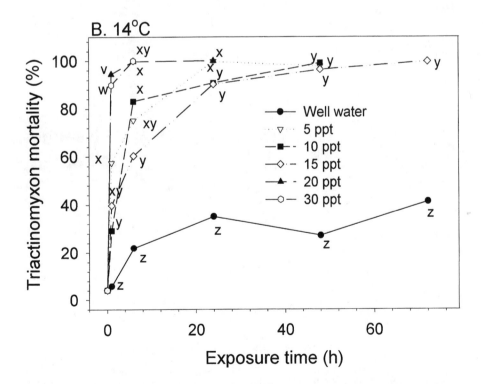

Figure 2b. Mean percentage of dead TAMs at reported saline concentrations over time, at 14°C. Significant differences ($P \leq 0.05$) between saline concentrations within a sampling time are noted by a different letter.

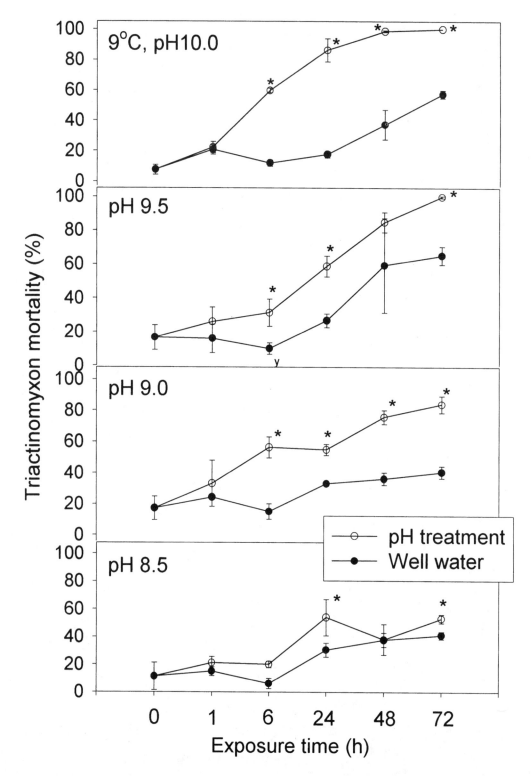

Figure 3. The effects of time and high pH on TAMs viability at 9°C, reported as mean percent of TA dead ± SD. Significant differences ($P \leq 0.05$) between well water and pH groups within a sampling time are noted by a asterisk.

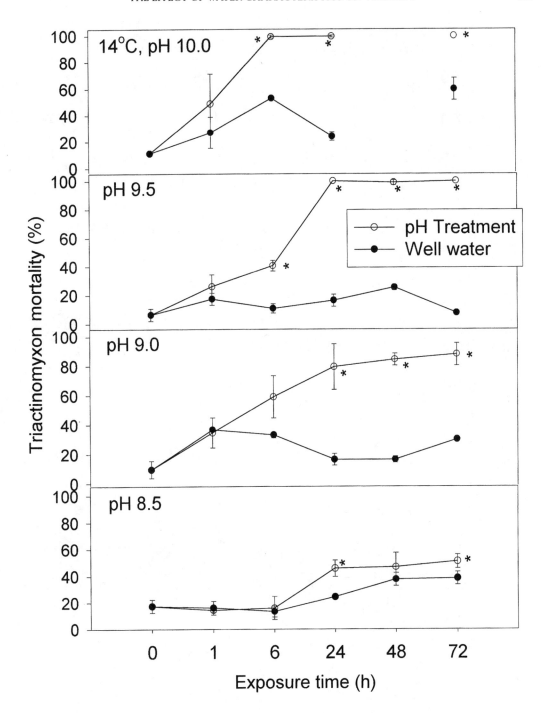

Figure 4. The effects of time and high pH on TAMs viability at 14°C, reported as mean percent of TA dead ± SD. Significant differences ($P \leq 0.05$) between well water and pH groups within a sampling time are noted by an asterisk.

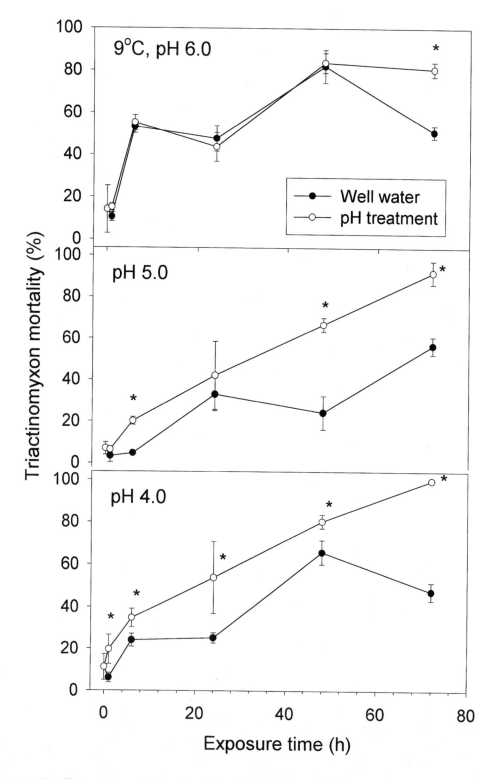

Figure 5. The effects of time and low pH on TAM viability at 9°C, reported as mean percent of TAMs dead ± SD. Significant differences ($P \leq 0.05$) between well water and pH groups within a sampling time are noted by an asterisk.

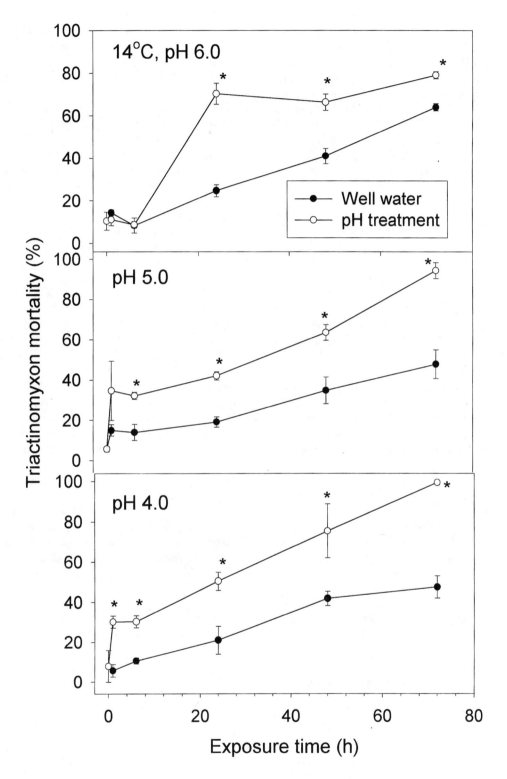

Figure 6. The effects of time and low pH on TAMs viability at 14°C, reported as mean percent of TA dead ± SD. Significant differences ($P \leq 0.05$) between well water and pH groups within a sampling time are noted by an asterisk.

Total Hardness

Total hardness was tested at extreme levels, evaluating both water considered very hard (650 mg/L) and soft (10 mg/L) at exposure times from 24 to 72 h. Even at these extreme levels of water hardness, there was no significant effect of either hard or soft water on the viability of TAMs ($P \leq 0.05$; Table 1). At 72 h, nearly 50% of observed TAMs died in the soft-water treatments, while approximately 30 and 45% mortality, respectively, occurred in the hard water and control treatments.

Dissolved Oxygen

Dissolved oxygen was tested at a very low level (0–0.5 mg/L) for a period of 72 h. At both 9°C and 14°C, DO decreased from 0.5 to 0.0 mg/L over the time period. In both exposures, low DO had no significant effect on the viability of TAMs (Table 1). In both well water and low DO treatments, 30–40% of TAMs were dead after the 72 h.

DISCUSSION

The epidemiology of whirling disease is still far from being fully understood. Factors that influence the viability and abundance of the alternate host, susceptibility of the fish host, and infection severity are still being explored. It is known that dose (i.e., the number of TAMs a fish is exposed to) has a significant impact on the development of clinical signs and mortality (Markiw 1992c). The research conducted in this study examined the effect of water quality characteristics on TAM viability, which affects the number of TAMs available to infect fish. Conditions that reduce viability would logically reduce the number of viable TAMs passing by a given point on a stream.

Vital staining using fluorochromes has been a useful method for viability assessment (Jones and Senft 1985). It has been used previously with M. cerebralis (Markiw 1992b) and viability has also been correlated with survival of Giardia cysts by Schupp and Erlandsen (1987). These authors noted that Giardia muris cysts that were stained red by PI were incapable of infecting mice, whereas green, FDA-stained cysts resulted in infection. Later work (Smith and Smith 1989; Labatiuk et al. 1991) has indicated that the dyes are conservative, overestimating actual viability of Giardia cysts. In this study, the TAMs staining both red and green were considered as potentially viable, but in reality may not be viable. Interpretation of the lethal effects of the variables tested are based on the vital staining result of 100% dead (all red).

In this study, pH was the variable most likely to influence TAM viability in freshwater. Both high and low values influenced viability in ranges that would not be deadly to fish. Some fishes have been found in certain lakes and streams at pH levels ranging from 4 to 10 (Cleary 1955). However, lethal levels may vary because sensitivity varies with species, presence of other ions such as Al+, and the stage of development (Doudoroff and Katz 1950). For example, Woodward et al. (1991) noted survival of greenback cutthroat trout Oncorhynchus clarki stomias alevins was reduced by 68% at pH 5.0. The data in this study suggest that low pH conditions typical in the eastern U.S. (Schofield 1982; Patrick 1996) would tend to reduce the lifespan of any TAMs released. This would be in agreement with the lack of severe population impacts noted for this region (Hulbert 1996). However, other variables such as relative

Table 1. Mean percentage (± SD) of dead TAMs for low dissolved oxygen, soft (10mg/L total hardness) and hard (650 mg/L total hardness) water exposed groups at given temperature and time. There were no significant ($P \leq 0.05$) differences between treatments or temperatures.

Water quality variable	Concentration (mg/L)	Temperature(°C)	Time (h)	% Mortality ± SD Test	Control
Hardness[a]	650	9	24	23.3 ± 1.9	17.6 ± 2.0
Hardness	650	9	48	31.0 ± 2.2	28.3 ± 6.5
Hardness	650	9	72	27.1 ± 3.8	44.2 ± 4.3
Hardness	10	9	72	51.0 ± 1.4	38.4 ± 6.1
Hardness	10	14	72	61.7 ± 4.3	57.7 ± 0.5
DO	0	9	72	27.1 ± 3.8	44.2 ± 4.3
DO	0	14	72	27.7 ± 4.0	28.0 ± 5.1

[a]total hardness as $CaCO_3$

abundance of tubificid worms and temperature regimes may also act to limit the severity of the disease in the region. Also, not all eastern streams are acidic. Effects on TAMs were noted on the basic end of the pH scale as well, at levels that are not considered lethal to salmonids ($\leq$ pH 9.5–9.8; Daye and Garside 1975; Murray and Ziebell 1984). Data from this study also indicated that high pH levels typically found in eutrophic waters in summer could reduce the lifespan of the infective stage. Studies of the effect of pH on *T. tubifex* indicate that lethal levels are more extreme than they are for salmonids; 96-h LC$_{50}$ values were pH 3.6 and 10.5 without sediment or 2.5 and 10.7 with sediment (Chapman et al. 1982). Therefore, direct effects of pH on the worm host would not be expected, but the combined effect of parasitism and pH on TAM release has yet to be studied. The ultimate impact of the parasite under high or low pH conditions must also take into account worm population dynamics, temperature, natural predation, and a variety of other variables.

The effect of temperature on the disease has been examined in a few different studies. R. Vincent (Montana Fish, Wildlife, and Parks, personal communication) noted that the peak infection rates of sentinel fish occurred at about 12–13°C, from mid-May to Mid-July. Research with *T. tubifex* indicated that temperature played a key role in the development and production of TAMs. TAM production from worms transferred from 15°C to 25°C or 30°C stopped after 4 d, and after 15 d at 20°C (El-Matbouli et al. 1999). Worms transferred to 5°C, 10°C, or left at 15°C continued to produce TAMs. Markiw (1992b) noted that temperature influenced the lifespan of the TAM, ranging from only 2 to 3 d at 19–24°C to 7–8 d at 7°C. Similarly, we found that nearly 50% of TAMs from control groups at 9°C and 14°C were dead after 72 h. The lifespan data from this study may be conservative due to the collection of TAMs from cultures only three times a week (Monday, Wednesday, Friday). Therefore some TAMs may be 2–3 d old at temperatures from 13°C to 15°C, before initiating the test. This would mean the half-life of the TAMs could be from 3 to 6 d at 9°C or 14°C. El-Matbouli et al. (1999) found that TAMs may still be viable after 15 d at water temperatures up to 15°C.

The salinity data indicated that the TAMs are sensitive to salt, with higher concentrations causing significant mortality after only an hour. Lower concentrations at longer durations (e.g., 5–10 ‰ for 24 h or more) were similarly toxic. From a disinfection perspective, 20–30‰ salinity for at least 6 h would be required to effectively kill all TAMs. The sensitivity of TAMs to higher salinity is of interest relative to anadromous salmonids in important estuaries such as the mouth of the Columbia River. Salt may also be useful as a control measure in situations such as treatment of suspect water used in fish stocking trucks.

The lack of any effect of hardness at concentrations of 10 or 650 mg/L was not too surprising. Salmonids have a similar wide tolerance to hardness per se, with no effect noted until concentrations reach 1 mg/L (Brown and Lynam 1981). However, low hardness and alkalinity reduce the buffering capacity of the water, making any fish present very susceptible to drastic pH fluctuations. The lack of any effect of dissolved oxygen on viability was also not surprising. The worm host is typically found in sediments in which the DO can drop to zero. *Tubifex tubifex* is known to survive and even reproduce under anoxic conditions (Fox and Taylor 1955; Famme and Knudsen 1985). It would be a useful adaptation for the parasite to also survive such conditions, although this has yet to be evaluated *in vivo*.

The data presented here contribute to the understanding of the infection process in wild situations. Further research should elucidate other factors and relationships that affect disease severity. Once these are determined, these could be applied to fisheries management as it relates to whirling disease.

ACKNOWLEDGMENTS

We thank Grant White for the opportunity to collect infected worms from his farm for TAM production. We also thank Don Roberts at Utah State University for some of the TAMs used in the study. We thank Ronney Arndt and Quinn Cannon for their help in maintaining the worm cultures. The work was generously funded by the National Partnership for the Management of Wild and Native Cold Water Fisheries in cooperation with the Utah Division of Wildlife Resources. In coordination with matching funds from the Federal Aid in Sport Fish Restoration program, the Salt Lake County Fish and Game Association provided a generous donation used to purchase the epifluorescence microscope used in the study.

REFERENCES

APHA (American Public Health Association), American Water Works Association and Water Pollution Control Federation. 1989. Standard methods for the examination of water and wastewater, 17th edition. APHA, Washington, D.C.

Bauer, O. N. 1962. Relationships between host fishes and their parasites. Pages 84–103 in V. A. Dogiel, G. K. Petrushevski, and Y. I. Polyanski, editors. Parasitology of fishes. (English translation by Z. Kabata). Oliver and Boyd, London.

Brown, D. J. A., and S. Lynam. 1981. The effect of sodium and calcium concentrations on the hatching of eggs and survival of the yolk sac fry of brown trout, Salmo trutta L., at low pH. Journal of Fish Biology 19:205–211.

Chapman, P. M., M. A. Farrell, and R. O. Brinkhurst. 1982. Relative tolerances of selected aquatic oligochaetes to individual pollutants and environmental factors. Aquatic Toxicology 2:47–67.

Cleary, E. J. 1955. Stream pollution: aquatic life water quality criteria. Sewage and Industrial Wastes 27:321–331.

Daye, P. G., and E. T. Garside. 1975. Lethal levels of pH for brook trout, Salvelinus fontinalis (Mitchell). Canadian Journal of Zoology 53:639–641.

Doudoroff, P., and M. Katz. 1950. Critical review of literature on the toxicity of industrial wastes and their components to fish. Sewage and Industrial Wastes 22:1432–1458.

El-Matbouli, M., T. S. McDowell, D. B. Antonio, K. B. Andree, and R. P. Hedrick. 1999. Effect of water temperature on the development, release and survival of the triactinomyxon stage of Myxobolus cerebralis in its oligochaete host. International Journal for Parasitology 29:627–641.

Famme, P., and J. Knudsen. 1985. Anoxic survival, growth and reproduction by the freshwater annelid, Tubifex sp., demonstrated using a new simple anoxic chemostat. Comparative Biochemistry and Physiology 81A:251–253.

Fox, H. M., and A. E. R. Taylor. 1955. The tolerance of oxygen by aquatic invertebrates. Proceedings Royal Society B 143:214–225.

Hulbert, P. J. 1996. Whirling disease: a resource stewardship challenge. Fisheries 21(6):26–27.

Jones, K. H., and J. A. Senft. 1985. An improved method to determine cell viability by simultaneous staining with flourescein diacetate-propidium iodide. Journal of Histochemistry and Cytochemistry 33:77–79.

Labatiuk, C. W., F. W. Schaefer III, G. R. Finch, and M. Belosevic. 1991. Comparison of animal infectivity, excystation, and fluorogenic dye as measures of Giardia muris cyst inactivation by ozone. Applied and Environmental Microbiology 57:3187–3192.

Markiw, M. E. 1992a. Salmonid Whirling Disease. U.S. Fish and Wildlife Service, Leaflet 17. Washington, D.C.

Markiw, M. E. 1992b. Experimentally induced whirling disease II. Determination of longevity of the infective triactinomyxon stage of Myxobolus cerebralis by vital staining. Journal of Aquatic Animal Health 4:44–47.

Markiw, M. E. 1992c. Experimentally induced whirling disease I. Dose response of fry and adults of rainbow trout to the triactinomyxon stage of Myxobolus cerebralis. Journal of Aquatic Animal Health 4:40–43.

Markiw, M. E., and K. Wolf. 1983. Myxosoma cerebralis (Myxozoa; Myxosporea) etiologic agent of salmonid whirling disease requires tubificid worm (Annelida: Oligochaeta) in its life cycle. Journal Protozoology 30:561–564.

Murray, C. A., and C. D. Ziebell. 1984. Acclimation of rainbow trout to high pH to prevent stocking mortality in summer. Progressive Fish-Culturist 46:176–179.

Nehring, R. B., and P. G. Walker. 1996. Whirling disease in the wild: the new reality in the Intermountain West. Fisheries 21(6):28–30.

Patrick, R. 1996. Rivers of the United States, volume III, The eastern and southeastern states. Wiley and Sons, New York.

Schofield, C. L. 1982. Historical fisheries changes in the United States related to decreases in surface water pH. Pages 57–67 in T. A. Haines and R. E. Johnson, editors. Acid Rain/Fisheries. American Fisheries Society, Bethesda, Maryland.

Schupp, D. G., and S. L. Erlandsen. 1987. A new method to determine Giardia cyst viability: correlation of fluorescein diacetate and propidium iodide staining with animal infectivity. Applied and Environmental Microbiology 53:704–707.

Smith, A. L., and H. V. Smith. 1989. A comparison of fluorescein diacetate and propidium iodide staining and in vitro excystation for determining Giardia intestinalis cyst viability. Parasitology 99:329–331.

SPSS Inc. 1996. SPSS® base 7.0 for Windows™ user's guide. SPSS Inc., Chicago.

Vincent, E. R. 1996. Whirling disease and wild trout: the Montana experience. Fisheries 21(6):32–33.

Woodward, D. F., A. M. Farag, E. E. Little, B. Steadman, and R. Yancik. 1991. Sensitivity of greenback cutthroat trout to acidic pH and elevated aluminum. Transactions of the American Fisheries Society 120:34–42.

American Fisheries Society Symposium 29:239–247, 2002
© 2002 by the American Fisheries Society

Efficacy of Fumagillin and TNP-470 to Prevent Experimentally Induced Whirling Disease in Rainbow Trout Oncorhynchus mykiss

LINDA STATON

U.S. Fish & Wildlife Service, Bozeman Fish Health Center, Bozeman, Montana 59718, USA
Phone: (406) 582-8656; Fax: (406) 587-3998; email: linda_staton@fws.gov

DAVE ERDAHL

U.S. Fish & Wildlife Service, Bozeman Fish Technology Center/NIO, Bozeman, Montana 59715, USA
Phone: (406)587-9265; Fax: (406)582-0242; email: dave_erdahl@fws.gov

MANSOUR EL-MATBOULI

Institute of Zoology, Fish Biology, and Fish Disease, University of Munich, Munich, Germany
email: elmatbouli@zoofisch.vetmed.uni-muenchen.de

ABSTRACT Two potential therapeutants, fumagillin, an antibiotic derived from the fungus *Aspergillus fumagatus*, and TNP-470, a superactive analog of fumagillin, were tested for efficacy to prevent *Myxobolus cerebralis* infection in rainbow trout. The study was conducted at the Wild Trout Research Laboratory in Bozeman, Montana utilizing rainbow trout *Oncorhynchus mykiss* fry (~2.0 g). Treatment groups included negative controls (no treatment, no exposure), positive controls (no treatment, exposure), fumagillin top-dressed on feed, fumagillin incorporated in feed, and TNP-470 incorporated in feed. The exposure dose was 1000 triactinomyxons (TAMs) per fish for 2 h. Medicated feed treatment was initiated 24 h after exposure to TAMs. All treated groups received medicated feed for 10 d, with the exception of a single TNP-470 treatment group that was fed for 26 d. Five replicate tanks of fish were used for each treatment group. Presence and level of *Myxobolus cerebralis* infection were determined by histology, spore counts, and electron microscopy evaluation conducted 150 d postexposure. Hematology samples were also collected to evaluate potential toxic effects of treatment. Spore count and histological evaluation indicated that fumagillin and TNP-470 treatment was not efficacious in preventing or reducing *Myxobolus cerebralis* infection. Although fumagillin and TNP-470 administered for 10 d or 26 d did result in a reduction in spore numbers, results were not significantly different than observed in positive controls. Furthermore, histological scores were similar for all treatment groups and the positive controls. Although electron microscopy revealed spore deformation in both fumagillin and TNP-470 treated groups, no treatment group was effective in preventing *Myxobolus cerebralis* infection. Fumagillin treatment and TNP-470 fed for 10 d did not appear to negatively impact fish performance. However, toxicity was observed in fish fed TNP-470 for 26 d. These fish were observed to be lethargic 30 d pe, and blood samples revealed low hematocrits, severely decreased lymphocytes, and reduced numbers of blast cells. Histological evaluation revealed abnormal cytology in both the kidney and thymus. Based on study results, fumagillin and TNP-470 did not appear to be effective therapeutants for use in the prevention or control of *Myxobolus cerebralis* infection in rainbow trout.

Whirling disease of salmonids is a disease that although endemic to Eurasia, has been present in North America only since the 1950s. It is caused by the myxosporean parasite *Myxobolus cerebralis*. Although whirling disease is currently one of the most widely discussed fish diseases in the United States, it is obviously not a new disease to wild and hatchery fish in this country. Since it was first detected in Pennsylvania in 1956, it has been found in 23 states. However, whirling disease has never been so notorious as it has become following its discovery in Montana's blue-ribbon trout streams in late 1994. In short order, steering committees were organized and workshops and taskforces assembled to vigorously address this divisive, complex "problem" to which, unfortunately, only limited previous research had been directed. Out of these recent efforts has come a wealth of information including data on the life history of *Myxobolus cerebralis* and pathology associated with it, pathogen distribution, and the development of new diagnostic tools. Unfortunately, this new information has provided no obvious solutions to controlling/mitigating the effects of whirling disease, particularly with respect to incidence in wild populations. However, there is little doubt that stocking of infected fish may greatly exacerbate the problem. If whirling disease is ever going to be controlled or managed, it is imperative that stocking of infected fish be minimized, if not terminated entirely.

Until recently, the only practical mechanism for the control of whirling disease in aquaculture facilities was thought to be by control/isolation of hatchery water supplies or by control of the age at which fish are exposed to *Myxobolus cerebralis*. The former option carries no guarantees of a disease-free environment and is tremendously expensive, while the latter is often not possible. If drug or therapeutant treatment of exposed fish could be used to reduce or eliminate incidence of whirling disease, it would be of tremendous benefit to fisheries programs dependent upon stock supplementation. It would also indirectly benefit all fisheries programs in the country by reducing overall system infectivity.

Research conducted by El-Matbouli and Hoffmann (1991) using the relatively long-known antibiotic fumagillin, indicated that this drug may be effective in the prevention of experimentally induced whirling disease in rainbow trout *Oncorhynchus mykiss* under laboratory conditions. Fumagillin is an antibiotic that is produced by certain strains of the fungus *Aspergillus fumigatus*. Although it has limited antibacterial or antifungal activity, and little or no effect against common animal viruses, interest in fumagillin has resulted mainly as a result of its amebicidal activity. The use of fumagillin as a therapeutant in aquaculture was developed originally in Europe for the treatment of proliferative kidney disease (PKD) in rainbow trout (Hedrick et al. 1988; le Gouvello et al. 1999), which is caused by *Tetracapsula bryosalmonae*. Higgins and Kent (1996) reported that a single 10 d treatment of fumagillin administered orally at 3 mg/kg body weight per d to be effective in controlling the severity of PKD infection in coho salmon *Oncorhynchus kisutch*. A preliminary investigation by Erdahl and Staton (unpublished data) found that administration of fumagillin top-coated feed was not efficacious in controlling whirling disease in rainbow trout and chinook salmon *Oncorhynchus tshawyscha*.

Although the mode of action of fumagillin in controlling whirling disease is presently not well understood, a reduction in spore numbers, as well as changes in spore morphology visible under light microscopy, suggest an effect on DNA or RNA synthesis (Hartwig and Przelacka 1971; Jaronski 1972), or an effect on the integrity of the spore wall (Liu 1973). TNP-470 was developed as a superactive analog of fumagillin. The compound is reported to have a higher bioreactivity and lower toxicity than fumagillin (Coyle et al. 1998).

OBJECTIVES

The overall objective of this study was to evaluate the efficacy of fumagillin and TNP-470 to control whirling disease caused by *Myxobolus cerebralis* in cultured rainbow trout. This study was conducted following a "pivotal" efficacy trial protocol. Pivotal study protocols reflect studies that are statistically defensible, scientifically valid, and meet the U.S. Food and Drug Administrations (FDA) requirements for new animal drug approval. It was anticipated that data from these studies would be submitted to FDA's public master file with the intention of fulfilling FDA efficacy requirements for the approved use of fumagillin and TNP-470 in the U.S. to prevent and control whirling disease. Specific study objectives were 1) Determine the efficacy of two different formulations of fumagillin ("Sanofi" and "Chinoin") to prevent experimentally induced whirling disease by reducing/inhibiting the number of viable *Myxobolus cerebralis* spores in rainbow trout; 2) Determine the efficacy of TNP-470 to prevent experimentally induced whirling disease by reducing/inhibiting the number of viable *Myxobolus cerebralis* spores in rainbow trout; 3) Compare the efficacy of feed top-coated with fumagillin versus the efficacy of feed with fumagillin incorporated in the feed pellet; and 4) Compare the efficacy of TNP-470 fed to fish for 10 d versus 26 d.

METHODS

Treatments

Two different formulations of fumagillin were used in this study. One formulation of fumagillin dicyclohexylamine (DCH) salt was obtained from Sanofi Santé Nutrition Animale (Z.I. La Ballastiere, 33501 Libourne Cedex, France). This fumagillin was a "pure" powder that contained approximately 65% fumagillin. The other formulation of fumagillin DCH salt was obtained from Chinoin Pharmaceutical, Hungary (from the University of Munich, Germany). This fumagillin pre-mix formulation was the same fumagillin that was used successfully to control whirling disease in rainbow trout by El-Matbouli and Hoffmann (1991). TNP-470 was obtained from the National Cancer Research Institute, Frederick, Maryland. Sanofi fumagillin was either top-coated on feed or incorporated in the feed. Chinoin fumagillin and TNP-470 were only incorporated in the feed. Feed was top-coated with a mixture of fumagillin DCH in ethanol and then overlaid with pure vegetable oil (Raphaëla le Gouvello, Sanofi Santé, personal communication; M. J. Higgins, Department of Fisheries and Oceans, Pacific Biological Station, Naniamo, B.C., personal communication). Incor-porated feed was prepared by macerating pelleted feed, adding the appropriate concentration of drug, and repelleting the feed. All feed with incorporat-ed fumagillin or TNP-470 was prepared by the University of Munich. All medicated feed was pre-pared from the same feed that was fed to control fish, which was Rangen size #1 crumbles. Medicat-ed and control feeds were stored at 4°C.

Experimental Design

The study was conducted at the Wild Trout Research Laboratory in Bozeman, Montana utiliz-ing 12 weeks old Arlee strain rainbow trout fry (~2.0 g). Treatment groups included negative con-trols, positive controls, Sanofi-fumagillin top-coat-ed on feed and fed for 10 d, Sanofi-fumagillin incorporated in feed and fed for 10 d, Chinoin-fumagillin incorporated in feed and fed for 10 d, TNP-470 incorporated in feed and fed for 10 d, and TNP-470 incorporated in feed and fed for 26 d. All fish were reared at a constant 13°C. Five replicate tanks of 38 fish per tank were used for each treat-ment group. Positive controls and all treated groups were exposed to 1,000 TAMs per fish for 2 h in a 2 L aerated bucket of source water for each tank. Medicated feed treatment was initiated to all treat-ed groups 24 h after exposure to TAMs. Fish were fed fumagillin-treated feed at a dosage of 7.5 mg/kg body weight per d. Fish were fed TNP-470 treated feed at a dosage of 0.75 mg/kg body weight per d. All treated groups received medicated feed for 10 d, with the exception of a single TNP-470 treatment group that was fed for 26 d. Following the treat-ment period, all fish received the same unmedicat-ed diet for the remainder of the trial. Mortalities were counted and removed daily. Fish were exam-ined visually weekly for clinical signs of whirling

disease (blacktail, skeletal and cranial deformities, whirling behavior, etc.).

Determination of Infection

Presence and level of Myxobolus cerebralis infection was determined by histology, spore counts, and elec-tron microscopy evaluation. Samples for histological evaluation were collected 6, 10, 26, 30, 61, and 150 d postexposure to TAMs. Histological evaluation was conducted using the MacConnell/Baldwin Scale, which evaluates lesion severity based on a 0–5 scoring system (Hedrick et al. 1999). Samples for spore counts were collected 150 d postexposure to TAMs. Spore counts were conducted after tissues were disrupted using the pepsin-trypsin digest method. Viable and damaged spores were enumerat-ed using a hemacytometer according to procedures outlined in the Blue Book (Thoesen 1994).

Sample Collection

Samples for scanning and transmission electron microscopy were collected 5 min and 1, 2, 4, 6, 10, 26, 30, 61, and 150 d postexposure to TAMs. All electron microscopy samples were evaluated by the University of Munich. Hematology samples were collected 30, 61, and 150 d postexposure to TAMs to determine potential toxic effects of treatment. Evaluation of these samples included histology, blood smears, and hematocrits.

Statistical Analysis

Statistical tests were performed using a two-sam-pled t-test, and a parametric or nonparametric analysis of variance (ANOVA). A Tukey's pair-wise multiple comparison test was performed when normality and equal variance tests were passed. A Dunn's pair-wise multiple comparison test was con-ducted when normality and equal variance tests failed. Where differences are stated to be signifi-cant, a level of P less than or equal to 0.05 is implied. All study data were analyzed by SigmaStat Version 2.03 (SPSS Inc. 1992).

Results

Results of this study indicated that treatment of Arlee strain rainbow trout fry exposed to Myxobolus cerebralis with fumagillin or TNP-470 treated feed was not efficacious in the prevention or control of whirling disease. Myxobolus cerebralis infection was confirmed in both the positive control and all treat-ed groups by spore counts and histological evalua-tion. Mean spore count for the positive control group 150 d postexposure was 29,516, as compared with mean spore counts for the treated groups that ranged from 8,066–13,533 (Table 1). Although spore counts were lower for the treated groups, these differences were not significant. Mean histology score for the positive control group 150 d postexposure was 3.4, as compared with mean histology scores for the treated groups that ranged from 3.0 to 3.4 (Table 2). Although histological evaluation identified a moderate level of Myxobolus cerebralis infection, differences were not significant between control and treated groups. Spore counts and histological scores did not indicate a difference with respect to either fumagillin formulation (Sanofi versus Chinoin) or method of medicated feed preparation (top-coated versus incorporated).

Mean percent total mortality was similar with respect to all treatment groups with the exception of TNP-470 fed for 26 d, in which mortality was significantly higher (11.05% as compared with a range of 0.00–3.68% in all other treatment and control groups; Table 3). Elevated mortality in the

Table 1. Spore count data collected from rainbow trout half-heads[1] 150 d pe to TAMs at the Wild Trout Research Lab, Bozeman, MT (Note: fish were exposed to 1,000 TAMs per fish for 2 h).

Treatment[2]	Spore count[3] (mean value)	Standard deviation	95% C.I	Coefficient of variation
Negative control	0	0	0	—
Positive control	29,516	15,922	19770	54%
Sanofi fumagillin (incorporated)	8,467	3,512	4,361	41%
Sanofi fumagillin (top-coated)	13,533	4,179	5,189	31%
Chinoin fumagillin (incorporated)	11,000	2,163	2,686	20%
TNP-470 incorporated (10-day treatment)	10,622	5,510	6,842	52%
TNP-470 incorporated (26-day treatment)	8,066	5,500	6,829	68%

[1]One half head used for spore count, other half head used for histological evaluation
[2]All treatment groups were conducted in quintuplet
[3]Overall mean from quintuplet treatment, 10 fish sampled per replicate

Table 2. Histology score of rainbow trout half-heads[1] 150 days post-exposure to TAMs at the Wild Trout Research Lab, Bozeman, MT (Note: fish were exposed to 1,000 TAMs per fish for 2 hours).

Treatment[2]	Histology score[3,4] (mean value)	Standard deviation	95% C.I	Coefficient of variation
Negative control	0.0	0.0	0.0	—
Positive control	3.4	0.4	0.7	11.8%
Sanofi fumagillin (incorporated)	3.1	0.7	1.4	22.6%
Sanofi fumagillin (top-coated)	3.4	0.2	0.5	5.9%
Chinoin fumagillin (incorporated)	3.2	0.2	0.5	6.3%
TNP-470 incorporated (10-day treatment)	3.3	0.5	0.9	15.2%
TNP-470 incorporated (26-day treatment)	3.0	0.4	0.8	13.3%

[1]One half head used for histological evaluation, other half head used for spore count
[2]All treatment groups were conducted in quintuplet
[3]Overall mean from quintuplet treatment, 5 fish sampled per replicate
[4]Histological score based on the MacConnell/Baldwin Scale

TNP-470 26 d group indicated a toxic effect of the TNP-470 treatment. Fish in all five replicates of this treatment group were observed to be lethargic, riding high in the water, and exhibited a loss of appetite 30 d pe.

Histololgical evaluation 30 d postexposure revealed abnormal cytology in both the kidney and thymus of fish from the TNP-470 26 d group. As compared with normal hematopoietic kidney tissue, which is predominately basophilic in composition, hematopoietic tissue in these fish was found to be very eosinophilic (Figure 1). Although kidney tubules remained intact, hematopoietic tissue in the kidney was replaced with connective tissue. In the thymus, normal basophilic thymocytes were replaced with connective tissue and supporting cells (Figure 2). Blood smears evaluated 30 d postexposure revealed no blast cells and no immature erythrocytes in the TNP-470 26 d treatment group, as compared with mean total blast cells and immature erythrocyte numbers of 14.5 and 34.0, respectively, for the negative control group (Table 4). Mean

Table 3. Mean total mortality, mean fish size (wt), and mean incidence of clinical signs of whirling disease.[1] Data collected from rainbow trout 150 d pe to TAMs at the Wild Trout Research Lab, Bozeman, MT (Note: fish were exposed to 1,000 TAMs per fish for 2 h).

Treatment[2]	Mean percent total mortality[3]	Mean fish weight (g)[3]	Percent mean incidence of clinical signs of WD[4]
Negative control	1.58	23.7	0
Positive control	1.58	21.4	32
Sanofi fumagillin (incorporated)	3.68	21.3	22
Sanofi fumagillin (top-coated)	2.63	19.7	12
Chinoin fumagillin (incorporated)	3.16	21.6	8
TNP-470 incorporated (10-day treatment)	0.00	21.7	6
TNP-470 incorporated (26-day treatment)	11.05	22.6	12

[1]Primary clinical signs of whirling disease evaluated included whirling behavior, cranial deformities, opercular deformities, exopthalmia, and skeletal deformities.
[2]All treatment groups were conducted in quintuplet
[3]Determination based on evaluation of 10 fish sample per replicate
[4]Determination based on evaluation of ~20 fish sample per replicate

Figure 1. Kidney tissue from positive control group and fish treated with TNP-470 for 26 d.

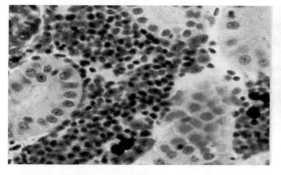

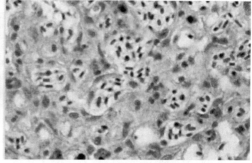

a. "Normal" kidney from positive controls.

b. Kidney from TNP-470 26 group showing decreased hematopoietic tissue and increased connective tissue.

number of lymphocytes 30 d postexposure for the TNP-470 26 d group was 0.5, as compared with 36.0 in the negative control group. An increase in the mean number of erythrocytic smudge cells and degenerate cells 30 d postexposure was also observed in the TNP-470 26 d treatment group as compared with the negative control group. Mean hematocrit values at 30 d postexposure were significantly lower for the TNP-470 26 d group (24.4) than for the negative control group (41.3; Table 4). However, 61 d postexposure all measured blood parameters were comparable for the TNP-470 26 d group and the negative control group, with the exception of lymphocyte numbers. Although lymphocyte numbers had recovered somewhat in the treated fish, cell counts remained ~40% lower than in negative controls.

Scanning and transmission electron micrographs (SEM and TEM) from positive controls (Figure 3a), fumagillin treated groups, and TNP-470 treated groups revealed a treatment effect with respect to spore ultrastructure. SEM revealed that TNP-470 treatment resulted in a disruptive effect on the valvogenic cells, causing a collapse of the spore shell (Figure 3b). SEM also showed that TNP-470 treatment resulted in perforation of the spore shell, and over time, total destruction of the shell (Figure 3c). TEM revealed that although both fumagillin and TNP-470 affected sporogenesis, the effects of each compound were different. Fumagillin treatment resulted in vacuolization of intracellular spore constituents and deformation of polar capsules, without affecting formation of spore valves (Figure 4a). Conversely, TNP-470 treatment appeared to disrupt the formation of both spore valves and polar capsules (Fibure 4b). Similar effects of TNP-470 treatment were also noted during the early developmental stages of sporogenesis.

DISCUSSION

It would be an understatement to say that the lack of efficacy observed following fumagillin treatment of fish exposed to *Myxobolus cerebralis* in this study was a disappointment. Based on the preliminary

Figure 2. Thymus tissue from positive control group and fish treated with TNP-470 for 26 d.

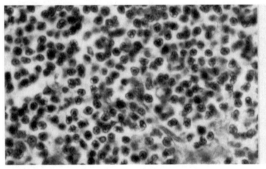

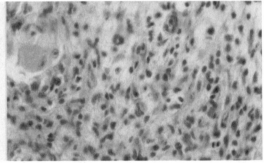

a. "Normal" thymus from positive controls.

b. Thymus from TNO-470 26 d group showing decreased thymocytes and increased connective tissue and supporting cells.

Table 4. Hematology data collected from rainbow trout 30 d pe to TAMs (All cell numbers are based on averaged counts of blood smears from 2 fish with 500 cells evaluated per fish).

Treatment[1]	Blast cells[2]	Immature erythrocytes[2]	Degenerative cells[2]	Lymphocytes[2]	Erythrocyte smudge cells[2]	Hematocrit[2]
Negative control	14.5	34.0	6.0	36.0	30.5	41.3%
TNP-470 incorporated (26-day treatment)	0.0	0.0	15.6	0.5	38.4	24.4%

[1]All treatment groups were conducted in quintuplet
[2]Overall mean from quintuplet treatment, 5 fish sampled per replicate

Figure 3. Scanning electron micrographs of *Myxobolus cerebralis* spores.

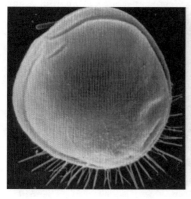 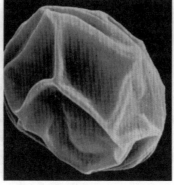 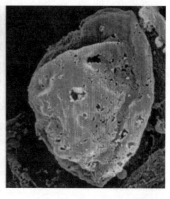

a. "Normal" spore from positive control fish.

b. Spore from TNP-470 treated fish showing deformation of spore shell.

c. Spore from TNP-470 treated fish showing perforation of spore shell.

Figure 4. Transmission electron micrographs of spores from Fumagillin and TNP-470 treated fish.

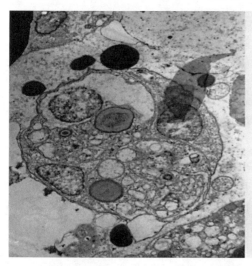 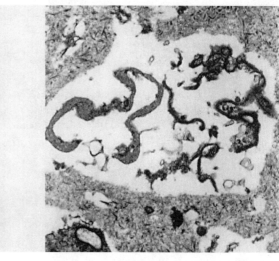

a. Spore from fumagillin treated fish showing vacuolization of spore.

b. Spore from TNP-470 treated fish showing irregularities in spore valves.

studies of El-Matbouli and Hoffmann (1991) it was expected that fumagillin treatment would result in a significant reduction in the level of *Myxobolus cerebralis* infection. This expectation was further heightened by results of Hedrick et al. (1988), Wishkovsky et al. (1990), Higgins and Kent (1996), and le Gouvello et al. (1999), who all reported some degree of efficacy with respect to the use of fumagillin to control proliferative kidney disease (PKD) in various salmonid species. As PKD is also caused by a myx-osporean parasite, it seemed probable to anticipate that fumagillin might also be efficacious when used to control *Myxobolus cerebralis*. Data from this study did not support such a conclusion. Although fumagillin and TNP-470 treatment in this study did result in somewhat lower spore counts as compared with positive control fish, this difference was not statistically significant. Furthermore, histological evaluation indicated no difference in level of infection between treated and positive control groups.

At this time, there are no clear hypotheses to explain the lack of efficacy of fumagillin or TNP-470 treatment to control/prevent whirling disease that was observed in this study, or in previous studies conducted by Erdahl and Staton (unpublished data). At one time it was hypothesized that by exposing very young fish to relatively high levels of *Myxobolus cerebralis*, the ability of fumagillin treatment to control level of infection was being "overwhelmed." However, this conclusion was not supported by the lack of efficacy observed in this study where somewhat older rainbow trout (12 weeks versus 8 weeks old fish) were exposed to lower levels of TAMs (1000 versus 1900 TAMs/fish for 2 h). During the course of these trials it was also hypothesized that a delay in time of treatment relative to time of exposure may have been contributing to a lack of observed efficacy, but this conclusion was not supported by results of this study or earlier trials. Following the initial studies by Erdahl and Staton (unpublished data), it was also hypothesized that fumagillin formulation and/or method of incorporation of fumagillin into feed may have had a bearing on the lack of observed efficacy. However, results from this study where two different formulations of fumagillin (including the one used successfully by El-Matbouli and Hoffmann (1991)) and a super-active analog of fumagillin were used and drug was both top-coated and incorporated in feed did not support this hypothesis.

Recently, it has been reported that "German" rainbow trout are more resistant to whirling disease than rainbow trout from the United States (Mansour El-Matbouli, personnel communication). If indeed so, resistance in these fish is likely an artifact of generations of exposure to the parasite in Europe. The development of resistance in these fish over time bodes well for the long-term survival of rainbow trout populations in the United States. It also may at least partially explain the lack of efficacy observed with respect to fumagillin treatment in studies conducted in the United States as compared with the work conducted in Germany by El-Matbouli and Hoffmann (1991). If the German strain rainbow trout used by El-Matbouli and Hoffmann (1991) were indeed "resistant" to whirling disease, the synergistic effects of natural resistance and fumagillin treatment may have combined to result in decreased infection level.

In addition to the fact that fumagillin and TNP-470 treatment did not appear to be efficacious, the elevated mean total mortality in replicate tanks fed TNP-470 for 26 d indicated toxicity with respect to this treatment regime. While this result was not entirely unexpected, it was also somewhat disappointing. Although previous studies have shown the fumagillin treatment at a high dosage (0.5–1.0 g/kg feed) for an extended period of time (4–8 weeks) can result in pathological changes in kidney and spleen tissue (Wishkovsky et al. 1990), it was not anticipated that TNP-470 (which is reported to be less toxic than fumagillin) treatment for 26 d would result in any adverse effects. However, extended TNP-470 treatment duration may be a moot point, as no differences were observed with respect to spore counts, histology, or electron microscopy between 10 d or 26 d treatment groups.

One potentially significant, positive finding from this study was that SEM and TEM revealed definitive spore deformities resulting from the treatment of *Myxobolus cerebralis*-exposed fish with either fumagillin or TNP-470 treatment. Although these two drugs appear to have different modes of action with respect to the formation of mature spores, treatment with either may result in a major disruption of spore ultrastructure. As it would seem logical to assume that disruption of spore structure would have at least some negative effect on spore function and ultimately on level of infection, some role may yet exist for the use of fumagillin and/or TNP-470 to help control whirling disease. Further research will be required to fully understand the mechanism of action and therapeutic potential of fumagillin and/or TNP-470.

SUMMARY

Results of this study did not indicate that fumagillin or TNP-470 treatment was effective in the prevention or control of whirling disease. Nonetheless, the development of a drug or therapeutant treatment remains a viable alternative to assist in the management of whirling disease in both captive and wild fisheries. Currently, whirling disease continues its rapid spread throughout the western United States where already the distribution of *Myxobolus cerebralis* in Montana and Colorado is a map of some of the finest trout waters in the world. The development of an effective therapeutant for the control of whirling disease would be of tremendous benefit to not only fisheries programs dependent upon stock supplementation, but would also indirectly benefit all fisheries programs by reducing overall system infectivity. To date, there are no obvious solutions to controlling/mitigating the effects of whirling disease, particularly with respect to incidence in wild populations. However, there is little doubt that stocking of infected fish may greatly exacerbate the problem. If whirling disease is ever going to be controlled or managed, it is imperative that stocking of infected fish be minimized, if not terminated entirely. Although preliminary studies indicated that fumagillin may be such a compound, unfortunately, the results of this study did not indicate that fumagillin or TNP-470 treatment is an effective management strategy to prevent or control whirling disease.

ACKNOWLEDGMENTS

The authors would like to thank the National Partnership on the Management of Wild and Native Cold Water Fisheries for providing the funding required to accomplish this study. We would also like to thank Beth MacConnell (Montana Department of Fish, Wildlife, and Parks) for her insight and advice with respect to study design and test procedures. Thank you to Andreas Schottes for all his EM work.

REFERENCES

Coyle, C., M. Kent, H. Tanowitz, M. Wittner, and L. M. Weiss. 1998. TNP-470 is an effective antimicrosporidial agent. Journal of Infectious Diseases 177:515–518.

El-Matbouli, M., and R. W. Hoffmann. 1991. Prevention of experimentally induced whirling disease in rainbow trout *Oncorhynchus mykiss* by fumagillin. Diseases of Aquatic Organisms 10:109–113.

le Gouvello, R., T. Pobel, and R. H. Richards. 1999. Field assessment of the efficacy of a ten day treatment of fumagillin against proliferative kidney disease in rainbow trout *Oncorhynchus mykiss*. Aquaculture 171:27–40.

Hartwig, A., and A. Przelacka. 1971. Nucleic acids in intestine of *Apis mellifica* infected with *Nosema apis* and treated with Fumagillin DCH: cytochemical and autoradiographic studies. Journal of Invertebrate Pathology 18:331–336.

Hedrick, R. P., J. M. Groff, P. Foley, and T. McDowell. 1988. Oral administration of fumagillin DCH protects chinook salmon (*Oncorhynchus tshawytscha*) from experimentally-induced proliferative kidney disease. Disease of Aquatic Organisms 4:165–168.

Hedrick, R. P., T. S. McDowell, M. Gay, G. D. Marty, M. P. Georgiadis, and E. MacConnell. 1999. Comparative susceptibility of rainbow trout *Oncorhynchus mykiss* and brown trout *Salmo trutta* to *Myxobolus cerebralis*, the cause of salmonid whirling disease. Diseases of Aquatic Organisms 37:173–183.

Higgins, M. J., and M. L. Kent. 1996. Field trials with fumagillin for the control of Proliferative Kidney Disease in coho salmon. The Progressive Fish-Culturist 58:268–272.

Jaronski, S. T. 1972. Cytochemical evidence for RNA synthesis inhibition by fumagillin. Journal of Antibiotics 25:327–331.

Liu, T. P. 1973. Effects of Fumidil B on the spore of *Nosema apis* and on lipids of the host cell as revealed by freeze-etching. Journal of Invertebrate Pathology 22:364–368.

Thoesen, J. C. 1994. Suggested procedures for the detection and identification of certain finfish and shellfish pathogens, 4th edition. American Fisheries Society, Fish Health Section, Bethesda, Maryland.

Wishkovsky, A., J. M. Groff, D. J. Lauren, R. J. Toth, and R. P. Hedrick. 1990. Efficacy of fumagillin against poliferative kidney disease and its toxic side effects in rainbow trout *Oncorhynchus mykiss* fingerlings. Fish Pathology 25:157–163.